An Illustrated Guide to
Veterinary Medical Terminology

An Illustrated Guide to
Veterinary Medical
Terminology

Janet Amundson Romich, DVM, MS

DELMAR

™

THOMSON LEARNING

Africa • Australia • Canada • Denmark • Japan • Mexico • New Zealand • Philippines
Puerto Rico • Singapore • Spain • United Kingdom • United States

NOTICE TO THE READER

Delmar Staff

Business Unit Director: Susan L. Simpfenderfer
Executive Editor: Marlene McHugh Pratt
Developmental Editor: Andrea Edwards Myers
Executive Marketing Manager: Donna J. Lewis

Executive Production Manager: Wendy A. Troeger
Project Editor: Amy E. Tucker
Production Editor: Carolyn Miller
Technology Project Manager: Kim Schryer

Printed in Canada
  8 9 10   08 07 06 05

For more information, contact Delmar, 3 Columbia Circle, PO Box 15015, Albany, NY 12212-0515; or find us on the World Wide Web at http://www.delmar.com

Library of Congress Cataloging-in-Publication Data
Romich, Janet Amundson.
    An illustrated guide to veterinary medical terminology / Janet Amundson Romich.
        p. cm.
    ISBN: 0-7668-0751-7
    1. Veterinary medicine—Terminology. I. Title.
SF610.R66   1999
636.089'014—dc21

Contents

Preface

TO THE STUDENT

Medical terminology may seem like a foreign language to most students. Many of the terms you have never encountered before, seem strange, or do not make sense. However, to communicate in the medical world a thorough understanding of the language is imperative.

Most medical terms are based on word parts that may already be familiar to you. Words like appendicitis, gastritis, or tonsillectomy may be words you have heard or used in the past. You may not realize the number of medical terms you know. Building on this foundation, new word parts will make learning medical terminology more logical.

This text and accompanying materials make the process of learning medical terminology as simple as possible. Review the introductory sections so that you are familiar with the organizational scheme of the textbook and flashcard technology. Once you get comfortable with the materials, you will find yourself learning medical terms faster than you have ever imagined.

Chapter Organization

The chapters in *An Illustrated Guide to Veterinary Medical Terminology* are organized in the following fashion:

► Introduction to medical terms
► Anatomical foundations
► Body systems
► Species-specific chapters

Chapter 1 provides the student with the basics of how medical terms are formed, analyzed, and defined. Chapter 2 provides the student with terms used in everyday dialogue regarding positioning of animals and relationships between body parts. Chapters 3 and 4 discuss anatomical landmarks both internally (musculoskeletal system) and externally (common terms for landmarks on an animal's body). Chapter 5 consists of terms used in the animal industry to describe males and females of selected species as well as terms for their young and groups of their species. Chapters 6 through 15 are organized by body systems. These chapters describe the anatomy of the body system and clinical terms used in reference to that body system and conclude with diagnostic tests, pathology, and procedures for that body system. Chapters 16 and 17 relate tests, procedures, and treatments used in the medical field to

care for animals. Chapters 18 through 23 are species-specific chapters that can be studied independently to enhance your knowledge of a particular species or that may be incorporated into other chapters to assess your progress. Appendix A consists of tables of abbreviations and Appendix B is plural forms of medical terms.

It should be noted that several puns appear in the text such as "the plane truth", "the hole truth", and "procede with caution". These play on words are designed to be entertaining and should not be used as a key to studying spelling and vocabulary of these sections.

Learning Supplements

FLASH!

An Illustrated Guide to Veterinary Medical Terminology is supplemented with Flash!, a flashcard program on CD-ROM. Students and instructors have been successfully using flashcards as a method of learning new terms and facts for many years. Instructors will find that using Flash! will allow students to actually study the medical terms immediately versus having students spend time making up flashcards. Instructors will find Flash! a useful teaching supplement to *An Illustrated Guide to Veterinary Medical Technology* since each chapter in the flashcard program correlates to the same chapter in the textbook. Flash! will have similar, if not identical, word definitions to the textbook. Instructors will find that the Flash! program is easy for students to use by choosing a chapter and having terms appear on the screen along with the pronounciation. The turn over key will then provide the definition. Students can manually turn over the definition or have the system on automatic turn over with the amount of time between term and definition selected by the user. The Flash! user also has the capability to set the program so that the definition appears first and the user must provide the term. This is a great way for the user to practice the spelling and recognition of these terms. The Flash! program also has a tag function to allow for tagging of difficult terms or to allow the instructor to tag only certain terms in the chapter that he/she wants to use. Lists of all the terms in the chapter are available and can be printed out in chart form or for classroom use. Instructors will appreciate Flash! as it can be used during review sessions (with projection equipment), for classroom activities (like

Jeopardy!®, since the categories with terms and definitions are already made), and for quizzes and tests (can test knowledge of definitions, terms, and spelling).

Students will like using the Flash! program along with the textbook because it is easy to use, can vary between asking for definitions or for medical terms, can display both question and answer at the same time, and allows for tagging of missed terms. Flash! allows for self-paced study, which is especially helpful for students with a limited veterinary medical background. Students can choose to randomize terms so that the terms do not appear in the same order in each chapter. At the end of the program students will find a chapter on word parts and a chapter on abbreviations that are valuable resources for many classes. Students can use Flash! for reviewing for other classes such as anatomy and physiology because basic structural terms are included as well as more common terms such as fetlock joint. Students can review chapter fifteen prior to entering laboratory courses and chapters sixteen and seventeen prior to entering pharmacology, surgery, or techniques courses. Flash! can also serve as a good review tool for board certification examinations. Since students can work individually with the Flash! program, self-paced and self-directed study is a key attribute to using the computer disc in conjunction with the textbook. Flash! also allows for fun and creative ways for group study of medical terms, definitions, and spelling of these terms.

Note: Instructors should expect students to master the terms in each section before progressing to the next section, since the word parts will not be repeated in subsequent chapters. For example, the prefix hypo- may first be seen in gastrointestinal chapter and not repeated in the endocrine chapter. However, words containing the prefix hypo- will be found in the endocrine chapter.

TO THE INSTRUCTOR

An instructor's guide for this text is available to meet the challenge of teaching. The instructor's guide provides answer keys for all exercises in the text, and teaching tips to enhance the teaching of medical terminology. It will provide both new and seasoned instructors a fresh outlook on medical terminology instruction.

ACKNOWLEDGMENTS

Special thanks to the following people who helped review this text and answered many questions regarding medical terminology throughout its development. Without their expertise the text would not have been as complete as it is.

Kay Bradley, BS, CVT
Instructor
Madison Area Technical College
Madison, WI

Kenneth Brooks, DVM, Diplomate ABVP
Lodi Veterinary Hospital, SC
Lodi, WI

Anne E. Chauvet, DVM, Diplomate ACVIM-Neurology
University of Wisconsin Veterinary Medical Teaching Hospital
Madison, WI

Michael T. Collins, DVM, PhD
University of Wisconsin School of Veterinary Medicine
Madison, WI

Thomas Curro, DVM, MS
Henry Doorly Zoo
Omaha, Nebraska

Deb Donohoe, LATG, Registered Laboratory Animal Technologist
Medical College of Wisconsin
Milwaukee, WI

Wendy Eubanks, CVT
Delafield Small Animal Hospital
Delafield, WI

Ron Fabrizius, DVM, Diplomate ACT
Poynette Veterinary Service, Inc
Poynette, WI

Kelly Gilligan, DVM
Four Paws Veterinary Clinic, LLC
Prairie du Sac, WI

John Greve, DVM, PhD
Iowa State University
Ames, IA

Mark Jackson, DVM, PhD, Diplomate ACVIM, MRCVS
North Carolina State University
Raleigh, NC

Linda Kratochwill, DVM
Duluth Business College
Duluth, MN

Amy Lang, RTR
University of Wisconsin Veterinary Medical Teaching Hospital
Madison, WI

Laura L. Lien, CVT
University of Wisconsin Veterinary Medical Teaching Hospital
Madison, WI

Carole Maltby, DVM
Maple Woods Community College
Kansas City, MO

Sheila McGuirk, DVM, PhD, Diplomate ACVIM
University of Wisconsin Veterinary Medical Teaching
    Hospital
Madison, WI

James Meronek, DVM
Sauk Prairie Veterinary Clinic
Prairie du Sac, WI

Karl Peter, DVM
Foothill College
Los Altos Hills, CA

Katherine Polzin, BA
University of Wisconsin Veterinary Medical Teaching
    Hospital
Madison, WI

Teri Raffel, CVT
Madison Area Technical College
Madison, WI

Linda Sullivan, DVM
University of Wisconsin Veterinary Medical Teaching
    Hospital
Madison, WI

Laurie Thomas, BA, MA
Clinicians Publishing Group/Partners in Medical
    Communications
Clifton, NJ

Beth Uldal Thompson, VMD
Veterinary Technician/Veterinary Learning Systems
Trenton, NJ

The author and Delmar Publishers wish to acknowledge the following reviewers of the Flash! CD-ROM:

**Deb Donohoe, LATG, Registered Laboratory Animal
    Technologist**
Medical College of Wisconsin
Milwaukee, WI

Ron Fabrizius, DVM, Diplomate ACT
Poynette Veterinary Service, Inc
Poynette, WI

Kelly Gilligan, DVM
Four Paws Veterinary Clinic
Prairie du Sac, WI

John H. Greve, DVM, PhD
Iowa State University
Ames, IA

**Mark Jackson, DVM, PhD, Diplomate ACVIM,
    MRCVS**
North Carolina State University
Raleigh, NC

Linda Kratochwill, DVM
Duluth Business College
Duluth, MN

Laura L. Lien, CVT
University of Wisconsin Veterinary Medical Teaching
    Hospital
Madison, WI

Sheila McGuirk, DVM, PhD, Diplomate ACVIM
University of Wisconsin Veterinary Medical Teaching
    Hospital
Madison, WI

James Meronek, DVM
Sauk Prairie Veterinary Clinic
Prairie du Sac, WI

Linda Sullivan, DVM
University of Wisconsin
Veterinary Medical Teaching Hospital
Madison, WI

I would also like to express my gratitude to Beth Thompson, VMD, and Laurie Thomas, BA, MA, of Veterinary Learning Systems for their determination in advancing my writing skills through the publication of journal articles for *Veterinary Technician Journal*. Without their guidance I would not have honed my writing skills. I would also like to thank the many veterinary technician and laboratory animal technician students at Madison Area Technical College for their support and continued critique of the Veterinary Terminology course. A special thank you goes to the 1998 veterinary technician and laboratory animal technician students at Madison Area Technical College who learned terminology through my rough draft of this text. Lastly, I would like to thank the excellent staff at Delmar Publishing/Thomson Learning and my family for their continued support and understanding.

HOW TO USE THIS TEXT

An Illustrated Guide to Veterinary Medical Terminology helps you learn and retain medical terminology using a logical approach to medical word parts and associations. The keys to learning from this text include:

Illustrations

Complete with detailed labeling, the text's line drawings clarify key concepts and contain important information of their own. In addition to line drawings, actual photos are included to enhance the visual perception of medical terms and to allow for better retention of medical terms and usage of these terms in the "real" world. Review each illustration and photo carefully for easy and effective learning.

Charts and Tables

Charts and tables condense material into a visually appealing and organized fashion to allow for rapid learning. Some tables include terms organized by opposites or body systems to enhance relating the information to various situations.

New Terms

New terms appear in bold type, with the pronunciation and definition following.

Pronunciation System

The pronunciation system is an easy to learn and follow approach to learning the sounds of medical terms. This system is not laden with a lot of linguistic marks and variables, so that the student does not get bogged down in understanding the key. Once students become familiar with the key it is very easy for them to progress in speaking the medical language.

Pronunciation Key

PRONUNCIATION GUIDE

Pronunciation guides for common words are omitted.

Any vowel that has a dash above it represents the long sound, as in

$\bar{a}$	hay
$\bar{e}$	we
$\bar{\imath}$	ice
$\bar{o}$	toe
$\bar{u}$	unicorn

Any vowel followed by an "h" represents the short sound, as in

ah	apple
eh	egg
ih	igloo
oh	pot
uh	cut

Unique vowel combinations are as follows

oo	boot
ər	higher
oy	boy
aw	caught
ow	ouch

OTHER PRONUNCIATION GUIDELINES

Word parts are represented in the text as prefixes, combining forms, and suffixes. The notation for a prefix is a word part followed by a hyphen. The notation for a combining form (word root and its vowel to ease pronunciation) is the root followed by a / and its vowel, as in nephr/o. The notation for a suffix is a hyphen followed by the word part. The terms prefix, combining form, and suffix will not appear in the definitions.

Learning Objectives

The beginning of each chapter lists learning objectives to help students know what is expected of them as they read the text and complete the exercises.

Review Exercises

At the end of each chapter are exercises to help you interact with and review the chapter's information. The exercises include several formats: multiple choice, matching, case studies, word building, and diagram labeling. The answers to these exercises are found in the instructor's manual.

 # Ready, Set, Go

► Identify and recognize the parts of a medical term
► Define commonly used prefixes, combining forms, and suffixes presented in this chapter
► Analyze and understand basic medical terms
► Recognize the importance of spelling medical terms correctly
► Practice pronunciation of medical terms
► Recognize the importance of medical dictionary use
► Practice medical dictionary use

INTRODUCTION TO MEDICAL TERMINOLOGY

Medical terms are used everyday in such varying environments as medical offices, newspapers, television, and conversational settings. Most of us are familiar with many medical terms; however, the appearance of other medical terms seems complicated and foreign. Learning and understanding how medical terminology developed can aide in the understanding of these terms.

Current medical vocabulary is based on terms of Greek and Latin origin, **eponyms** (words formed from a person's name), and modern language terms. The majority of medical terms are derived from word parts based on Greek and Latin words. Becoming familiar with these Greek and Latin terms and identifying word parts will enable one to learn common medical terms and to recognize unfamiliar medical terms by word analysis. Medical terminology may seem daunting at first because of the length of medical words and seemingly curious spelling rules, but once you learn the basic rules of breaking a word down into its constituents the words become easier to read and understand.

ANATOMY OF A MEDICAL TERM

Many medical terms are composed of word part combinations. Recognizing these word parts and their meanings simplifies learning medical terminology. These word parts are

prefix: word part found at the beginning of a word. Usually indicates number, location, time, or status.

root: word part that gives the essential meaning of the word.

combining vowel: single vowel, usually an "o," that is added to the end of a root to make the word easier to pronounce.

combining form: combination of the root and combining vowel.

suffix: word part found at the end of a word. Usually indicates procedure, condition, disease, or disorder.

Understanding the meaning of the word parts allows one to dissect medical terms in a logical way. It also allows for greater expansion of one's medical vocabulary by the process of breaking down unfamiliar terms into recognizable word parts.

Prefixes

Prefixes are added to the beginning of a word or root to modify its meaning. For example, the term operative can be modified using various prefixes.

▶ The prefix **pre-** means before. **Preoperative** means before or preceding an operation.
▶ The prefix **peri-** means around. **Perioperative** means pertaining to the period around an operation or the period before, during, and after an operation.
▶ The prefix **post-** means after. **Postoperative** means after an operation.

Many prefixes have another prefix whose meaning is opposite of its own. Initially, when learning prefixes it is helpful to learn them in these pairs or in similar groups (Table 1–1 and Figure 1–1).

Combining Vowels

A combining vowel is sometimes used to make the medical term easier to pronounce. The combining vowel is used when the suffix begins with a consonant as in the suffix -scope. An **arthroscope** is an instrument to visually examine the joint. Because the suffix -scope begins with a consonant, the combining vowel "o" is used. "o" is the most commonly used combining vowel; however, "i" and "e" may be used as well. A combining vowel is not used when the suffix begins with a vowel as in the suffix -itis. **Gastritis** is inflammation of the stomach. Because the suffix -itis begins with a vowel, the combining vowel "o" is not used.

Combining Forms

A word root plus a vowel is the combining form. Combining forms usually describe a part of the body. New words are created by adding combining forms with prefixes, other combining forms, and suffixes. For example, the term panleukopenia is composed of the word parts:

▶ **pan-** (pahn), a prefix meaning all
▶ **leuk/o** (loo-kō), a combining form meaning white
▶ **-penia** (pē-nē-ah), a suffix meaning deficiency or reduction in number

Panleukopenia is a deficiency of all types of white blood cells.

TABLE 1–1
Contrasting Prefixes

Without a prefix the root traumatic means pertaining to injury	**a-** (ah or ā) means without or no. **atraumatic** means without injury.
Without a prefix the root uria means urination	**an-** (ahn) means without or no. **anuria** means absence of urine.
ab- (ahb) means away from. **abduction** means to take away from midline.	**ad-** (ahd) means towards. **adduction** means move toward the midline.
Without a prefix the root emetic means pertaining to vomiting.	**anti-** (ahn-tī or ahn-tih) means against **antiemetics** work against or prevent vomiting.
dys- (dihs) means difficult, painful, or bad. **dysphagia** means difficulty eating or swallowing.	**eu-** (yoo) means good, easy, or normal. **euthyroid** means having a normally functioning thyroid gland.
endo- (ehn-dō) means within or inside. **endocrine** means to secrete internally.	**ex-** (ehcks) or **exo-** (ehcks-ō) means without, out of, outside, or away from. **exocrine** means to secrete externally (via a duct).
endo- means within or inside. **endoparasite** is an organism that lives within the body of the host.	**ecto-** (ehck-tō) means outside. **ectoparasite** is an organism that lives on the outer surface of the host.
hyper- (hī-pər) means increased or more than normal. **hyperglycemia** means increased amounts of blood glucose.	**hypo-** (hī-pō) means decreased or less than normal. **hypoglycemia** means decreased amounts of blood glucose.
inter- (ihn-tər) means between. **intercostal** means between the ribs.	**intra-** (ihn-trah) means within. **intramuscular** means within the muscle.
poly- (pohl-ē) means many. **polyuria** means frequent or increased amount of urination.	**oligo-** (ohl-ih-gō) means scant or little. **oliguria** means infrequent or decreased amount of urination.
pre- (prē) means before. **preanesthetic** means pertaining to before anesthesia.	**post-** (pōst) means after. **postanesthetic** means pertaining to after anesthesia.
sub- (suhb) means below, under, or less. **sublingual** means under the tongue.	**super-** (soo-pər) and **supra-** (soo-prah) mean above, beyond, or excessive. **supernumerary** means in excess of the regular number. **suprascapular** means above the shoulder blade.

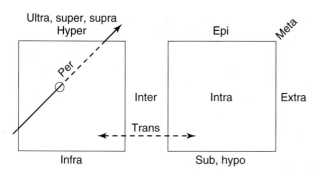

FIGURE 1–1 Directional prefixes

Suffixes

Suffixes are attached to the end of a word part to modify its meaning. For example, the combining form gastr/o means stomach and can be modified using the various suffixes.

▶ The suffix **-tomy** means cutting into or incision. **Gastrotomy** is incision into the stomach.
▶ The suffix **-stomy** means to surgically create a new opening. **Gastrostomy** is surgically creating an opening between the stomach and the body surface.
▶ The suffix **-ectomy** means to surgically remove or excise. **Gastrectomy** is surgical removal of the stomach.

Many suffixes can be grouped together by meaning or by the category they modify. Initially, when learning suffixes it is easiest if the learner groups them by meaning or category.

"PERTAINING TO" SUFFIXES

▶ **-ac** (ahck), as in the example **cardiac** (pertaining to the heart)
▶ **-al** (ahl), as in the example **renal** (pertaining to the kidney)
▶ **-ar** (ahr), as in the example **lumbar** (pertaining to the loin)
▶ **-ary** (ahr-ē), as in the example **alimentary** (pertaining to the gastrointestinal tract)
▶ **-an** (ahn), as in the example **ovarian** (pertaining to the ovary)
▶ **-eal** (ē-ahl), as in the example **laryngeal** (pertaining to the larynx)
▶ **-ic** (ihck), as in the example **enteric** (pertaining to the intestines)
▶ **-ine** (ihn), as in the example **uterine** (pertaining to the uterus)
▶ **-ous** (uhs), as in the example **cutaneous** (pertaining to the skin)
▶ **-tic** (tihk), as in the example **nephrotic** (pertaining to the kidneys)

SURGICAL SUFFIXES

▶ **-ectomy** (ehck-tō-mē) = surgical removal
mastectomy is surgical removal of the breast or mammary glands
▶ **-stomy** (stō-mē) = surgically create a new opening
colostomy is surgically creating a new opening between the colon and body surface
▶ **-tomy** (tō-mē) = cutting into
laparotomy is to cut into the abdomen
▶ **-pexy** (pehck-sē) = suture to stabilize
gastropexy is to surgically stabilize the stomach to the abdominal wall
▶ **-plasty** (plahs-tē) = surgical repair
rhinoplasty is surgical repair of the nose

PROCEDURAL SUFFIXES

▶ **-centesis** (sehn-tē-sihs) = surgical puncture to remove fluid or gas (either for diagnosis or to remove excess fluid or gas)
cystocentesis is surgical puncture of the urinary bladder with a needle to remove fluid (urine)
▶ **-gram** (grahm = record of
electrocardiogram is the hard copy record
▶ **-graph** (grahf) = instrument that records (or used as record)
electrocardiograph is the machine that records the electrical activity of the heart
▶ **-graphy** (grahf-ē) = procedure that records
electrocardiography is the procedure used to test the electrical activity of the heart
▶ **-lysis** (lī-sihs) = separation or breakdown
urinalysis is separation of the urine into its constituents
▶ **-scope** (skōp) = instrument to visually examine
endoscope is an instrument to visually examine inside the body
▶ **-scopy** (skōp-ē) = procedure to visually examine
endoscopy is the procedure of visually examining inside the body
▶ **-therapy** (thehr-ah-pē) = treatment
chemotherapy is treatment with chemical substances or drugs

DOUBLE "R" SUFFIXES

▶ **-rrhagia** or **-rrhage** (rā-jē-ah or rihdj) = bursting forth
hemorrhage is bursting forth of blood from the vessels
▶ **-rrhaphy** (rahf-ē) = to suture
enterorrhaphy is suturing the intestines
▶ **-rrhea** (rē-ah) = flow, discharge
diarrhea is complete discharge of the bowels
▶ **-rrhexis** (rehck-sihs) = rupture
myorrhexis is rupture of the muscle

What is the difference between human and veterinary medical terminology?
Most times, the medical terms used in human medical settings are identical to the ones used in veterinary medical settings. The increased number of species in veterinary medicine, and the addition of terms used in animal production, greatly expand the vocabulary of veterinary professionals. Species-specific anatomical differences will also negate or augment the terms used in a specific area. Do you know where the calf muscle is located on a person? Where is the calf muscle in a calf?

CONDITIONAL SUFFIXES

▶ **-algia** and **-dynia** (ahl-jē-ah or dihn-ē-ah) = pain
arthralgia and **arthrodynia** are joint pain
▶ **-itis** (ī-tihs) = inflammation
hepatitis is inflammation of the liver
▶ **-malacia** (mah-lā-shē-ah) = abnormal softening
osteomalacia is abnormal softening of bone
▶ **-megaly** (mehg-ah-lē) = enlargement
cardiomegaly is enlargement of the heart
▶ **-osis** (ō-sihs) = abnormal condition
cardiosis is an abnormal condition of the heart
▶ **-pathy** (pahth-ē) = disease
enteropathy is disease of the intestines
▶ **-sclerosis** (skleh-rō-sihs) = abnormal hardening
arteriosclerosis is abnormal hardening of the arteries
▶ **-um** (uhm) = structure
pericardium is the structure surrounding the heart

Suffixes may change the part of speech of a word. Different suffixes may change the word from a noun (naming people, places, or things) to an adjective (descriptor) (Figure 1–2). Examples of this include:

cyan*osis* is a noun meaning condition of blue discoloration, while **cyano*tic*** is an adjective meaning pertaining to blue discoloration
an***emia*** is a noun meaning a blood condition of deficient red blood cells or hemoglobin, while **anem*ic*** is an adjective meaning pertaining to a blood condition of deficient red blood cells or hemoglobin
muc***us*** is a noun meaning a slime-like substance that is composed of glandular secretion, salts, cells, and leukocytes, while **muc*ous*** is an adjective meaning pertaining to mucus
ili***um*** is a noun meaning a part of the hip, while **ili*ac*** is an adjective pertaining to the hip

ANALYZING MEDICAL TERMS

Medical terminology can be more easily understood if the following objectives are adhered to:

1. ***Dissect:*** First analyze the word structurally by dividing it into its basic components
2. ***Begin at the end:*** After dividing the word into its basic parts, define the suffix first, the prefix second, and then the roots. If there are two roots, divide each and read them from left to right.
3. ***Anatomic order:*** Where body systems are involved, the words are usually built in the order in which the organs occur in the body. For example, **gastroenteritis** is the proper term for inflammation of the stomach and intestine. Since food passes from the stomach into the intestine, the medical term for stomach appears before the medical term for intestine. The order of medical word parts in a medical term may also represent the order of blood flow through organs. The exception to this involves some diagnostic procedures in which tools or substances are passed retrograde or in the opposite direction of anatomic order. In these cases the words are built in the order in which the equipment passes the body part.

Using the above guidelines, analyze the term ovariohysterectomy. First divide the term into its basic components: ovari/o/hyster/ectomy. Defining from back to front, the suffix -ectomy is surgical removal, one combining form ovari/o means ovary, and the other combining form hysteri/o means uterus. Together the term **ovariohysterectomy** means surgical removal of the ovaries and uterus. This term is built based upon the order in which the ovaries and uterus are found in the body.

WHAT DID YOU SAY?

Proper pronunciation of medical terms takes time and practice. Listening to how words are pronounced by medical professionals and using medical dictionaries and textbooks are your best sources for learning pro-

NOUN	SUFFIX	ADJECTIVE	SUFFIX
cyanosis	- osis	cyanotic	- tic
anemia	- emia	anemic	- ic
mucus	- us	mucous	- ous
ilium	- um	iliac	- ac
condyle	- e	condylar	- ar
carpus	- us	carpal	- al

FIGURE 1–2 Suffix variation depending on usage

nunciation. There will be individual variations in using medical terms either due to geographic location or personal preference. Medical dictionaries also vary on how they present pronunciation of medical terms. Some sources mark the syllable receiving the greatest emphasis with a primary accent (′) and the syllable receiving the second most emphasis with a secondary accent (″). Other sources will boldface and capitalize the syllable receiving the most emphasis, while other sources do not emphasize syllables. Consult with your reference prior to pronouncing the word.

General Pronunciation Guidelines

In general, all vowels in scientific words are pronounced. Vowels can either be short or long (Table 1–2). Consonants are generally pronounced as in other English words.

DOES SPELLING COUNT?

Be aware of spelling when using medical terminology. Changing one or two letters can change the meaning of a word. **Hepatoma** is a liver mass, while **hematoma** is a mass or collection of blood. The **urethra** takes urine from the urinary bladder to the outside of the body, while **ureters** collect urine from the kidney and transport it to the urinary bladder. Medical terms may be pronounced the same but have different meanings, therefore, spelling is important. For example, ileum and ilium are pronounced the same. However, **ileum** is the distal part of the small intestine (e = enter/o or e = eating) while **ilium** is part of the hip bone (h**i**p has I in it). Some medical terms actually have the same spelling as terms used for other body parts. For example, the combining form myel/o represents the spinal cord and bone marrow (it originates from the term meaning white substance). Other terms have different spellings depending on how the term is used grammatically. For example, when used as a noun mucus (that slimy stuff secreted from mucous membranes) is spelled differently than when it is used as an adjective (as in mucous membrane).

When looking a medical term up in the dictionary, spelling plays an important role. However, the term may not be spelled the way it sounds. The following guidelines can be used to find a word in the dictionary:

▶ If it sounds like f, it may begin with f or ph.
▶ If it sounds like j, it may begin with g or j.
▶ If it sounds like k, it may begin with c, ch, k, or qu.
▶ If it sounds like s, it may begin with c, ps, or s.
▶ If it sounds like z, it may begin with x or z.

TABLE 1–2
Pronunciation of Vowels

Vowel	Sound	Example
"a" at the end of a word	ah	idea
"ae" followed by r or s	ah	aerobic
"i" at the end of a word	ī	bronchi
"oe"	eh	oestrogen (old English form)
"oi"	oy	sarcoid
"eu"	ū	euthanasia
"ei"	ī	Einstein
"ai"	ay	air
"au"	aw	auditory
Exceptions to Consonant Pronunciations		
Consonant	Sound	Example
"c" before e, i, and y	s	cecum
"c" before a, o, and u	k	cancer
"g" before c, i, and y	j	genetic
"g" before a, o, and u	g	gall
"ps" at beginning of word	s	psychology
"pn" at beginning of word	n	pneumonia
"c" at end of word	k	anemic
"cc" followed by i or y	first c = k second c = s	accident
"ch" at beginning of word	k	chemistry
"cn" in middle of word	both c (pronounce k) and n (pronounce ehn)	gastrocnemius
"mn" in middle of word	both m and n	amnesia
"pt" at beginning of word	t	pterodactyl
"pt" in middle of word	both p and t	optical
"rh"	r	rhinoceros
"x" at beginning of word	z	xylophone

REVIEW EXERCISES

Match the word parts in column I with the correct definition in column II

Column I	Column II
_____ -itis	A. incision or cutting into
_____ -gram	B. before
_____ post-	C. surgical puncture to remove fluid
_____ -tomy	D. difficult, painful, or bad
_____ pre-	E. enlargement
_____ -centesis	F. excision or surgical removal
_____ -therapy	G. combining form for liver
_____ dys-	H. combining form for kidney
_____ peri-	I. inflammation
_____ ren/o	J. record
_____ hepat/o	K. after
_____ -megaly	L. treatment
_____ -ectomy	M. around

Supply the medical term that represents the following definitions

Pertaining to the stomach	_____
Inflammation of the liver	_____
Abnormal softening of bone	_____
Joint pain	_____
Procedure to visually examine inside the body	_____
Heart enlargement	_____
Pertaining to the kidney	_____
Bursting forth of blood from vessels	_____
Suturing of stomach to body wall	_____
Treatment with chemicals or drugs	_____

Choose the correct answer

1. The prefix _____ means away from midline.
   a. ad-
   b. ab-
   c. ex-
   d. endo-

2. The suffix _____ means instrument to visually examine.
   a. -ectomy
   b. -scope
   c. -scopy
   d. -graphy

3. The prefix _____ means increased, while the prefix _____ means decreased.

   a. pre-, post-
   b. endo-, exo-
   c. hyper-, hypo-
   d. inter-, intra-

4. The suffix _____ means pertaining to.

   a. -al
   b. -ary and -ar
   c. -ic
   d. all of the above

5. The suffix _____ means incise or cut into.

   a. -ex
   b. -tomy
   c. -ectomy
   d. -graphy

6. The suffix _____ means abnormal condition.

   a. -osis
   b. -rrhea
   c. -rrhagia
   d. -uria

7. The suffix _____ means separation or breaking into parts.

   a. -gram
   b. -pexy
   c. -um
   d. -lysis

8. The prefix _____ means below.

   a. supra-
   b. super-
   c. inter-
   d. sub-

9. The prefix(es) _____ means many.

   a. olig-
   b. a-, an-
   c. poly-
   d. eu-

10. The prefix(es) _____ means without or no.

    a. a-, an-
    b. olig-
    c. dys-
    d. hyper-

Case Study—Fill in the blanks to complete the case history of patients

A five year old male, neutered cat is presented to a veterinary clinic with _____ (painful urination) and _____ (scant urine production). Upon examination the abdomen is <u>palpated</u> and _____ (an enlarged urinary bladder) is noted. After completing the examination, the veterinarian suspects an <u>obstruction</u> of the _____ (the anatomic tube that carries urine from the urinary bladder to outside the body). Blood is taken for analysis and the cat is admitted to the clinic. The cat is anesthetized and a urinary <u>catheter</u> is passed. Urine is collected for _____ (breakdown of urine into its components). In addition to the obstruction, the cat is treated for _____ (inflammation of the urinary bladder).

In the above case study, the meanings of some unfamiliar medical terms (underlined in the above text) cannot be understood by breaking up the term into its basic components. Using a dictionary or dictionary Web site, define the following medical terms

1. palpated _____

2. obstruction _____

3. catheter _____

2 Where, Why, and What?

Objectives *In this chapter, you should learn to:*

▶ Identify and recognize body planes, positional terms, directional terms, and body cavities
▶ Identify terms used to describe the structure of cells, tissues, and glands
▶ Define terms related to body cavities and structure
▶ Recognize, define, spell, and pronounce medical terms related to pathology and procedures
▶ Identify body systems by their components, functions, and combining forms
▶ Identify prefixes that assign numeric value

IN POSITION

Positional terms are important for accurately and concisely describing body locations and relationships of one body structure to another. The terms forward and backward, up and down, in and out, and side to side are not clear enough descriptions by themselves to have universal understanding in the medical community. Therefore, very specific terms were developed so that there would not be confusion amongst people as to what was trying to be conveyed. Listed in Table 2-1 and illustrated in Figures 2–1 and 2–2 are directional terms used in veterinary settings.

THE PLANE TRUTH

Planes are imaginary lines that are used descriptively to divide the body into sections:

▶ **midsagittal** (mihd-sahdj-ih-tahl) **plane** is the plane that divides the body into *equal* right and left halves. It is also called the **median** (mē-dē-ahn) plane and **midline** (Figure 2–3).
▶ **sagittal** (sahdj-ih-tahl) **plane** is the plane that divides the body into unequal right and left halves (Figure 2–4).

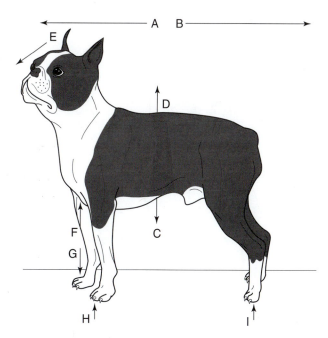

FIGURE 2–1 Directional terms. The arrows on this Boston terrier represent the following directional terms: A = cranial, B = caudal, C = ventral, D = dorsal, E = rostral, F = proximal, G = distal, H = palmar, I = plantar.

9

TABLE 2–1
Terms Used to Describe Direction

ventral (vehn-trahl) refers to the belly or underside of a body or body part (ventr/o in Latin means belly) (venture means to <u>under</u>take) (ventral fin is on the belly)	**dorsal** (dōr-sahl) refers to the back (dors/o in Latin means back) (endorse means sign on the back) (dorsal fin is on the back)
cranial (krā-nē-ahl) means towards the head (crani/o in Latin means skull)	**caudal** (kahw-dahl) means towards the tail (cauda in Latin means tail)
anterior (ahn-tēr-ē-ər) means front of the body (anteri/o in Latin means before) Used more in description of organs or body parts because front and rear are confusing terms in quadrupeds. A quadruped's belly is oriented downward, not forward as in humans.	**posterior** (pohs-tēr-ē-ər) means rear of the body (posteri/o in Latin means behind)
rostral (rohs-trahl) means nose end of the head (rostrum in Latin means beak) **cephalic** (seh-fahl-ihck) means pertaining to the head (kephale in Greek means head)	**caudal** (kahw-dahl) means towards the tail (cauda in Latin means tail)
medial (mē-dē-ahl) means toward midline (medi/o in Latin means middle)	**lateral** (laht-ər-ahl) means away from midline (later/o in Latin means side)
superior (soo-pēr-ē-ər) means uppermost, above, or toward the head (super in Latin means above)	**inferior** (ihn-fēr-ē-ər) means lowermost, below, or toward the tail (inferi in Latin means lower)
proximal (prohck-sih-mahl) means nearest midline or beginning of a structure (proxim/o in Latin means next)	**distal** (dihs-tahl) means farthest from midline or beginning of a structure (dist/o in Latin means distant)
superficial (soop-ər-fihsh-ahl) means near the surface; also called external (super in Latin means above)	**deep** (dēp) means away from the surface; also called internal (deep means beneath surface)
palmar (pahl-mahr) means bottom of the front foot or hoof (refers to anything distal to carpus) (palmar in Latin means hollow of hand)	**plantar** (plahn-tahr) means bottom of the rear foot or hoof (refers to anything distal to tarsus) (plantar in Latin means sole of foot)

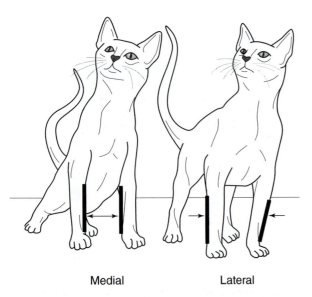

Medial   Lateral

FIGURE 2–2 Medial versus lateral. The arrows on these cats represent the directional terms medial and lateral.

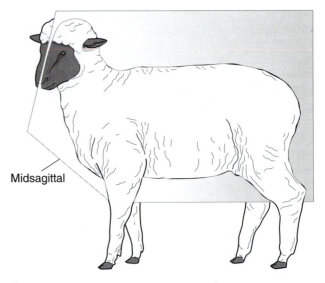

Midsagittal

FIGURE 2–3 Planes of the body. The midsagittal or median plane divides the body into *equal* left and right portions.

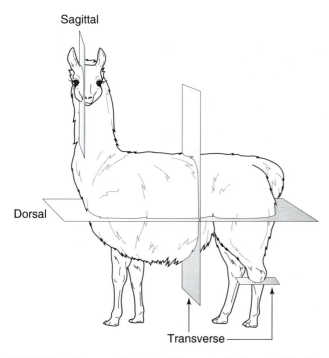

FIGURE 2–4 Planes of the body. The sagittal plane divides the body into *unequal* right and left parts, the dorsal plane divides the body into back and belly parts, and the transverse plane divides the body into cranial and caudal parts. The transverse plane also describes a perpendicular transection to the long axis of an appendage.

▶ **dorsal** (dōr-sahl) **plane** is the plane that divides the body into dorsal (back) and ventral (belly) parts (Figure 2–4). It is also called the **frontal** (frohn-tahl) **plane** or **coronal** (kō-roh-nal) **plane.** In humans, the frontal plane is a vertical plane due to the fact that people stand erect.

▶ **transverse** (trahnz-vərs) **plane** is the plane that divides the body into cranial and caudal parts. It is also called the **horizontal plane** or **cross-sectional plane.** The transverse plane may also be used to describe a perpendicular transection to the long axis of an appendage.

STUDYING

The suffix **-logy** means the study of. There are specific terms used to describe specific branches of study. The study of body structure is termed **anatomy** (ah-naht-ō-mē). **Physiology** (fihz-ē-ohl-ō-jē) is the study of body function(s). **Pathology** (pahth-ohl-ō-jē) is the study of the nature, causes, and development of abnormal conditions. Combining physiology and pathology results in the term **pathophysiology** (pahth-ō-fihz-ē-ohl-ō-jē)

which is the study of changes in function caused by disease. The study of disease causes is **etiology** (ē-tē-ohl-ō-jē).

YOU HAVE SAID A MOUTHFUL

Describing positions in the mouth has become increasingly important with the rise of veterinary dentistry. The dental **arcade** (ahr-kād) is the term used to describe how individual teeth are arranged in the mouth. Arcade means a series of arches, which is how the teeth are arranged in the oral cavity. Surfaces of the teeth are named for the area in which they contact (Figure 2–5). The **lingual** (lihng-gwahl) **surface** is the aspect of the tooth that faces the tongue. Remember linguistics is the study of language, and the tongue is used to make sounds. Some people use lingual surface to describe the tooth surface that faces the tongue on both the maxilla and mandible. More correctly the **palatal** (pahl-ah-tahl) **surface** is the tooth surface of the maxilla that

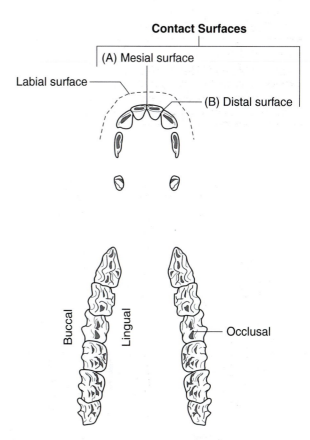

FIGURE 2–5 Teeth surfaces of the maxilla. Teeth surfaces are identified by the area they are near. In the mandible, the lingual surface is called the palatal surface.

faces the tongue, and the lingual surface is the tooth surface of the mandible that faces the tongue. The **buccal** (būk-ahl) **surface** is the aspect of the tooth that faces the cheek. Bucca is Latin for cheek. The buccal surface is sometimes called the **vestibular** (vehs-tih-buh-lahr) **surface.** Vestibule in Latin means space or cavity at an entrance. The **occlusal** (ō-klū-zahl) **surfaces,** are the aspects of the teeth that meet when you chew. Think of the teeth occluding or stopping things from passing between them when you clinch them. The **labial** (lā-bē-ahl) **surfaces** is the tooth surface facing the lips. **Labia** is the medical term for lips. **Contact** (kohn-tahct) **surfaces** are the aspects of the tooth that touch other teeth. Contact surfaces are divided into **mesial** (mē-zē-ahl) and **distal** (dihs-tahl). The mesial contact surface is the one that is closest to the midline of the dental arcade or arch. The distal contact surface is the one that is furthest from the midline of the dental arcade (think distance). Each tooth has both contact surfaces, even the last molar which only touches one tooth surface.

THE HOLE TRUTH

A body **cavity** (kahv-ih-tē) is a hole or hollow space within the body that contains and protects internal organs. The **cranial** (krā-nē-ahl) **cavity** is the hollow space that contains the brain in the skull. The **spinal** (spī-nahl) **cavity** is the hollow space that contains the spinal cord within the spinal column. The **thoracic** (thō-rahs-ihck) **cavity** or **chest cavity** is the hollow space that contains the heart and lungs within the ribs between the neck and diaphragm. The **abdominal** (ahb-dohm-ih-nahl) **cavity** is the hollow space that contains the major organs of digestion between the diaphragm and pelvic cavity. The abdominal cavity is commonly referred to as the **peritoneal** (pehr-ih-tohn-ē-ahl) **cavity,** but this is not quite accurate. The peritoneal cavity is the hollow space within the abdominal cavity between the parietal peritoneum and the visceral peritoneum. The **pelvic** (pehl-vihck) **cavity** is the hollow space that contains the reproductive and some excretory system organs formed by the pelvic bones.

Cavities are just one way to segregate the body. Regional terms are also used to describe areas of the body. The **abdomen** (ahb-dō-mehn) is the portion of the body between the thorax and the pelvis containing the abdominal cavity. The **thorax** (thō-rahcks) is the chest region located between the neck and diaphragm. The **groin** (groyn) is the lower region of the abdomen between itself and the thigh (it is also known as the **inguinal** (ihng-gwih-nahl)) **area.**

Membranes (mehm-brānz) are thin layers of tissue that cover a surface, line a cavity, or divide a space or organ. The **peritoneum** (pehr-ih-tō-nē-uhm) is the membrane lining the walls of the abdominal and pelvic cavities and covers some organs in this area. The peritoneum may be further divided in reference to its location. The **parietal** (pah-rī-eh-tahl) **peritoneum** is the outer layer of the peritoneum, while the **visceral** (vihs-ər-ahl) **peritoneum** is the inner layer of the peritoneum. Inflammation of the peritoneum is termed **peritonitis** (pehr-ih-tō-nī-tihs).

Other terms associated with the abdomen and peritoneum include the umbilicus mesentery and retroperitoneal. The **umbilicus** (uhm-bihl-ih-kuhs) is the pit in the abdominal wall marking the point where the umbilical cord entered the fetus. In veterinary terminology the umbilicus is also referred to as the **navel** (nā-vuhl). **The mesentery** (mehs-ehn-tehr-ē or mehz-ehn-tehr-e), is the layer of the peritoneum that suspends parts of the intestine within the abdominal cavity. **Retroperitoneal** (reh-trō-pehr-ih-tō-nē-ahl) pertains to behind the peritoneum.

Other membranes of the body will be covered in the specific body region in which they are found.

LYING AROUND

Lay, lie, laid, and lying are confusing words in English, much less learning a new medical term for the same thing. However, the only medical term for lying down is **recumbent** (rē-kuhm-behnt). Recumbent is then modified depending upon which side is facing down (Figure 2–6).

dorsal recumbency (dōr-sahl rē-kuhm-behn-sē) is lying on the back
ventral recumbency (vehn-trahl rē-kuhm-behn-se) is lying on the belly = **sternal** (stər-nahl) recumbency
left lateral recumbency (laht-ər-ahl rē-kuhm-behn-se) is lying on the left side
right lateral recumbency is lying on the right side

There are two less commonly used terms derived from human medical terminology that relate to lying down. **Prone** (prōn) means lying in ventral or sternal recumbency; **supine** (soo-pīn) means lying in dorsal recumbency.

To clarify the lies spread about recumbency terms remember

► lay = to put, place, or prepare
► laid = past tense of lay
► laying = active form of lay
► lie = to recline or be situated
► lain = past tense of lie
► lying = active form of lie

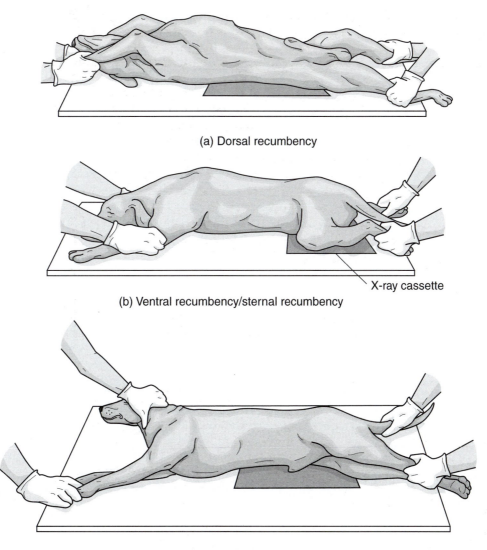

(a) Dorsal recumbency

X-ray cassette

(b) Ventral recumbency/sternal recumbency

(c) Lateral recumbency

FIGURE 2–6 Recumbency positions. The position in which an animal lies is important in veterinary medicine especially when radiographing an animal. (a) This dog is in dorsal recumbency. (b) This dog is in ventral or sternal recumbency. (c) This dog is in right lateral recumbency.

MOVING RIGHT ALONG

Medical terms used to describe movement may involve changing prefixes and/or suffixes to change direction. The terms adduction and abduction look very similar yet have opposite meanings (Figure 2–7). **Adduction** (ahd-duhck-shuhn) means to move toward the midline (think addition to something) while **abduction** (ahb-duhck-shuhn) means to move away from midline (think child abduction means to take away).

Flexion (flehck-shuhn) means to bend a joint or reduce the angle between two bones. Flexing your biceps involves bending your elbow. **Extension** (ehcks-tehn-shuhn)

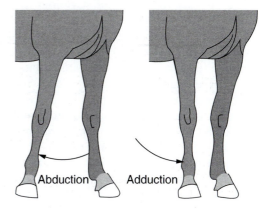

Abduction   Adduction

FIGURE 2–7 Adduction versus abduction

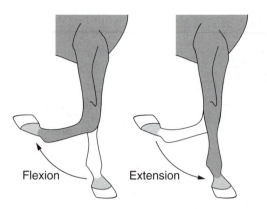

Flexion   Extension

FIGURE 2–8 Flexion versus extension

means to straighten a joint or increase the angle between two bones (Figure 2–8). You extend your hand for a handshake. **Hyperflexion** (hī-pər-flehcks-shuhn) and **hyperextension** (hī-pər-ehcks-tehn-shuhn) are when a joint is flexed or extended too far.

Supination and pronation are two less commonly used terms in veterinary settings. **Supination** (soo-pihnā-shuhn) is the act of rotating the limb or body part so that the ventral surface is turned upward, while **pronation** (prō-nā-shuhn) is the act of rotating the limb or body part so that the ventral surface is turned downward. Think of supination as in the movement involved with eating soup. **Rotation** (rō-tā-shuhn) is another term of movement that means circular movement around an axis.

SETTING OUR CYTES AHEAD

Cells are the structural units of the body (Figure 2–9). The combining form for cell is cyt/o (sī-tō). Cells are specialized and grouped together to form tissues and organs. **Cytology** (sī-tohl-ō-jē) is the study of cells. The suffix -logy means the study of. Cytology involves studying cell origin, structure, function, and pathology.

The cell membrane, cytoplasm, and nucleus are collectively called the **protoplasm** (prō-tō-plahzm). The suffix -plasm (plahzm) means formative material of cells and the combining form prot/o means first. The cell membrane is the structure lining the cell that protects the cell's contents. **Cytoplasm** (sī-tō-plahzm) is the material located within the cell membrane that is not part of the nucleus. The **nucleus** (nū-klē-uhs) is the structure within a cell that contains nucleoplasm, chromosomes, and the surrounding membrane. **Nucleoplasm** (nū-klē-ō-plahzm) is the material within the nucleus, and **chromosomes** (krō-mō-sōmz) are the structures in the nucleus composed of DNA, which transmits genetic information.

It's in the Genes

Genetic is a term used to denote something that pertains to genes or heredity. A **genetic** (jehn-eh-tihck) **disorder** is any disease or condition caused by defective genes. This term is different from **congenital** (kohn-jehn-ih-tahl), which denotes something that is present at birth. A genetic defect may be congenital, but a congenital defect only implies that something faulty is present at birth. An **anomaly** (ah-nohm-ah-lē) is a deviation from what is regarded as normal. Anomaly may be used instead of defect.

GROUPING THINGS TOGETHER

A group of specialized cells that are similar in structure and function is a **tissue** (tihsh-yoo). The study of the structure, composition, and function of tissue is **histology** (hihs-tohl-ō-jē). Hist/o is the combining form for tissue.

There are many different types of tissue (Figure 2–10). **Epithelial tissue** (ehp-ih-thē-lē-ahl tihsh-yoo) covers internal and external body surfaces. Epithelial tissue is further divided into epithelium and endothelium. **Epithelium** (ehp-ih-thē-lē-uhm) is the cellular covering that forms the outer layer of the skin and covers the external surfaces of the body. Epi- is a prefix that means above, thel/o is a combining form that means nipple but is now used to denote any thin membrane, and -um is a suffix that means structure. **Endothelium** (ehn-dō-thē-lē-uhm) is the cellular covering that forms the lining of the internal organs including the blood vessels. Endo- is a prefix meaning within. **Mesothelium** (mēs-ō-thē-lē-uhm) is a word similar to epithelium and endothelium and means the cellular covering that forms the lining of serous membranes like the peritoneum. The prefix meso- means middle.

Connective tissue is another tissue type. Connective tissue holds the organs in place and binds body parts together. Bones, cartilage, tendons, and ligaments are all types of connective tissue. **Adipose** (ahd-ih-pohs) tissue, another form of connective tissue, is also known as fat. Adip/o is the combining form for fat.

Tissue can form normally or abnormally. The suffix -plasia (plā-zē-ah) is used to describe formation, development, and growth of tissue and cell *numbers*. The suffix -trophy (trō-fē) means formation, development, and increased *size* of tissue and cells. The use of different prefixes describes problems with tissue formation.

- **aplasia** (ā-plā-zē-ah) is lack of development of an organ or tissue or cell
- **hypoplasia** (hī-pō-plā-zē-ah) is incomplete or less than normal development of an organ or tissue or cell
- **hyperplasia** (hī-pər-plā-zē-ah) is an abnormal increase in the number of normal cells in normal arrangement in an organ or tissue or cell

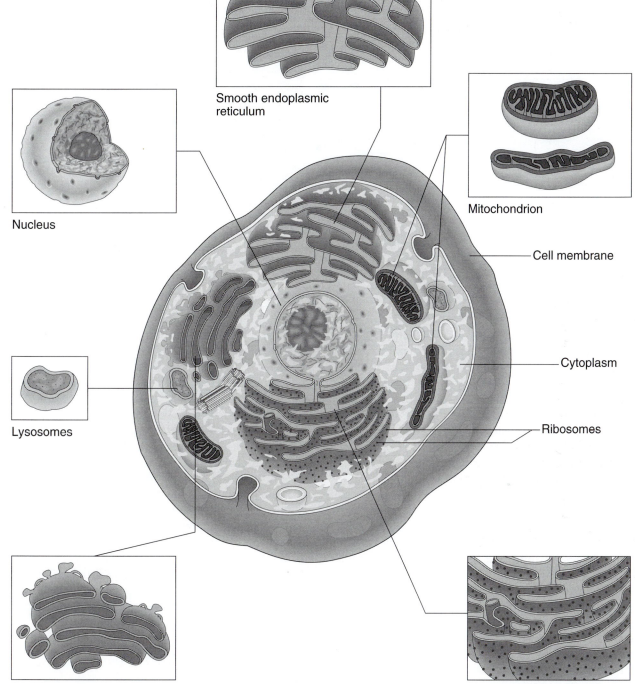

Smooth endoplasmic reticulum

Mitochondrion

Nucleus

Cell membrane

Lysosomes

Cytoplasm

Ribosomes

Golgi apparatus

Rough endoplasmic reticulum

FIGURE 2–9 Parts of the cell. Parts of the cell include the nucleus (brain of the cell), cytoplasm (semifluid medium containing organelles) mitochondrion (energy of the cell), Golgi apparatus (chemical processor of the cell), endoplasmic reticulum (synthesizers of protein, lipids, and some carbohydrates), ribosomes (protein synthesis), and lysosomes (digestive system of the cell).

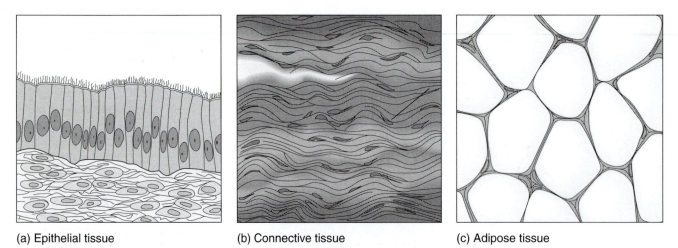

(a) Epithelial tissue   (b) Connective tissue   (c) Adipose tissue

FIGURE 2–10 Tissue types. Some examples of tissue types include (a) epithelial tissue, (b) connective tissue, and (c) adipose tissue.

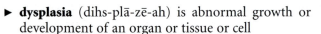

▶ **dysplasia** (dihs-plā-zē-ah) is abnormal growth or development of an organ or tissue or cell

▶ **anaplasia** (ahn-ah-plā-zē-ah) is a change in the structure of cells and their orientation to each other

▶ **neoplasia** (nē-ō-plā-zē-ah) is any abnormal new growth of tissue in which multiplication of cells is uncontrolled, more rapid than normal, and progressive. Neoplasms usually form a distinct mass of tissue called a **tumor** (too-mər). Tumors may be **benign** (beh-nīn) meaning not recurring or **malignant** (mah-lihg-nahnt) meaning tending to spread and life threatening. The suffix -oma (ō-mah) means tumor or neoplasm.

The prefix a- means without, hypo- means less than normal, hyper- means more than normal, dys- means bad, ana- means without, and neo- means new.

Glands (glahndz) are groups of specialized cells that secrete material used elsewhere in the body. Aden/o is the combining form for gland. Glands are divided into two categories: exocrine and endocrine (Figure 2–11). **Exocrine** (ehck-soh-krihn) **glands** are groups of cells that secrete their chemical substances into ducts that lead out of the body or to another organ. Exocrine glands may contain ducts. Examples of exocrine glands are sweat glands, sebaceous glands, and the portion of the pancreas that secretes digestive chemicals. **Endocrine** (ehn-dō-krihn) **glands** are groups of cells that secrete their chemical substances directly into the bloodstream, which transports them throughout the body. Endocrine glands are ductless. Examples of endocrine glands are the thyroid gland, pituitary gland, and the portion of the pancreas that secretes insulin.

An **organ** (ohr-gahn) is a part of the body that performs a special function or functions. Each organ has its own combining form or forms as listed in Table 2–2. The combining forms have either Latin or Greek origins. If a body part has two combining forms that are

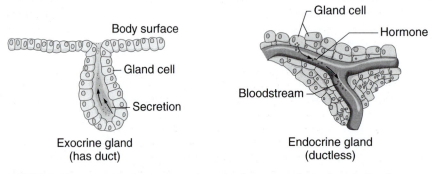

FIGURE 2–11 Types of glands. Exocrine glands secrete their chemical substances into ducts that lead out of the body or to another organ. Endocrine glands secrete their chemical substances directly into the bloodstream.

TABLE 2–2
Combining Forms for Organs

Body System	Combining Form	Major Functions
Skeletal system	bones = **osteo/o** (ohs-tē-ō), **oss/e** (ohs-ē), or **oss/i** (ohs-ih) joints = **arthr/o** (ahr-thrō) cartilage = **chondr/o** (kohn-drō)	Support and shape, protective, hematopoietic, store minerals
Muscular system	muscles = **my/o** (mī-ō) fascia = **fasc/i** (fahs-ē) or **fasci/o** (fahs-ē-ō) tendons = **ten/o** (tehn-ō), **tend/o** (tehn-dō), or **tendin/o** (tehn-dih-nō)	Locomotion, movement of body fluids, generate body heat
Cardiovascular system	heart = **cardi/o** (kahr-dē-ō) arteries = **arteri/o** (ahr-tē-rē-ō) veins = **ven/o,** (vēn-ō) or **phleb/o** (fleh-bō) blood = **hem/o** (hē-mō) or **hemat/o** (hē-maht-ō)	Provides delivery of oxygen and nutrients to tissue, transport cellular waste from body, immune function, and endocrine function
Lymphatic and immune system	lymph vessels, fluid, and nodes = **lymph/o** (lihm-fō) tonsils = **tonsill/o** (tohn-sih-lō) spleen = **splen/o** (spleh-nō) thymus = **thym/o** (thī-mō)	Provide nutrients to and remove waste from tissues, protect the body from harmful substances
Respiratory system	nose or nares = **nas/o** (nā-zō) or **rhin/o** (rī-nō) pharynx = **pharyng/o** (fahr-ihn-gō) trachea = **trache/o** (trā-kē-ō) larynx = **laryng/o** (lahr-ihng-gō) lungs = **pneum/o** (nū-mō) or **pneumon/o** (nū-mohn-ō)	Bring oxygen into the body for transportation to the cells, remove carbon dioxide and some water waste from the body
Digestive system	mouth = **or/o** (ōr-ō) or **stomat/o** (stō-maht-ō) esophagus = **esophag/o** (eh-sohf-ah-gō) stomach = **gastr/o** (gahs-trō) small intestine = **enter/o** (ehn-tər-ō) large intestine = **col/o** (kō-lō) or **colon/o** (kō-lohn-ō) liver = **hepat/o** (hehp-ah-tō) pancreas = **pancreat/o** (pahn-krē-ah-tō)	Digestion of ingested food, absorption of digested food, elimination of solid waste
Urinary system	kidneys = **ren/o** (rē-nō) or **nephr/o** (nehf-rō) ureters = **ureter/o** (yoo-rē-tər-ō) urinary bladder = **cyst/o** (sihs-tō) urethra = **urethr/o** (yoo-rē-thrō)	Filtration of blood to remove waste, maintain electrolyte balance, regulate fluid balance within the body
Nervous system	nerves = **neur/o** (nū-rō) or **neur/i** (nū-rē) brain = **encephal/o** (ehn-sehf-ah-lō) spinal cord = **myel/o** (mī-eh-lō) eyes = **ophthalm/o** (ohf-thahl-mō), **ocul/o** (ohck-yoo-lō), **opt/o** (ohp-tō), or **opt/i** (ohp-tē) sight = **optic/o** (ohp-tih-kō) ears = **ot/o** (ō-tō), **aur/i** (awr-ih), or **aur/o** (awr-ô)	Coordinating mechanism, reception of stimuli, transmission of messages
Integumentary system	skin = **dermat/o** (dər-mah-tō), **derm/o** (dər-mō), or **cutane/o** (kyoo-tā-nē-ō)	Protection of body, temperature, and water regulation
Endocrine system	adrenals = **adren/o** (ahd-reh-nō) gonads = **gonad/o** (gō-nahd-ō) pineal = **pineal/o** (pī-nē-ahl-ō) pituitary = **pituit/o** (pih-too-ih-tō) thyroid = **thyroid/o** (thī-royd-ō) or **thyr/o** (thī-rō)	Integrating body functions, homeostasis, growth
Reproductive system	testes = **orch/o** (ōr-kō), **orchi/o** (ōr-kē-ō), **orchid/o** (ōr-kihd-ō), or **testicul/o** (tehst-tihck-yoo-lō) ovaries = **ovari/o** (ō-vā-rē-ō), **oophor/o** (ō-ohff-ehr-ō) uterus = **hyster/o** (hihs-tehr-ō), **metr/o** (mē-trō), **metr/i** (mē-trē), **metri/o** (mē-trē-ō), or **uter/o** (yoo-tər-ō)	Production of new life

TABLE 2–3

Prefixes Assigning Number Value

Number Value	Latin Prefix	Greek Prefix	Examples
1	uni-	mono-	unicorn, unilateral monochromatic, monocyte
2	duo	dyo	duet, dyad
3	tri-	tri-	trio, triceratops, triathlon
4	quadri- or quadro-	tetr- or tetra-	quadruplet, tetralogy, tetroxide
5	quinqu-	pent- or penta-	quintet, pentagon
6	sex-	hex- or hexa-	sexennial, hexose, hexagon
7	sept- or septi-	hept- or hepta-	septuple, heptarchy
8	octo-	oct-, octa-, or octo-	octave, octopus
9	novem- or nonus	ennea-	nonuple, ennead
10	deca- or decem-	dek- or deka-	decade, dekanem

used to describe it, how do you know which form to use? In general, the Latin term is used to describe or modify something as in renal disease or renal tubule. The Greek term is generally used to describe a pathological finding as in nephritis or nephropathy.

1, 2, 3, GO

Medical terms can be further modified by the use of prefixes to assign number value (Table 2–3), numerical order, or parts of a whole. The following prefixes are also used in everyday English, so some of them may be familiar. For example, unicorns are animals with one horn (uni = one, corn = horn). It would make sense then that a **bicornuate uterus** (bi = two, corn = horn) is a uterus with two horns. Knowing that lateral means pertaining to the side, it would make sense that **unilateral** (yoo-nih-lah-tər-ahl) would mean pertaining to one side. **Bilateral** (bī-lah-tər-ahl) means pertaining to two sides.

REVIEW EXERCISES

For Figures 2–12, 2–13, 2–14, 2–15, and 2–16 follow the instructions in the captions.

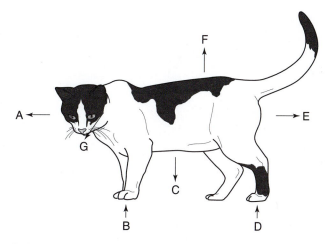

FIGURE 2–12 Label the arrows with the proper directional term.

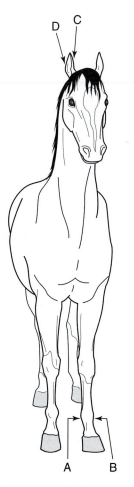

FIGURE 2–13 Label the arrows with the proper directional term.

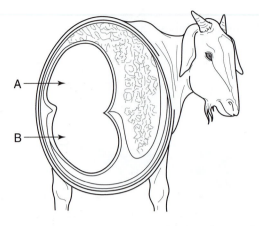

FIGURE 2–14 Label the sacs of the rumen. Through what plane is this goat sectioned?

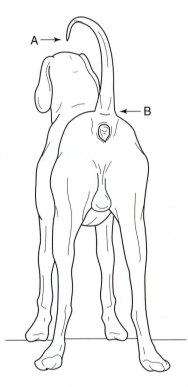

FIGURE 2–15 Is point A the more proximal or more distal end of the tail? Is point B the more proximal or more distal end of the tail?

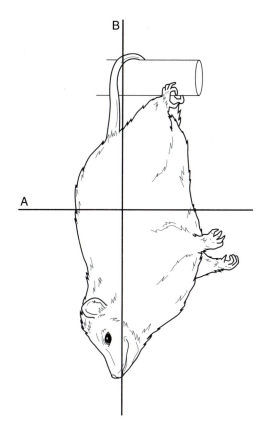

FIGURE 2–16   What plane of the body is plane A? What plane of the body is plane B?

Match the number with the prefix that describes it. Each number may have more than one correct answer.

_____	1	hept-
_____	2	mono-
_____	3	tri-
_____	4	deka-
_____	5	duo-
_____	6	uni-
_____	7	penta-
_____	8	octo-
_____	9	quadri-, quadro-
_____	10	tetra-
		sex-
		nonus-
		deca-
		quinqu-

Choose the correct response

1. Lateral means
   a. near the beginning
   b. near the front
   c. toward the side
   d. toward midline

2. The sagittal plane divides the body into
   a. cranial and caudal portions
   b. left and right portions
   c. equal left and right halves
   d. dorsal and ventral portions

3. The paw is _____ to the shoulder.
   a. caudal
   b. cranial
   c. proximal
   d. distal

4. The transverse plane divides the body into
   a. cranial and caudal portions
   b. left and right portions
   c. equal left and right halves
   d. dorsal and ventral portions

5. The lining of the abdominal cavity and some of its organs is called the
   a. mesentery
   b. peritoneum
   c. thoracum
   d. membrane

6. The study of structure, composition, and function of tissues is called
   a. cytology
   b. histology
   c. pathology
   d. organology

7. The _____ plane divides the body into dorsal and ventral portions.
   a. sagittal
   b. midsagittal
   c. dorsal
   d. transverse

8. The medical term for lying down is
   a. lateral
   b. sternal
   c. recumbent
   d. surface

9. The medical term for lack of development of an organ or tissue is
   a. anaplasia
   b. aplasia
   c. dysplasia
   d. hyperplasia

10. The medical term for the ventral surface of the rear paw, hoof, or foot is
    a. ventral
    b. dorsal
    c. palmar
    d. plantar

Fill in the blank with the correct answer

a. The _____ is also known as the navel.

b. _____ glands secrete chemical substances directly into the bloodstream.

c. A/An _____ is any new growth of tissue in which multiplication of cells is uncontrolled, more rapid than normal, and progressive.

d. A/An _____ is a deviation from what is regarded as normal.

e. The _____ cavity contains the heart and lungs.

f. The ventral surface of the front paw, foot, or hoof is the _____ surface.

g. The shoulder is _____ to the pelvis.

h. A/An _____ is the basic structural unit of the body.

i. The stomach is located _____ to the heart.

j. A/An _____ is a hollow space within a body or organ.

k. Another term for groin is _____.

l. The _____ is a layer of the peritoneum that suspends parts of the intestine within the abdominal cavity.

m. _____ is the suffix for formative material of cells.

n. Not malignant is _____.

o. The five combining forms for uterus are _____ , _____ , _____ , _____, and _____.

Meat and Bones

Objectives *In this chapter, you should learn to:*

► Identify and describe the major structures and functions of the musculoskeletal system
► Describe bone anatomy terms
► Differentiate between the axial and appendicular skeletons
► Recognize, define, spell, and pronounce terms related to the diagnosis, pathology, and treatment of the musculoskeletal system
► Construct musculoskeletal terms from word parts

FUNCTIONS OF THE SKELETAL SYSTEM

The **musculoskeletal** (muhs-kyoo-lō-skehl-eh-tahl) **system** consists of two systems that work together to support the body and allow for movement of the animal. The skeletal system consists of bones, joints, and cartilage. The bones of the skeletal system form the framework that supports and protects an animal's body. Within bone is the red bone marrow which functions to form red blood cells, white blood cells, and clotting cells. Joints aid in the movement of the body. Cartilage protects the ends of bones where they contact each other. Cartilage is also found in the ear and nose. The muscular system will be covered later in this chapter.

STRUCTURES OF THE SKELETAL SYSTEM

Make the Connection

The skeleton is made up of various forms of connective tissues. Connective tissue is a type of tissue in which the proportion of cells to extracellular matrix is small (less cells and more matrix). Connective tissue binds together and is the support of various structures of the body. Bone, tendons, ligaments, and cartilage are all connective tissues associated with the skeletal system.

Bone

Bone, a form of connective tissue, is one of the hardest tissues in the body. Embryonically, the skeleton is made of cartilage and fibrous membranes that harden into bone before birth. **Ossification** (ohs-ih-fih-kā-shuhn), the formation of bone from fibrous tissue, continues until maturity (maturity varies with the species of animal). Normal bone goes through a continuous process of building up and breaking down throughout an animal's life. This process allows bone to heal and repair itself. Bone growth is balanced between the actions of **osteoblasts** (ohs-tē-ō-blahsts) and **osteoclasts** (ohs-tē-ō-klahsts). Osteoblasts (**osteo** = bone, **-blasts** = immature) are immature bone cells that produce bony tissue, while osteoclasts (**osteo** = bone, **-clasts** = break) eat away bony tissue from the medullary cavity of bone. The combining forms for bone are **oste/o, oss/e,** and **oss/i.**

Red bone marrow, located within cancellous bone, is **hematopoietic** (hēm-ah-tō-poy-eht-ihck). The

---

epi- = above, **physis** = growth, **dia-** = between, **peri-** = surrounding, **oste/o** = bone, **-um** = structure,

endo- = within or inner, **meta-** = beyond

combining form **hemat/o** means blood and the suffix -**poietic** means pertaining to formation. Thus, red bone marrow produces red blood cells, white blood cells, and clotting cells. The **medullary** (mehd-yoo-lahr-ē) **cavity** of bone, or the inner space of bone, contains yellow bone marrow. In adult animals, yellow bone marrow replaces red bone marrow. Yellow bone marrow is composed mainly of fat cells and serves as a fat storage area.

Bone is divided into different categories based on bone types, bone shapes, and bone functions (Table 3–1 and Figure 3–1).

> **Soft versus hard:** Bone diseases can cause abnormal changes. Bones can become softer than normal or harder than normal. To describe these changes the suffixes -**malacia** (abnormal softening) and -**sclerosis** (abnormal hardening) are used.

Cartilage

Cartilage (kahr-tih-lihdj) is another form of connective tissue that is more elastic than bone. The elasticity of cartilage makes it useful in the more flexible portions of the skeleton. **Articular** (ahr-tihck-yoo-lahr) cartilage, a specific type of cartilage, covers the joint surfaces of bone. The **meniscus** (meh-nihs-kuhs) is a curved fibrous cartilage found in some joints, such as the canine stifle, that serves as a cushion to forces applied to the joint. The combining form for cartilage is **chondr/o.**

Joints

Joints or **articulations** (ahr-tihck-yoo-lā-shuhns) are connections between bones. Articulate means to join in a way that allows motion between the parts. The combining form for joint is **arthr/o.** There are different types of joints based on their function and degree of movement.

Joints are classified based on their degree of movement (Figure 3–2). **Synarthroses** (sihn-ahrth-rō-cēz) allow for no movement, **amphiarthroses** (ahm-fih-ahrth-rō-cēz) allow slight movement, and **diarthroses** (dē-ahrth-rō-cēz) allow free movement.

Synarthroses are immovable joints usually united with fibrous connective tissue. An example of a synarthrosis is a suture. A **suture** (soo-chuhr) is a jagged line where bones join and form a nonmovable joint. Sutures are typically found in the skull. A **fontanelle**

TABLE 3–1
Terminology Applied to Bone

Types of Bone	
cortical bone (kōr-tih-kahl)	hard, dense, strong bone that forms the outer layer of bone; also called compact bone **cortex** = bark or shell in Latin
cancellous bone (kahn-sehl-uhs)	lighter, less strong bone that is found in the ends and inner portions of long bones; also called spongy bone **cancellous** = latticework in Latin
Bone Anatomy Terms	
epiphysis (eh-pihf-ih-sihs)	wide end of bone, which is covered with articular cartilage and is composed of cancellous bone **proximal epiphysis** = located nearest midline of the body **distal epiphysis** = located farthest away from midline of the body
diaphysis (dī-ahf-ih-sihs)	shaft of a long bone that is mainly composed of compact bone
physis (fī-sihs)	segment of bone that involves growth of the bone; also called the **growth plate** or **epiphyseal cartilage**
metaphysis (meh-tahf-ih-sihs)	wider part of long bone shaft located adjacent to the physis; in adult animals is considered part of the epiphysis
periosteum (pehr-ē-ohs-tē-uhm)	tough, fibrous tissue the forms the outer covering of bone
endosteum (ehn-dohs-tē-uhm)	tough, fibrous tissue that forms the lining of the medullary cavity
Bone Classification	
long bones	consist of shaft, two ends, and a marrow cavity (i.e., femur)
short bones	cube shaped with no marrow cavity (i.e., carpal bones)
flat bones	thin, flat bones (i.e., pelvis)
pneumatic bones	sinus containing (i.e., frontal bone)
irregular bones	unpaired bones (i.e., vertebrae)
sesamoid bones	small bones attached to tendons (i.e., patella)

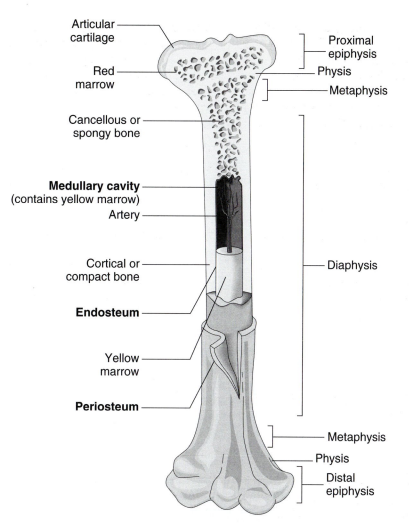

Articular cartilage

Red marrow

Cancellous or spongy bone

Medullary cavity
(contains yellow marrow)

Artery

Cortical or compact bone

Endosteum

Yellow marrow

Periosteum

Proximal epiphysis

Physis

Metaphysis

Diaphysis

Metaphysis

Physis

Distal epiphysis

FIGURE 3–1 Anatomy of a long bone

(fohn-tah-nehl) is a soft spot remaining at the junction of sutures that usually closes after birth.

Amphiarthroses are semimovable joints. An example of an amphiarthorsis is a symphysis. A **symphysis** (sihm-fih-sihs) is a joint where two bones join and are held firmly together so that they function as one bone. Another term for symphysis is **cartilaginous joint.** The halves of the mandible fuse at a symphysis to form one bone. This fusion is the **mandibular symphysis.** The halves of the pelvis also fuse at a symphysis, which is called the **pubic symphysis.**

Diarthroses are freely movable joints. Examples of diarthroses are synovial joints. **Synovial** (sih-no-vē-ahl) **joints** are further classified as ball and socket joints (also called **enarthrosis** (ehn-ahr-thrō-sihs) or spheroid joints, arthrodial (ahr-thrō-dē-ahl) or condyloid (kohn-dih-loyd) joints, trochoid (trō-koyd) or pivot (pih-voht) joints, ginglymus (jihn-glih-muhs) or hinge joints, and gliding joints. **Ball and socket joints** allow for a wide range of motion in many directions, such as the hip and shoulder joints. **Arthrodial** or **condyloid**

joints are joints with oval projections that fit into a socket, like the carpal joints (where radius meets carpus). **Trochoid joints** include pulley-shaped joints like the connection between the atlas to the axis. **Hinge joints** allow motion in one plane or direction, such as canine stifle and elbow joints. **Gliding joints** move or glide over each other as in the radioulnar joint or the articulating process between successive vertebrae. Primates have an additional joint called the saddle joint. The only **saddle joint** is located in the carpometacarpal joint of the thumb. This saddle joint allows primates to flex, extend, abduct, adduct, and circumduct the thumb.

Ligaments and Tendons

A **ligament** (lihg-ah-mehnt) is a band of fibrous connective tissue that connects one bone to another bone. **Ligament/o** is the combining form for ligament. A ligament is different from a tendon. A **tendon** (tehn-dohn) is a band of fibrous connective tissue that connects

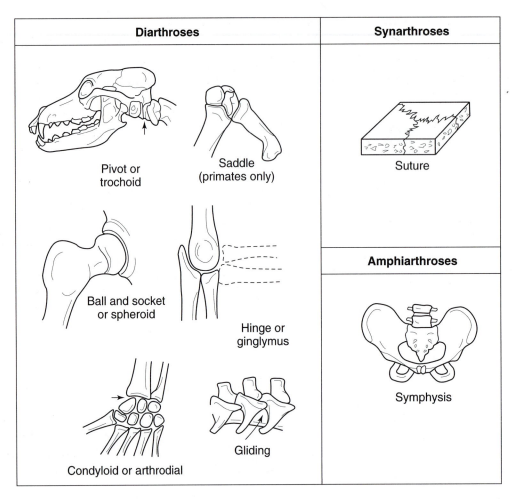

Diarthroses	Synarthroses
Pivot or trochoid	Suture
Saddle (primates only)	
Ball and socket or spheroid	**Amphiarthroses**
Hinge or ginglymus	
Condyloid or arthrodial	
Gliding	Symphysis

FIGURE 3–2 Types of joints

muscle to bone. The combining forms for tendon are **ten/o, tend/o,** and **tendin/o.**

> A way to remember that a tendon connects a muscle to bone is that both tendon and muscle have the same number of letters. Or that your Achilles tendon attaches your calf muscle to a bone.

Bursa

A **bursa** (bər-sah) is a fibrous sac that acts as a cushion to ease movement in areas of friction. Within the shoulder joint is a bursa where a tendon passes over bone. The combining form for bursa is **burs/o.** More than one bursa are **bursae** (bər-sē).

Synovial Membrane and Fluid

Bursae and synovial joints have an inner lining called the **synovial** (sih-nō-vē-ahl) **membrane.** The synovial membrane secretes synovial fluid, which acts as a lubricant to make joint movement smooth. **Synovi/o** is the combining form for synovial membrane and synovial fluid.

BONING UP

The skeleton is descriptively divided into two parts: the axial skeleton and the appendicular skeleton. The **axial** (ahcks-ē-ahl) **skeleton** is the framework of the body that includes the skull, hyoid bones, vertebral column, ribs, and sternum. The **appendicular** (ahp-ehn-dihck-yoo-lahr) **skeleton** is the framework of the body that consists of the extremities, shoulder, and pelvic girdle. Append means to add or hang, so think of the appendages or extremities as structures that hang from the axial skeleton.

The Axial Skeleton

TAKE IT FROM THE TOP

The **cranium** (krā-nē-uhm) is the portion of the skull that encloses the brain. The combining form **crani/o** means skull. The cranium consists of the following bones (Figure 3–3):

frontal (frohn-tahl) = forms the roof of the cranial cavity or "front" or cranial portion of the skull

parietal (pah-rī-ih-tahl) = paired bones that form the roof of the caudal cranial cavity

occipital (ohck-sihp-ih-tahl) = forms the caudal aspect of the cranial cavity where the foramen magnum or opening for the spinal cord is located. **Foramen** (fō-rā-mehn) is an opening in bone through which tissue passes. **Magnum** (māg-nuhm) means large.

temporal (tehm-pohr-al) = paired bones that form the sides and base of the cranium

sphenoid (sfeh-noyd) = paired bones that form part of the base of the skull and parts of the floor and sides of the bony eye socket

ethmoid (ehth-moyd) = forms the rostral part of the cranial cavity

incisive (ihn-sīs-ihv) = forms the rostral part of the hard palate and lower edge of nares

pterygoid (tahr-ih-goyd) = forms the lateral wall of the nasopharynx

In addition to bones the skull also has air- or fluid-filled spaces. These air- or fluid-filled spaces are called **sinuses** (sīn-uhs-ehz).

---

brachycephalic (brā-kē-ceh-fahl-ihck) dogs have short, wide heads like pugs and Pekingese

dolichocephalic (dō-lih-kō-ceh-fahl-ihck) dogs have narrow, long heads like collies and greyhounds

---

LET'S FACE IT

The bones of the face consist of the following:

zygomatic (zī-gō-mah-tihck) = projections from the temporal and frontal bones to form the cheekbone

maxilla (mahck-sih-lah) = forms the upper jaw

palatine (pahl-ah-tihn) = forms part of the hard palate

lacrimal (lahck-rih-mahl) = forms medial part of the orbit

incisive (ihn-sī-sihv) = forms the rostral part of the hard palate and lower edge of nares

nasal (nā-sahl) = forms the bridge of the nose

vomer (vō-mər) = forms the base of the nasal septum. The **nasal septum** (nay-sal sep-tum) is the cartilaginous structure that divides the two nasal cavities.

mandible (mahn-dih-buhl) = forms the lower jaw

hyoid (hī-oyd) = bone suspended between the mandible and the laryngopharynx

BACK TO BASICS

The **vertebral** (vər-teh-brahl) **column** (also called the **spinal column** or **backbone**) supports the head and body and provides protection for the spinal cord. The vertebral column is comprised of individual bones called **vertebra** (vər-teh-brah). The combining forms for vertebra are **spondyl/o** and **vertebr/o.** More than one vertebra are called **vertebrae** (vər-teh-brā).

Vertebrae are divided into parts, and the parts may vary depending upon the location of the vertebra and

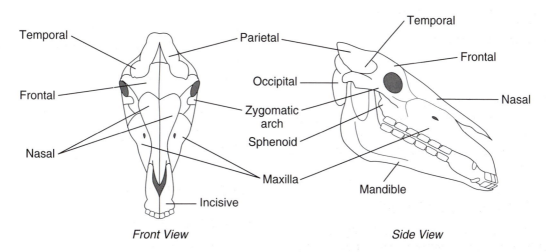

Front View   *Side View*

FIGURE 3–3 Selected bones of the skull and face

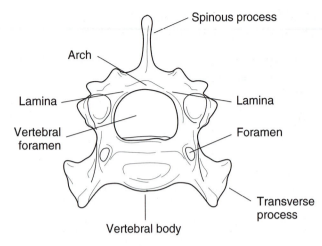

FIGURE 3–4 Parts of a vertebra

its function (Figure 3–4). The **body** is the solid portion ventral to the spinal cord. The **arch** is the dorsal part of the vertebra that surrounds the spinal cord. The **lamina** (lahm-ih-nah) is the left or right half dorsal part of the arch. Processes project from the vertebrae. The term process means projection. A **spinous process** is a single projection from the dorsal part of the arch. **Transverse processes** project caudolaterally from the right and left sides of a vertebra. **Articular processes** are paired cranial and caudal projections located on the dorsum of the arch.

Foramen (fō-rā-mehn) is a term meaning opening. The opening in the middle of the vertebra through which the spinal cord passes is the **vertebral foramen.**

The vertebrae are separated and cushioned from each other by cartilage discs called **intervertebral discs.**

Vertebrae are organized and named by region. The regions are named as follows (Figure 3–5):

Cervical	*Thoracic*	*Lumbar*	*Sacral*	*Coccygeal*
(sehr-vih kahl)	(thō-rahs -ihck)	(luhm- bahr)	(sā-krahl)	(kohck-sihd-jē-ahl) (also called **caudal**)
Neck area "C"	Chest area "T"	Loin area "L"	Sacrum area "S"	Tail area "Cy" or "Cd"

In addition, the first two vertebrae have individual names. C1 or cervical vertebra one is called the **atlas** and C2 or cervical vertebra two is called the **axis** (remember that they follow alphabetical order).

> 💡 **The vertebral formulas for different species are listed below:**
>
> dogs and cats: C-7, T-13, L-7, S-3, Cy-6–23
> equine: C-7, T-18, L-6 (or L-5 in some Arabians), S-5, Cy-15–21
> bovine: C-7, T-13, L-6, S-5, Cy-18–20
> pigs: C-7, T-14–15, L-6–7, S-4, Cy-20–23
> sheep and goats: C-7, T-13, L-6-7, S-4, Cy-16–18
> chicken: C-14, T-7, LS-14, Cy-6 (lumbar and sacral vertebrae are fused)

STICK TO YOUR RIBS

Ribs are paired bones that attach to thoracic vertebrae. The combining form for rib is **cost/o.** Ribs are sometimes called **costals.**

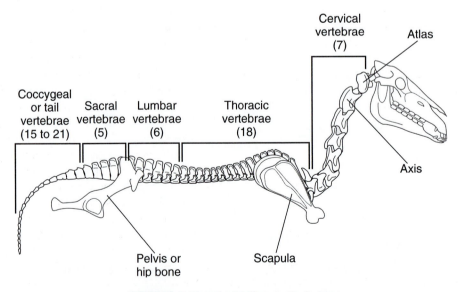

FIGURE 3–5 Vertebral column of a horse

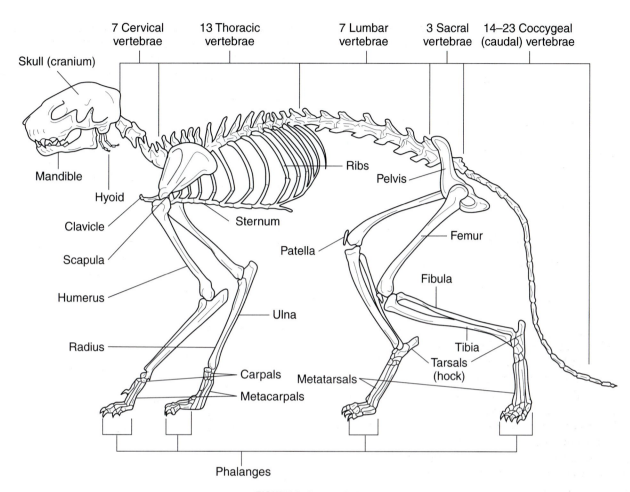

7 Cervical vertebrae   13 Thoracic vertebrae   7 Lumbar vertebrae   3 Sacral vertebrae   14–23 Coccygeal (caudal) vertebrae

Skull (cranium)

Mandible

Hyoid

Clavicle

Scapula

Humerus

Radius

Ribs

Pelvis

Sternum

Patella

Femur

Fibula

Ulna

Tibia

Tarsals (hock)

Carpals

Metatarsals

Metacarpals

Phalanges

FIGURE 3–6 Cat skeleton

The **sternum** (stər-nuhm) or breastbone forms the midline ventral portion of the rib cage. The sternum is divided into three parts: the manubrium, body, and xiphoid process. The **manubrium** (mah-nū-brē-uhm) is the cranial portion of the sternum. The **body** of the sternum is the middle portion. The caudal portion of the sternum is known as the **xiphoid** (zī-foyd) **process.**

The ribs, sternum, and thoracic vertebrae make up the boundaries of the thoracic cavity. The **thoracic cavity** or **rib cage** serves to protect the heart and lungs (Figure 3–6).

The Appendicular Skeleton

FROM THE FRONT

The bones of the front limb from proximal to distal consist of the scapula, clavicle, humerus, radius, ulna, carpus, metacarpals, and phalanges.

The **scapula** (skahp-yoo-lah) or **shoulder blade** is a large triangular shaped bone on the side of the thorax. The **clavicle** (klahv-ih-kuhl) or **collarbone** is a slender bone that connects the sternum to the scapula. Some animal species only have a **vestigial** (vehs-tihj-ahl) or **rudimentary** clavicle, while some animal species such as swine, ruminants, and equine do not have a clavicle.

The **humerus** (hū-mər-uhs) is the long bone of the proximal front limb. The humerus is sometimes referred to as the **brachium.** The radius and ulna are the two bones of the forearm or distal front limb. This region is called the **antebrachium. Ante-** means before. The **radius** (rā-dē-uhs) is the cranial bone of the front limb, while the **ulna** (uhl-nah) is the caudal bone of the front limb. The ulna has a proximal projection called the **olecranon** (ō-lehck-rah-nohn) that forms the point of the elbow. Some species have a fused radius and ulna.

The **carpal** (kahr-pahl) **bones** are irregularly shaped bones in the area known as the **wrist** in people. In small animals this joint is called the **carpus,** while in large animals this joint is called the **knee.** The **metacarpals** (meht-ah-kahr-pahlz) are bones found distal to the carpus (**meta-** = beyond). The metacarpals are identified by numbers from medial to lateral. In some species like the horse certain metacarpals do not articulate with the phalanges. In the horse, metacarpals (and metatarsals)

II and IV do not articulate with the phalanges and are commonly called **splint bones.** Splint bones are attached by an **interosseous** (ihn-tər-ohs-ē-uhs) ligament to the large third metacarpal (or metatarsal) bone, which is commonly called the **cannon bone** (Figure 3–7).

The **phalanges** (fā-lahn-jēz) are the bones of the digit. One bone of the digit is called a **phalanx.** Phalanges are numbered from proximal to distal. Most digits have three phalanges, but the most medial phalanx (digit I) has only two phalanges. **Digits** are the bones that relate to the human finger and vary in number in animals (Figure 3–8). Digit I of dogs is commonly called the **dewclaw,** and may be removed shortly after birth. **Ungulates** (uhng-yoo-lātz) or animals with hooves also have digits that are numbered in the same fashion. Animals with a cloven hoof or split hoof have digits III and IV, while digits II and V are vestigial. The vestigial digits of cloven hoofed animals are also called dewclaws.

Cloven hoofed animals, like ruminants and swine, have three phalanges in their digits with the distalmost phalanx (P3) encased in the hoof. Equine species have one digit (digit III). Within that digit are three phalanges. In livestock, the joints between the phalanges or between the phalanges and other bones have common names. The joint between metacarpal (metatarsal) III and the digit is the **fetlock joint.** The joint between PI and P2 is known as the **pastern joint.** The joint between P2 and P3 is known as the **coffin joint.** The phalangeal bones also have common names in livestock. P1 is the **long pastern bone,** P2 is the **short pastern bone,** and P3 is the **coffin bone.** (See Chapter 4 for illustration.)

Phalanx 3 or P3 may also called a **claw** or **nail** in nonhooved animals. The combining form for claw or nail is **onych/o.** In cats a common surgical procedure is removal of the claw. A common term for that procedure is a **declaw;** a medical term for that is **onychectomy** (ohn-ih-kehk-tō-mē).

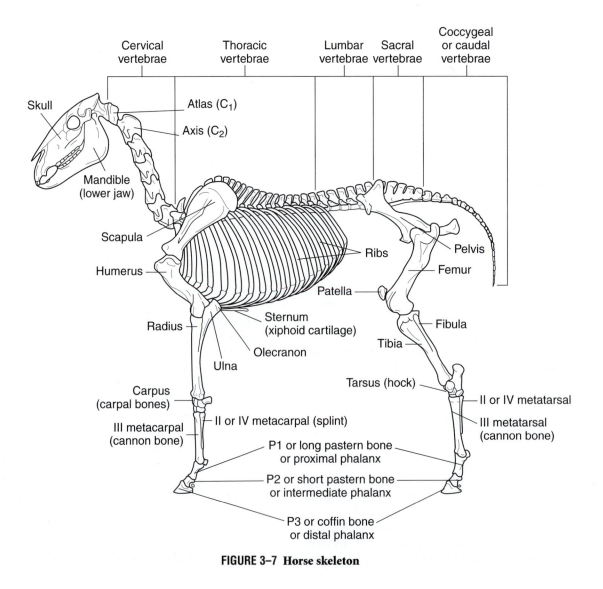

FIGURE 3–7 Horse skeleton

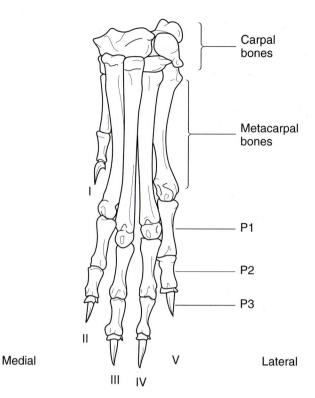

FIGURE 3–8 Digits versus phalanges. The digits of the hoof or paw are numbered medially to laterally. The phalanges are numbered from proximal to distal.

Sesamoid (sehs-ah-moyd) **bones** are small nodular bones embedded in a tendon or joint capsule. There are multiple sesamoid bones in animals. Some sesamoid bones will also have a common name. The **navicular bone** of horses is the common name for the sesamoid bone located inside the hoof on the palmar or plantar surface of P3.

TO THE BACK

The bones of the rear limb include the pelvis, femur, tibia, fibula, tarsals, metatarsals, and phalanges.

The **pelvis** (pehl-vihs) or **hip** consists of three pairs of bones: ilium, ischium, and pubis. The **ilium** (ihl-ē-uhm) is the largest pair that is blade shaped. The ilium articulates with the sacrum to form the **sacroiliac** (sā-krō-ihl-ē-ahck) **joint.** The **ischium** (ihs-kē-uhm) is the caudal pair of bones. The **pubis** (pehw-bihs) is the ventral pair of bones that are fused on midline by a cartilaginous joint called the **pubic symphysis** (pehw-bihck sihm-fih-sihs). The **acetabulum** (ahs-eh-tahb-yoo-luhm) is the large socket of the pelvic bone that forms where the three bones meet. The acetabulum forms the ball and socket joint with the femur (Figure 3–9).

The **femur** (fē-muhr) or **thigh bone** is the proximal long bone of the rear leg. The head of the femur ar-

 One way to remember the order of the ilium and ischium is that they follow alphabetical order from cranial to caudal. In cattle, the points of the ilium and ischium are called **hooks** and **pins.** They too follow alphabetical order from cranial to caudal.

ticulates proximally with the acetabulum. The **femoral** (fehm-ohr-ahl) **head** or head of the femur is connected to a narrow area, which is termed the **femoral neck.** Other structures found on the femur are the **trochanters** (trō-kahn-tehr), which means large, flat, broad projection on a bone and **condyles** (kohn-dīl-uhlz), which means rounded projection.

Sound Alikes: Ileum and ilium are both pronounced the same; yet they have different meanings. Ileum is the distal or aboral part of small intestine, while ilium is part of the hip bone. One way to keep these spellings straight is to remember ileum has an "e" in it like eating and enter/o, which involve the digestive tract. Ilium and hip both have an "i" in them and are part of the hip bone.

The **patella** (pah-tehl-ah) is a large sesamoid bone in the rear limb. In people it is called the kneecap and the joint is known as the knee. The knee is not a good term to use to describe the joint between the femur and tibia in animals, since in large animals the knee is commonly used to describe the carpus. The joint that houses the patella is termed the **stifle** (stī-fuhl) **joint** in animals. Another sesamoid bone in the rear limb of some animals is the **popliteal** (pohp-liht-ē-ahl). The popliteal sesamoid is located on the caudal surface of the stifle.

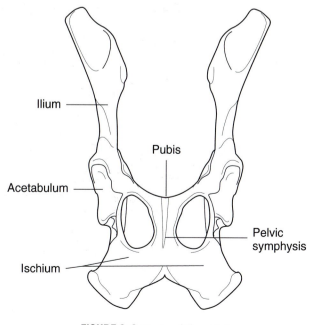

FIGURE 3–9 Parts of the pelvis

Knee Deep in Trouble: The term knee can be a confusing term in veterinary medicine. Lay people may use the term knee to refer to the stifle joint of dogs and cats; however, the term knee in large animals is used to describe the carpal joint. Most people in veterinary medicine use the term stifle for the joint located in the rear leg between the femur and tibia/fibula and reserve the term knee for the carpal joint in large animals.

The tibia and fibula are the distal long bones of the rear limb. The **tibia** (tihb-ē-ah) is the larger and more weight bearing bone of the two. The **fibula** (fihb-yoo-lah) is a long, slender bone. Some animals do not have a fibula that extends to the distal end, while others may have the tibia and fibula fused. The area of the rear limb between the stifle and hock is termed the **crus** (kruhs).

The tarsal bones are irregularly shaped bones found in the area known as the ankle in people. In small animals this joint is called the **tarsus** (tahr-suhs), while in large animals it is called the **hock** (hohck). One of the tarsal bones is the **talus** (tahl-uhs). The talus is the shorter, medial tarsal bone located in the proximal row of tarsal bones. Talus and tarsus both begin with "t" and sound similar, which makes associating them together easy. The long, lateral tarsal bone located in the proximal row of tarsal bones is the **calcaneus** (kahl-kā-nē-uhs). Calcaneus and carpus both begin with "c" and sound similar, but they are not located in the same area. Calcaneus was named because it reminded someone of a piece of chalk, which comes from the same source as calcium. The metatarsals are bones found distal to the tarsus (**meta-** = beyond). The metatarsals are numbered and have similar terminology as the metacarpals.

The phalanges are the bones of the digit (both front and rear limbs). The terminology used for the phalanges in the front limb is also used for the rear limb.

STRUCTURAL SUPPORT

Bones are not structurally smooth and many times have bumps or grooves or ridges. All of these structures have a medical term to describe them. By knowing what these descriptive terms mean, learning bone parts may be easier.

aperture (ahp-ər-chər) = opening
canal (kahn-ahl) = tunnel
condyle (kohn-dîl) = rounded projection (that articulates with another bone)
crest (krehst) = high projection or border projection = ridge = **crista** (kris-tah)
dens (dehnz) = toothlike structure
eminence (ehm-ih-nehns) = surface projection
facet (fahs-eht) = smooth area

foramen (fō-rā-mehn) = hole
fossa (fohs-ah) = trench or hollow depressed area
fovea (fō-vē-ah) = small pit
head = major protrusion
lamina (lahm-ih-nah) = thin, flat plate
line = low projection or ridge
malleolus (mah-lē-ō-luhs) = rounded projection (distal end of tibia and fibula)
meatus (mē-ā-tuhs) = passage or opening
process (proh-sehs) = projection
protuberance (prō-too-bər-ahns) = projecting part
ramus (rā-muhs) = branch or smaller structure given off by a larger structure
sinus (sīn-uhs) = space or cavity
spine (spīn) = sharp projection
sulcus (suhl-kuhs) = groove
suture (soo-chuhr) = seam
trochanter (trō-kahn-tehr) = broad, flat projection (on femur)
trochlea (trōck-lē-ah) = pulley-shaped structure in which other structures pass or articulate
tubercle (too-behr-kuhl) = small, rounded surface projection
tuberosity (too-beh-rohs-ih-tē) = projecting part

▶ **Test Me: Skeletal System** ◀

Diagnostic procedures performed on the skeletal system include
- **arthrocentesis** (ar-throh-sen-tee-sis) is the surgical puncture of a joint to remove fluid for analysis
- **arthrography** (ar-throg-rah-fee) is the injection of a joint with contrast material for radiographic examination
- **arthroscopy** (ar-thros-koh-pee) is the visual examination of the joint using a fiberoptic scope (when used in the joint is termed an arthroscope)
- **radiology** (rā-dē-ohl-ō-jē) = study of internal body structures after exposure to ionizing radiation; used to detect fractures and diseases of bones (Figure 3–10).

▶ **The Path of Destruction: Skeletal System** ◀

Pathologic conditions of the skeletal system include
- **arthropathy** (ahr-throhp-ah-thē) = joint disease
- **arthralgia** (ahr-thrahl-jē-ah) = joint pain
- **arthrodynia** (ahr-thrō-dihn-ē-ah) = joint pain
- **ankylosis** (ahng-kih-lō-sihs) = loss of joint motility due to disease, injury, or surgery
- **luxation** (luhck-sā-shuhn) = dislocation or displacement of a bone from its joint
- **subluxation** (suhb-luhck-sā-shuhn) = partial dislocation or displacement of a bone from its joint (see Figure 3–11).

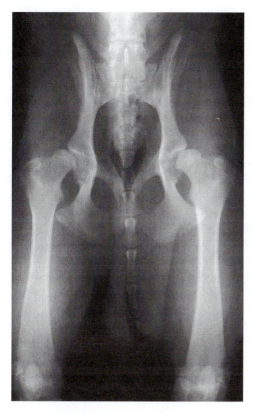

FIGURE 3–10 Radiograph of the canine hip. *Source:* Lodi Veterinary Hospital, Lodi, WI.

- **chondropathy** (kohn-drohp-ah-thē) = cartilage disease
- **chondromalacia** (kohn-drō-mah-lā-shē-ah) = abnormal cartilage softening
- **bursitis** (bər-sī-tihs) = inflammation of the bursa
- **arthritis** (ahr-thrī-tihs) = inflammatory condition of joint(s)
- **synovitis** (sihn-ō-vī-tihs) = inflammation of the synovial membrane of joints
- **osteoarthritis** (ohs-tē-ō-ahr-thrī-tihs) = degenerative joint disease commonly associated with aging or wear and tear on the joints; also called **degenerative joint disease** or **DJD**
- **rheumatoid arthritis** (roo-mah-toyd ahr-thrī-tihs) = autoimmune disorder of the connective tissues and joints
- **gouty arthritis** (gow-tē ahr-thrī-tihs) or **gout** = joint inflammation associated with the formation of uric acid crystals in the joint (seen more commonly in birds)
- **spondylosis** (spohn-dih-lō-sihs) = any degenerative disorder of the vertebrae
- **spondylosis deformans** (spohn-dih-lō-sihs dē-fōrm-ahnz) = chronic degeneration of the articular processes and the development of bony outgrowths around the ventral edge of the vertebrae
- **spondylitis** (spohn-dih-lī-tihs) = inflammation of the vertebrae

- **intervertebral disc disease** (ihn-tər-vər-tē-brahl dihsk dih-zēz) = rupture or protrusion of the cushioning disc found between the vertebrae that results in pressure on the spinal cord and/or spinal nerve roots; also called **herniated disc, ruptured disc,** or **IVDD**
- **lordosis** (lōr-dō-sihs) = position in which the vertebral column is abnormally curved ventrally, seen in cats in heat; commonly called **swayback**
- **discospondylitis** (dihs-kō-spohn-dih-lī-tihs) = inflammation of the intervertebral disc and vertebrae
- **periostitis** (pehr-ē-ohs-tī-tihs) = inflammation of the fibrous tissue that forms the outermost covering of bone
- **ostealgia** (ohs-tē-ahl-jē-ah) = bone pain
- **osteitis** (ohs-tē-ī-tihs) = inflammation of bone
- **osteomyelitis** (ohs-tē-ō-mī-eh-lī-tihs) = inflammation of bone and bone marrow
- **exostosis** (ehck-sohs-tō-sihs) = benign growth on the bone surface
- **osteonecrosis** (ohs-tē-ō-neh-krō-sihs) = death of bone tissue
- **sequestrum** (sē-kwehs-truhm) = piece of dead bone that is partially or fully detached from the adjacent healthy bone
- **osteomalacia** (ohs-tē-ō-mah-lā-shē-ah) = abnormal softening of bone
- **osteosclerosis** (ohs-tē-ō-skleh-rō-sihs) = abnormal hardening of bone
- **myeloma** (mī-eh-lō-mah) = tumor composed of cells derived from hematopoietic tissues of bone marrow
- **osteoporosis** (ohs-tē-ō-pō-rō-sihs) = abnormal condition of marked loss of bone density and an increase in bone porosity
- **osteochondrosis** (ohs-tē-ō-kohn-drō-sihs) = degeneration or necrosis of bone and cartilage followed by regeneration or recalcification

FIGURE 3–11 Radiograph of atlas axis subluxation in a dog. *Source:* Photo by Anne E. Chauvet, DVM, Diplomate ACVIM–Neurology, University of Wisconsin School of Veterinary Medicine.

- **osteochondrosis dissecans** (ohs-tē-ō-kohn-drō-sihs dehs-ih-kahns) = degeneration or necrosis of bone and cartilage followed by regeneration or recalcification with dissecting flap of articular cartilage and some inflammatory joint changes. Detached pieces of articular cartilage are referred to as **joint mice** or **osteophytes** (ohs-tē-ō-fītz)
- **Legg-Calvé-Perthes disease** (lehg-cah-veh-pər-thehz dih-zēz) = idiopathic necrosis of the femoral head and neck of small breed dogs; also called avascular necrosis of the femoral head and neck
- **hip dysplasia** (dihs-plā-zē-ah) = abnormal development of the pelvic joint causing the head of the femur and the acetabulum not to be aligned properly; most commonly seen in large breed dogs

Fracture Terminology
See Figure 3–12.

- **fracture** (frahk-suhr) = broken bone
- **crepitation** (krehp-ih-tā-shuhn) = cracking sensation that is felt and heard when broken bones move together

- **manipulation** (mahn-ihp-yoo-lā-shuhn) = the attempted realignment of the bone involved in a fracture or dislocation; also known as **reduction**
- **immobilization** (ihm-mō-bihl-ih-zā-shuhn) = the act of holding, suturing, or fastening a bone in a fixed position, usually with a bandage or cast
- **callus** (kahl-uhs) = bulging deposit around the area of a bone fracture that may eventually become bone
- **closed fracture** = broken bone in which there is no open wound in the skin; also known as a **simple fracture**
- **open fracture** = broken bone in which there is an open wound in the skin; also known as a **compound fracture**
- **comminuted** (kohm-ih-noot-ehd) **fracture** = broken bone that is splintered or crushed into multiple pieces
- **compression** (kohm-prehs-shuhn) **fracture** = broken bone produced when the bones are pressed together

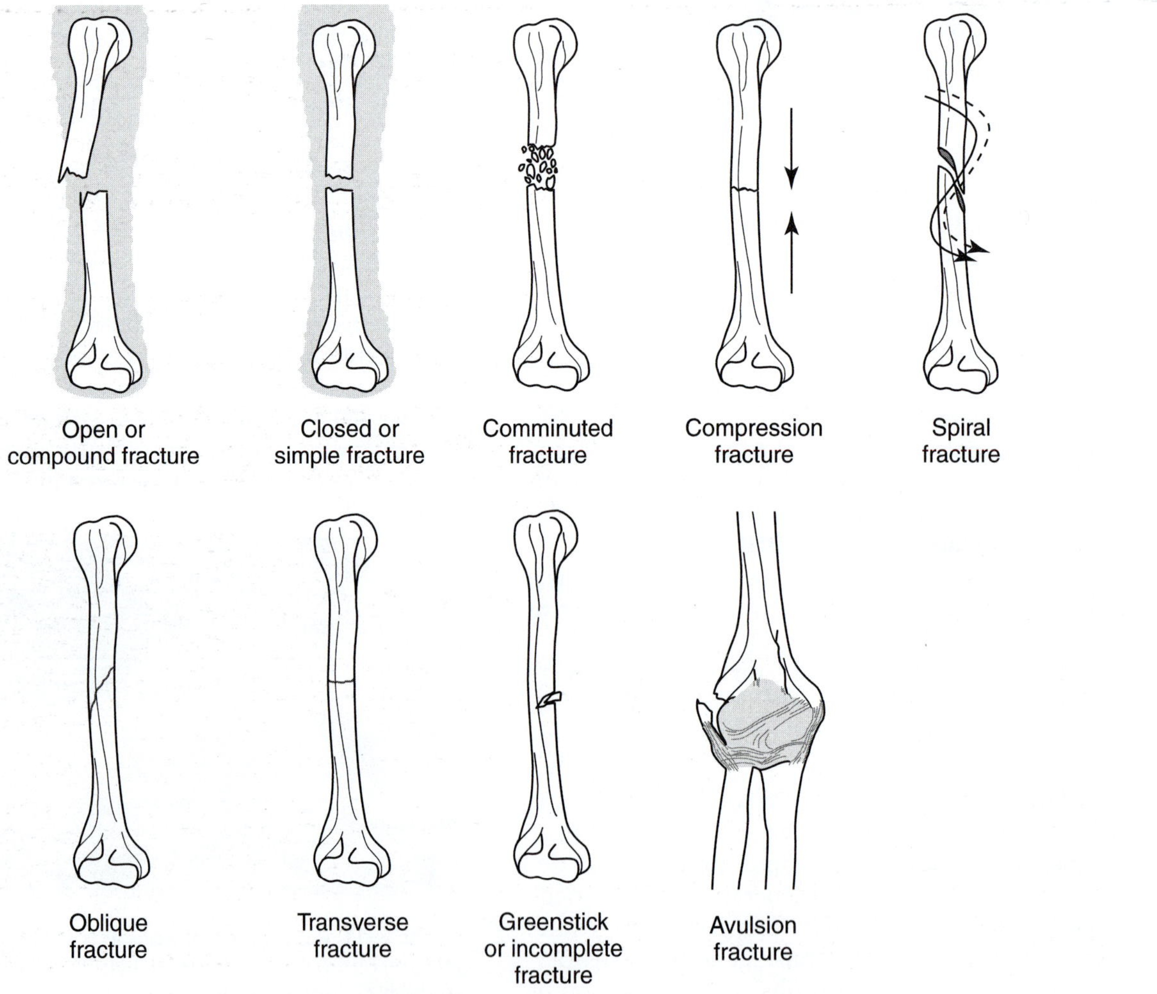

   Comminuted Compression  Spiral

Open or compound fracture / Closed or simple fracture / Comminuted fracture / Compression fracture / Spiral fracture

Oblique fracture / Transverse fracture / Greenstick or incomplete fracture / Avulsion fracture

FIGURE 3–12 Fracture types

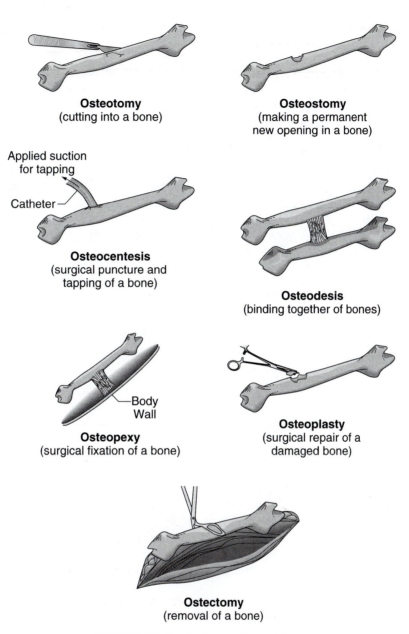

Osteotomy
(cutting into a bone)

Osteostomy
(making a permanent new opening in a bone)

Applied suction for tapping

Catheter

Osteocentesis
(surgical puncture and tapping of a bone)

Osteodesis
(binding together of bones)

Body Wall

Osteopexy
(surgical fixation of a bone)

Osteoplasty
(surgical repair of a damaged bone)

Ostectomy
(removal of a bone)

FIGURE 3–13 Surgical procedures of bone

- **spiral** (spī-rahl) **fracture** = broken bone in which the bone is twisted apart or spiraled apart
- **transverse** (trahnz-vərs) **fracture** = broken bone that is broken at right angles to its axis or straight across the bone
- **oblique** (ō-blēck) **fracture** = broken bone that has an angular break diagonal to the long axis
- **greenstick fracture** = bone that is broken only on one side and the other side is bent; also called **incomplete fracture**
- **physeal** (fī-sē-ahl) **fracture** = bone that is broken at the epiphyseal line or growth plate; these fractures are further categorized as **Salter-Harris I-V fractures**

- **avulsion** (ā-vuhl-shuhn) **fracture** = broken bone in which the site of muscle, tendon, or ligament insertion is detached due to a forceful pull

In or Out

Bones may abnormally bend in or bend out. Medical terms for this condition in bones are **varus** (var-uhs) and **valgus** (vahl-guhs). Valgus means bend out (think bend laterally; both valgus and lateral have an "l") while varus means bend in.

▶ Procede with Caution: Skeletal System ◀

Procedures performed on the skeletal system (see Figure 3–13) include:

- **arthrodesis** (ahr-thrō-dē-sihs) = fusion of a joint or the spinal vertebrae by surgical means
- **chemonucleolysis** (kē-mō-nū-klē-ō-lī-sihs) = process of dissolving part of the center of an intervertebral disc by injecting a foreign substance
- **laminectomy** (lahm-ih-nehck-tō-mē) = surgical removal of the dorsal arch of the vertebrae
- **ostectomy** (ohs-tehck-tō-mē) = surgical removal of bone
- **osteotomy** (ohs-tē-oht-ō-mē) = to surgically incise bone or sectioning of bone
- **osteostomy** (ohs-tē-ohs- tō-mē) = making a permanent new opening in bone
- **osteocentesis** (ohs-tē-ō-sehn-tē-sihs) = surgical puncture of a bone
- **osteodesis** (ohs-tē-ō- dē-sihs) = fusion of bones
- **osteopexy** (ohs-tē-ō-pehck-sē) = surgical fixation of a bone to the body wall
- **osteoplasty** (ohs-tē-ō-plahs-tē) = surgical repair of bone
- **craniotomy** (krā-nē-oht-ō-mē) = surgical incision or opening into the skull
- **amputation** (ahmp-yoo-tā-shuhn) = removal of all or part of a body part
- **trephination** (trē-fīn-nā-shuhn) = process of cutting a hole into a bone using a **trephine** (trē-fīn) (circular, sawlike instrument used to remove bone or tissue)

FUNCTIONS OF THE MUSCULAR SYSTEM

Muscles are organs that contract to produce movement. Muscles make movement possible. One type of movement is **ambulation** (ahm-bū-lā-shuhn) or walking, running, or otherwise moving from one place to another. Another type of movement is contraction of organs or tissues that result in normal functioning of the body. For example, contraction of sections of the gastrointestinal tract allow food to move through the digestive system, while contraction of vessels allow for movement of fluids such as blood. Movement also results in heat generation to keep the body warm.

STRUCTURES OF THE MUSCULAR SYSTEM

Making Another Connection

Like the skeletal system, the muscular system is made up of various forms of connective tissue. The connective tissues of the muscular system are the following.

Muscle Fibers

Muscles are made up of long, slender cells called muscle fibers. Each muscle consists of a group of muscle fibers encased in a fibrous sheath. The combining form for muscle is **my/o**; the combining forms for fibrous tissue are **fibr/o** and **fibros/o**.

There are three types of muscle cells based on their appearance and function. The three types are skeletal, smooth, and cardiac (Table 3–2 and Figure 3–14).

Fascia

Fascia (fahsh-ē-ah) is a sheet of fibrous connective tissue that covers, supports, and separates muscles. The combining forms for fascia are **fasci/o** and **fasc/i**.

TABLE 3–2
Muscle Types

Muscle Type	Other Name	Microscopic Appearance	Function
skeletal	**striated** (strī-āt-ehd) **voluntary**	long, cylindric multinucleated cells with dark and light bands to create a striated or striped look	attach bones to the body and make motion possible
smooth	**nonstriated unstriated involuntary visceral**	spindle shaped without stripes or striations	produce relatively slow contractions to allow unconscious functioning of internal organs
cardiac	**myocardium** (mī-ō-kahr-dē-uhm)	elongated branched cells that lie parallel to each other and have dark and light bands	involuntary contraction of heart muscle

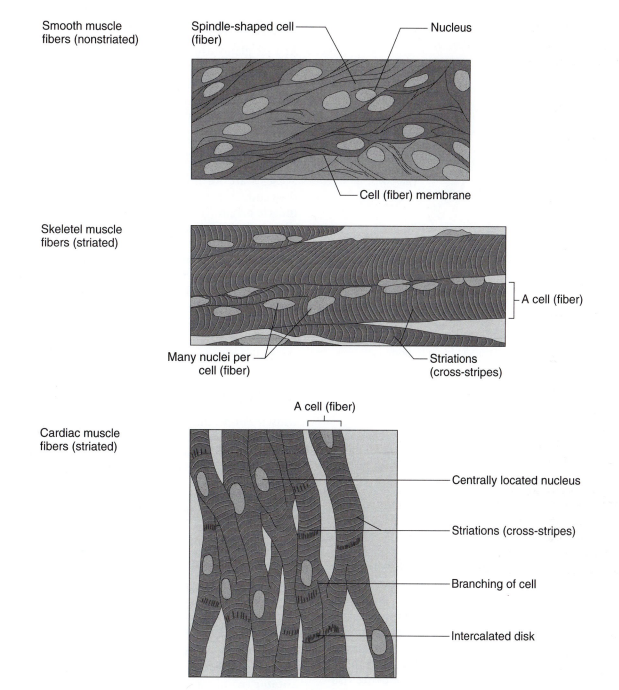

Smooth muscle fibers (nonstriated)

Spindle-shaped cell (fiber)

Nucleus

Cell (fiber) membrane

Skeletel muscle fibers (striated)

A cell (fiber)

Many nuclei per cell (fiber)

Striations (cross-stripes)

A cell (fiber)

Cardiac muscle fibers (striated)

Centrally located nucleus

Striations (cross-stripes)

Branching of cell

Intercalated disk

FIGURE 3–14 Muscle types

Tendons

A **tendon** (tehn-dohn) is a narrow band of connective tissue that attaches muscle to bone. The combining forms for tendon are **tend/o, tendin/o,** and **ten/o.** Remember to make the distinction between a tendon and a ligament (Figure 3–15).

Technically, tendons connect muscles to bones or other structures. One example of this is the linea alba. Linea alba means white line in Latin. The **linea alba** is a fibrous band of connective tissue on the ventral abdominal wall that is the center attachment of the abdominal muscles.

Aponeurosis

An **aponeurosis** (ahp-ō-nū-rō-sihs) is a fibrous sheet that gives attachment to muscular fibers and serves as a means of origin or insertion of a flat muscle. The combining form for aponeurosis is **aponeur/o** and the plural form of aponeurosis is **aponeuroses.**

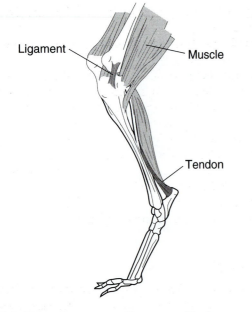

Ligament

Muscle

Tendon

FIGURE 3–15 Tendons versus ligaments. Tendons connect muscle to bone; ligaments connect bone to bone

SHOW SOME MUSCLE

One of the functions of muscle is to allow movement. The combining form **kinesi/o** and the suffix **-kinesis** mean movement. **Kinesiology** (kih-nē-sē-ohl-ō-jē) is the study of movement.

Some muscles of the body are arranged in pairs that work opposite or against each other. One muscle pair may produce movement in one direction, while another muscle pair produces movement in the opposite direction. Muscles that work against or opposite each other are termed **antagonistic** (ahn-tā-gohn-ihs-tihck). **Anti-** means against; **agon** means struggle. Other muscles of the body are arranged to work with another muscle. **Synergists** (sihn-ər-gihsts) are muscles that contract at the same time as another muscle to help movement or support movement. **Syn-** means together; **erg** means work. Antagonistic muscles work by producing contraction of one pair of muscles while the other pair relaxes. **Contraction** (kohn-trahck-shuhn) means tightening. **Relaxation** (rē-lahk-sā-shuhn) means lessening of tension. During contraction the muscle becomes shorter and thicker. During relaxation the muscle returns to its original shape. Muscles are signaled to contract or relax by nerve impulses. A **neuromuscular** (nū-rō-muhs-kū-lahr) **junction** is the point at which nerve endings come in contact with the muscle cells.

Tonus (tō-nuhs) or muscle tone is balanced muscle tension. The combining form for tone, tension, or stretching is **ton/o.**

WHAT'S IN A NAME?

At first glance the names of muscles and the task of learning them may seem impossible. Dividing muscle names into their basic components or taking a closer look at how the names were derived may help in learning their names and functions.

In the Beginning (and Ending)

Muscles are formed by where they begin and where they end. Terms used to denote these two locations are **muscle origin** (ōr-ih-gihn) and **muscle insertion** (ihn-sihr-shuhn). Muscle origin is the place where a muscle begins or originates and is the more fixed attachment or the end of the muscle closest to midline. Muscle insertion is the place where a muscle ends or inserts and is the more movable end or portion of the muscle farthest from midline. Muscles may be named according to where they originate and end. Brachioradialis muscles are connected to the brachium (humerus) and to the radius.

Are They Moving?

Muscles move in a variety of ranges. **Range of motion** is a term used to describe the types of muscle movements. Range of motion is sometimes abbreviated ROM. Muscles may be named for the manner in which they move:

abductor (ahb-duhck-tər) = muscle that moves a part away from midline

adductor (ahd-duhck-tər) = muscle that moves a part towards midline

flexor (flehck-sər) = muscle that bends a limb at its joint or decreases the joint angle

extensor (ehcks-tehn-sər) = muscle that straightens a limb or increases the joint angle

levator (lē-vā-tər) = muscle that raises or elevates a part

depressor (dē-prehs-sər) = muscle that lowers or depresses a part

rotator (rō-tā-tər) = muscle that turns a body part on its axis

supinator (soo-pih-nā-tər) = muscle that rotates the palmar or plantar surface upward

pronator (prō-nā-tər) = muscle that rotates the palmar or plantar surface downward

Where Do They Live?

Muscles are also named for their location on the body or the organ they are near. **Pectoral muscles** are located on the chest (**pector** = chest). Muscles may also be named for their location in relation to something else. **Epaxial** (ehp-ahcks-ē-ahl) **muscles** are located above the pelvic axis (**epi-** = above, **axis** = line about where rotation can occur), **intercostal muscles** are located between the ribs (**inter-** = between, **cost/o** = rib), **infraspinatus muscles** are located beneath the spine of the scapula (**infra-** = beneath or below), and **supraspinatus muscles** are located above the spine of the scapula (**supra-** = above). Muscle names may also indicate their location within a group, such as **inferior** (below or deep), **medius** (middle), and **superior** (above). Other terms indicating depth of muscles are **externus** (outer) and **internus** (inner). **Orbicularis** are muscles surrounding another structure.

Which Way Do They Go?

Muscles may also be named according to the direction of the muscle fibers.

▶ **rectus** (rehck-tuhs) means straight. Rectus muscles align with the vertical axis of the body.
▶ **oblique** (ō-blēck) means slanted. Oblique muscles slant outward away from midline.
▶ **transverse** means crosswise. Transverse muscles form crosswise to midline.
▶ **sphincter** means tight band. Sphincter muscles are ringlike and constrict the opening of a passageway.

How Many Parts Are There?

Some muscles are named for the number of divisions they have.

▶ **biceps** (bī-sehpz) generally have two divisions; **bi-** means two
▶ **triceps** (trī-sehpz) generally have three divisions; **tri-** means three
▶ **quadriceps** (kwohd-rih-cehps) generally have four divisions; **quadri-** means four

Some muscles are not paired or divided. **Azygous** (ah-zī-guhs) means not paired (**a-** means without; **zygoto** means joined).

How Big Is It?

Muscles may also be named for their size. Muscles may either be small (**minimus**) or large (**maximus** or **vastus**), broad (**latissimus**) or narrow (**longissimus** or **gracilis**). **Major** and **minor** are terms also used to describe larger and smaller parts, respectively.

How Is It Shaping Up?

Some muscles are shaped like a familiar object and have been named accordingly.

▶ **deltoid** (dehl-toyd) muscles look like the Greek letter delta (Δ)
▶ **quadratus** (kwohd-rā-tuhs) muscles are square or four-sided in shape
▶ **rhomboideus** (rohm-boy-dē-uhs) muscles are diamond shaped (rhomboid is a four-sided figure which may have unequal adjoining sides, but have equal opposite sides)
▶ **scalenus** (skā-lehn-uhs) muscles are unequally three-sided (skalenos is Greek for uneven)
▶ **serratus** (sihr-ā-tuhs) muscles are saw-toothed (serratus is Latin for notched)
▶ **teres** (tər-ēz) muscles are cylindrical (teres is Latin for smooth and round or cylindrical)

No Rules

Sometimes muscles are just named for what they look like or how they relate to something else. **Sartorius muscle** (one muscle of the thigh area) is named because this muscle flexes and adducts the leg of a human to that position assumed by a tailor sitting cross-legged at work (sartorius means tailor). The **gemellus** is named because it is a twined muscle (gemellus means twin). The **gastrocnemius muscle** is the leg muscle that resembles the shape of the stomach (**gastr/o** means stomach, **kneme** means leg).

▶ **Test Me: Muscular System** ◀

A diagnostic procedure performed on the muscular system is
- **electromyography** (ē-lehck-trō-mī-ohg-rah-fē) = process of recording the strength of muscle contraction due to electrical stimulation; abbreviated EMG. An **electromyogram** (ē-lehck-trō-mī-ō-grahm) is the record of the strength of muscle contraction due to electrical stimulation.

▶ **Path of Destruction: Muscular System** ◀

Pathologic conditions of the muscular system include:
- **tendinitis** (tehn-dih-nī-tihs) = inflammation of the band of fibrous connective tissue that connects muscle to bone
- **fibroma** (fī-brō-mah) = tumor composed of fully developed connective tissue; also called **fibroid** (figh-broyd)
- **fasciitis** (fahs-ē-ī-tihs) = inflammation of the sheet of fibrous connective tissue that covers, supports, and separates muscles (fascia)
- **ataxia** (ā-tahck-sē-ah) = inability of voluntary control of muscle movement; "wobbly"

- **atonic** (ā-thohn-ihck) = pertaining to lacking muscle control
- **myotonia** (mī-ō-tō-nē-ah) = delayed relaxation of a muscle after contraction
- **myoclonus** (mī-ō-klō-nuhs) = spasm of muscle
- **myositis** (mī-ō-sī-tihs) = inflammation of voluntary muscles
- **myopathy** (mī-ohp-ah-thē) = abnormal condition or disease of muscle
- **myasthenia** (mī-ahs-thē-nē-ah) = muscle weakness
- **adhesion** (ahd-hē-shuhn) = band of fibers that hold structures together in an abnormal fashion
- **hernia** (hər-nē-ah) = protrusion of a body part through tissues that normally contain it
- **dystrophy** (dihs-trō-fē) = defective growth
- **tetany** (teht-ahn-ē) = muscle spasms or twitching
- **laxity** (lahx-ih-tē) = looseness, not tight

▶ Procede With Caution: Muscular System ◀

Procedures performed on the muscular system include
- **tenectomy** (teh-nehck-tō-mē) = surgical removal of a part of a tendon (fibrous connective tissue that connects muscle to bone)
- **tenotomy** (teh-noht-ō-mē) = surgical division of a tendon (fibrous connective tissue that connects muscle to bone)
- **myectomy** (mī-ehck-tō-mē) = surgical removal of muscle or part of a muscle
- **myoplasty** (mī-ō-plahs-tē) = surgical repair of muscle
- **myotomy** (mī-oht-ō-mē) = surgical incision into or dividing a muscle

REVIEW EXERCISES

Multiple Choice—Choose the correct answer

1. A common name for the tarsus is the
   a. elbow
   b. calcaneus
   c. hock
   d. wrist

2. The _____ joints are the freely movable joints of the body.
   a. suture
   b. synovial
   c. symphysis
   d. cartilaginous

3. The correct order of the vertebral segments is
   a. cervical, thoracic, lumbar, sacral, and coccygeal
   b. cervical, lumbar, thoracic, coccygeal, and sacral
   c. thoracic, lumbar, cervical, sacral, and coccygeal
   d. thoracic, cervical, lumbar, sacral, and coccygeal

4. A _____ is a fibrous band of connective tissues that connects bone to bone.
   a. fascia
   b. tendon
   c. synovial membrane
   d. ligament

5. The acetabulum is the
   a. patella
   b. cannon bone
   c. large socket in the pelvic bone
   d. crest of the scapula

6. The three parts of the pelvis are
   a. ileum, pubis, and acetabulum
   b. ilium, pubis, and sacrum
   c. ilium, sacrum, and coccyx
   d. ilium, ischium, and pubis

7. The digits contain bones that are called
   a. carpals
   b. phalanges
   c. tarsals
   d. tarsus

8. Components of the axial skeleton include
   a. scapula, humerus, radius, ulna, and carpus
   b. skull, hyoid, vertebrae, ribs, and sternum
   c. pelvic girdle, femur, tibia, fibula, and tarsus
   d. scapula, pelvis, humerus, femur, tibia, fibula, radius, and ulna

9. Another term for growth plate is
   a. physis
   b. shaft
   c. diaphysis
   d. trophic

10. The bones of the front limb include
    a. humerus, tibia, fibula, tarsal, metatarsal, and phalanges
    b. humerus, radius, ulna, carpal, metacarpal, and phalanges
    c. femur, tibia, fibula, tarsal, metatarsal, and phalanges
    d. radius, humerus, ulna, carpal, metatarsal, and phalanges

11. Rectus means
    a. ringlike
    b. straight
    c. angled
    d. rotating

12. Muscles may be classified as
    a. voluntary
    b. involuntary
    c. cardiac
    d. all of the above

13. A term for when a muscle becomes shorter and thicker is
    a. relaxation
    b. contraction
    c. rotation
    d. depression

14. Levator muscles _____ a muscle.
    a. decrease the angle of
    b. increase the angle of
    c. raise
    d. depress

15. A fibrous band of connective tissue that connects muscle to bone is
    a. cartilage
    b. tendon
    c. ligament
    d. aponeurosis

16. Looseness is termed
    a. laxity
    b. rigidity
    c. spasm
    d. tonus

17. Protrusion of a body part through tissues that normally contain it is called a
    a. projection
    b. hernia
    c. prominence
    d. myotonia

18. A muscle that forms a tight band is called a/an
    a. purse-string
    b. sartorius
    c. sphincter
    d. oblique

19. Surgical removal of a muscle or part of a muscle is called
    a. myositis
    b. myotomy
    c. myectomy
    d. myostomy

20. Abnormal condition or disease of muscle is termed
    a. myodynia
    b. myography
    c. myasthenia
    d. myopathy

Fill in the Blank

1. _____ and _____ are terms used for displacement of a bone from its joint.

2. The _____ is the tough, fibrous tissue that forms the outermost covering of bone.

3. A/An _____ is a curved fibrous cartilage found in some synovial joints.

4. Connections between two bones are called _____ or _____.

5. The cranial portion of the sternum is called the _____.

6. A/An _____ is removal of all or part of a limb or body part.

7. A _____ is a piece of dead bone that is partially or fully detached from the surrounding healthy bone.

8. Inward curvature of a bone is termed _____.

9. Visual examination of the internal structure of a joint using a fiberoptic instrument is _____.

10. _____ is loss of mobility of a joint.

11. _____ is abnormal softening of cartilage.

12. A muscle that straightens a limb at a joint is called a _____.

13. Extreme straightening of a limb beyond its normal limits is called _____.

14. A/An _____ is a band of fibers that holds structures together in an abnormal fashion.

15. Dogs with short, wide skulls are said to be _____.

16. Involuntary muscle is also called _____, _____, or _____.

17. Surgical removal of a claw is _____.

18. A/An _____ is a broken bone in which there is an open wound in the skin.

19. The _____ is the fibrous band of connective tissue on the ventral abdominal wall that is the center attachment of the abdominal muscles.

20. A/An _____ is the place where muscle ends that is the more movable end or portion away from midline.

Matching—Match the bone or joint with its common name.

_____P1
_____P2
_____P3
_____tarsus
_____splint bone
_____fetlock joint
_____pastern joint
_____coffin joint
_____knee
_____stifle
_____clavicle
_____cannon bone
_____dewclaw
_____sternum

a. carpus in large animals
b. hock
c. coffin bone
d. short pastern
e. long pastern
f. metacarpal/metatarsal III in equine and metacarpal/metatarsal III and IV in ruminants
g. collarbone
h. metacarpo-/metatarsophalangeal joint of equine and ruminants
i. metacarpal/metatarsal II and IV in equine
j. connection between phalanx I and II in equine and ruminants
k. distal interphalangeal joint of phalanx II and III in equine and ruminants
l. variable digit depending on species; digit I in dogs, digits II and V in ruminants
m. synovial joint located between the femur and tibia
n. breastbone

Matching—Match the bone with the area where it is located.

_____humerus   a. distal front limb

_____fibula   b. proximal front limb

_____tibia   c. proximal hind limb

_____ulna   d. distal hind limb

_____femur   e. joint in front limb

_____tarsus   f. joint in hind limb

_____radius   g. distal part of front and hind limbs

_____carpus

_____metacarpal

_____metatarsal

_____phalanx

Label the diagrams in Figures 3–16 and 3–17.

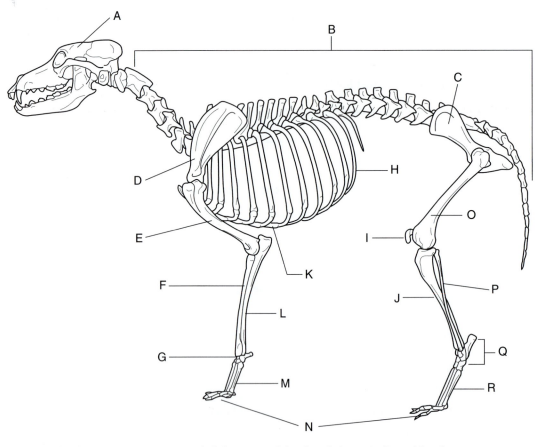

FIGURE 3–16 Dog skeleton. Label the parts of the dog skeleton indicated by the arrows.

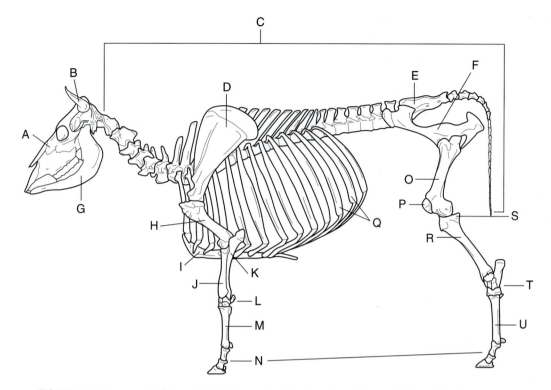

FIGURE 3–17 Bovine skeleton. Label the parts of the bovine skeleton indicated by the arrows.

4 Head to Toe (and All Parts in Between)

Objectives *In this chapter, you should learn to:*

▶ Identify and describe the common and anatomic term for the species provided
▶ Recognize, define, spell, and pronounce terms relating to their anatomic location

TWO WORDS; SAME MEANING

Medical terminology is a specific language used by health care professionals (both human and veterinary) to describe conditions in a concise manner. Lay people also have a language that is used to describe anatomy and medical conditions in a concise manner. In veterinary medicine, there are many different species and many different terms used to describe animal anatomy and diseases. In this chapter the anatomic lay terms used by many people in the veterinary community will be described. The lay terms for disease will be covered in the chapters on individual species.

Common Anatomic Terms For Equine Species

See Figure 4–1.

forelock (fōr-lohck)	in maned animals, the most anterior part of the mane, hanging down between the ears and onto the forehead
poll (pōl)	top of the head; the occiput
mane (mān)	region of long coarse hair at the dorsal border of the neck and terminating at the poll
crest (krehst)	the root of the mane
muzzle (muh-zuhl)	the two nostrils (including the skin and fascia) and the muscles of the upper and lower lip
cheek (chēk)	fleshy portion of either side of the face; forms the sides of the mouth and continues rostrally to the lips
shoulder (shōl-dər)	region around the large joint between the humerus and scapula
chest (chehst)	part of the body between the neck and abdomen; the thorax
knee (nē)	the carpus in ungulates (an ungulate is an animal with hooves)
cannon bone (kahn-nohn)	third metacarpal (metatarsal) of the horse; also called the shin bone
fetlock joint (feht-lohck)	metacarpophalangeal and metatarsophalangeal joint (joint between the cannon bone and long pastern bone [phalanx I]) in ungulates
pastern joint (pahs-tərn)	proximal interphalangeal joint (joint between the long and short pastern bones [phalanx I and II, respectively]) in ungulates
coffin joint (kawf-ihn)	distal interphalangeal joint (joint between the short pastern and coffin bones [phalanx II and III, respectively]) in ungulates (Figure 4–2)
coronary band (kohr-ō-nār-ē)	junction between the skin and the horn of the hoof; also called the **coronet** (kor-oh-net)
hoof wall (huhf wahl)	the hard, horny, outer layer of the covering of the digit in ungulates
hoof (huhf)	the hard covering of the digit in ungulates (Figure 4–3)
sole (sōl)	bottom of the hoof
frog (frohg)	V-shaped pad of soft horn between the bars on the sole of the equine hoof
bars (bahrz)	raised V-shaped structure on ventral surface of hoof
white line	fusion between the wall and sole of the hoof

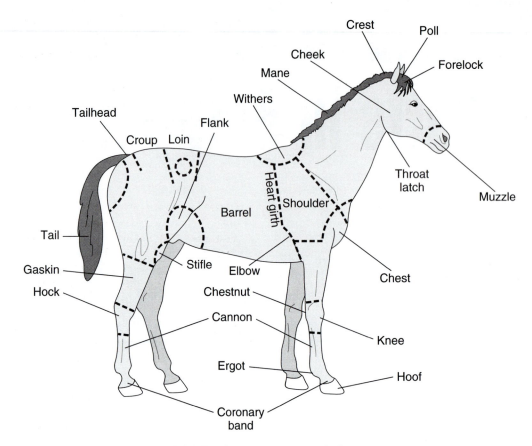

FIGURE 4–1 Anatomic parts of a horse

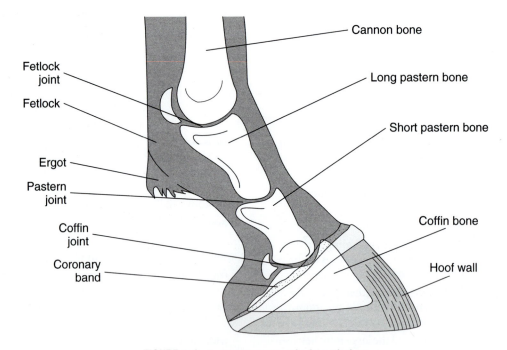

FIGURE 4–2 Anatomic parts of a horse's foot

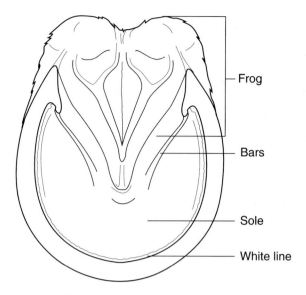

FIGURE 4–3 Anatomic parts of a horse's hoof

ergot (ār-goht) (or ər- goht) small mass of horn in a small bunch of hair on the palmar or plantar aspects of the equine fetlock

barrel (bār-ahl) capacity of the chest or trunk

flank (flānk) side of the body between the ribs and ilium

gaskin (gahs-kihn) muscular portion of the hindlimb between the stifle and hock; also called the **crus**

hock (hohck) tarsal joint

croup (kroop) muscular area around and above the tail base

withers (wih-thərz) region over the dorsum where the neck joins the thorax and where the dorsal margins of the scapula lie just below the skin

tail head (tā-uhl hehd) base of the tail where it connects to the body

dock (dohck) solid part of the equine tail

loin (loyn) lumbar region of the back, between the thorax and pelvis

nippers (nihp-pərz) central incisors of equine

cutters (kuht-ərz) second incisors of equine

corners (kōr-nərz) third incisors of equine

stifle joint (stī-fuhl) femorotibial and femoropatellar joint in quadrupeds

chestnuts (chehs-nuhtz) horny, irregular growths on the medial surface of the equine leg; in the front legs the chestnuts are just above the knee; in the rear legs the chestnuts are near the hock

elbow (ehl-bō) forelimb joint formed by distal humerus, proximal radius, and proximal ulna

udder (uh-dər) mammary gland

teat (tēt) nipple of mammary gland

tail (tā-uhl) caudal part of the vertebral column extending beyond the trunk

heart girth (hahrt gərth) circumference of the chest just caudal to the shoulders and cranial to the back

Common Anatomic Terms of Cattle

See Figure 4–4.

poll (pōl) top of the head; the occiput

crest (krehst) upper margin of the neck

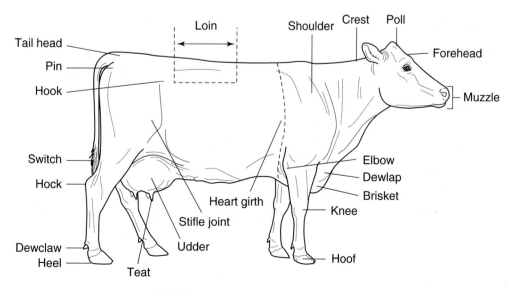

FIGURE 4–4 Anatomic parts of a bovine

muzzle (muh-zuhl)	the two nostrils (including the skin and fascia) and the muscles of the upper and lower lip
shoulder (shōl-dər)	region around the large joint between the humerus and scapula
forearm (fōr-ahrm)	part of the foreleg supported by the radius and ulna, between the elbow and knee
knee (nē)	the carpus in ungulates
pedal (pē-dahl)	pertaining to the foot
cannon (kahn-nohn) **bone**	third and fourth metacarpal (metatarsal) of the ruminant (not commonly used); also called the **shin bone**
fetlock (feht-lohck) **joint**	metacarpophalangeal and metatarsophalangeal joint (joint between the cannon bone and the long pastern bone [phalanx I]) in ungulates
pastern (pahs-tərn) **joint**	proximal interphalangeal joint (joint between the long and short pastern bones [phalanx I and II, respectively]) in ungulates
coffin (kawf-ihn) **joint**	distal interphalangeal joint (joint between the short pastern and coffin bones [phalanx II and III, respectively]) in ungulates
hoof (huhf wahl) **wall**	the hard, horny, outer layer of the covering of the digit in ungulates
hoof (huhf)	the hard covering of the digit in ungulates
flank (flānk)	side of the body between the ribs and ilium
hock (hohck)	tarsal joint
tail head (tā-uhl hehd)	base of the tail where it connects to the body
loin (loyn)	lumbar region of the back, between the thorax and pelvis
stifle (stī-fuhl) **joint**	femorotibial and femoropatellar joint in quadrupeds
elbow (ehl-bō)	forelimb joint formed by distal humerus, proximal radius, and proximal ulna
dewlap (doo-lahp)	loose skin under the throat and neck, which may become pendulous in some breeds
brisket (brihs-kiht)	mass of connective tissue, muscle, and fat covering the cranioventral part of the ruminant chest between the forelegs
dewclaw (doo-klaw)	accessory claw of the ruminant foot that projects caudally from the fetlock
heel (hēl)	caudal region of the hoof; also called the **bulb**

toe (tō)	cranial end of the hoof
sole (sōl)	bottom of the hoof
hooks (hookz)	protrusion of the wing of the ilium on the dorsolateral area of ruminants
pins (pihnz)	protrusion of the ischium bones just lateral to the base of the tail in ruminants
dock (dohck)	amputation of the tail
forehead (fōr-hehd)	region of the head between the eyes and ears
udder (uh-dər)	mammary gland
teat (tēt)	nipple of mammary gland
tail (tā-uhl)	caudal part of the vertebral column extending beyond the trunk
heart girth (hahrt gərth)	circumference of the chest just caudal to the shoulders and cranial to the back
switch (swihtch)	tuft of hair at the end of the tail

Common Anatomic Terms for Goats

See Figure 4–5.

poll (pōl)	top of the head; the occiput
crest (krehst)	upper margin of the neck
muzzle (muh-zuhl)	the two nostrils (including the skin and fascia) and the muscles of the upper and lower lip
shoulder (shōl-dər)	region around the large joint between the humerus and scapula
forearm (fōr-ahrm)	part of the foreleg supported by the radius and ulna, between the elbow and knee
knee (nē)	the carpus in ungulates
pedal (pē-dahl)	pertaining to the foot
cannon (kahn-nohn) **bone**	third and fourth metacarpal (metatarsal) of the ruminant (not commonly used); also called the **shin bone**
fetlock (feht-lohck) **joint**	metacarpophalangeal and metatarsophalangeal joint (joint between the cannon bone and the long pastern bone [phalanx I]) in ungulates
pastern (pahs-tərn) **joint**	proximal interphalangeal joint (joint between the long and short pastern bones [phalanx I and II, respectively]) in ungulates
coffin (kawf-ihn) **joint**	distal interphalangeal joint (joint between the short pastern and coffin bones [phalanx II and III, respectively]) in ungulates
hoof wall (huhf wahl)	the hard, horny, outer layer of the covering of the digit in ungulates

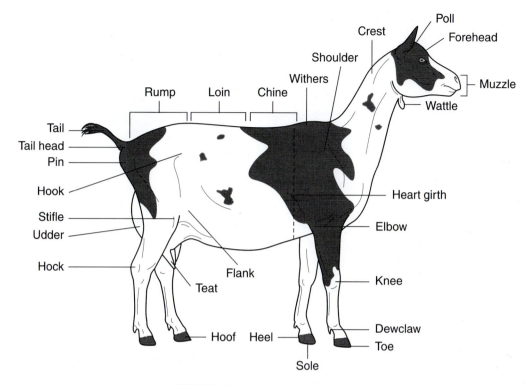

FIGURE 4–5 **Anatomic parts of a goat**

hoof (huhf)	the hard covering of the digit in ungulates
flank (flānk)	side of the body between the ribs and ilium
hock (hohck)	tarsal joint
tail head (tā-uhl hehd)	base of the tail where it connects to the body
loin (loyn)	lumbar region of the back, between the thorax and pelvis
chine (kīn)	thoracic region of the back
rump (ruhmp)	sacral to tailhead region of the back
stifle (stī-fuhl) **joint**	femorotibial and femoropatellar joint in quadrupeds
elbow (ehl-bō)	forelimb joint formed by distal humerus, proximal radius, and proximal ulna
brisket (brihs-kiht)	mass of connective tissue, muscle, and fat covering the cranioventral part of the ruminant chest between the forelegs
dewclaw (doo-klaw)	accessory claw of the ruminant foot that projects caudally from the fetlock
heel (hēl)	caudal region of the hoof; also called the bulb
toe (tō)	cranial end of the hoof
sole (sōl)	bottom of the hoof
hooks (hookz)	protrusion of the wing of the ilium on the dorsolateral area of ruminants
pins (pihnz)	protrusion of the ischium bones just lateral to the base of the tail in ruminants
forehead (fōr-hehd)	region of the head between the eyes and ears
udder (uh-dər)	mammary gland
teat (tēt)	nipple of mammary gland
horn butt (hōrn buht)	poll region between the eyes and ears of previous horn growth
withers (wih-thərz)	region over the dorsum where the neck joins the thorax and where the dorsal margins of the scapula lie just below the skin
wattle (waht-tuhl)	appendages suspended from the head (usually under the chin)
tail (tā-uhl)	caudal part of the vertebral column extending beyond the trunk
heart girth (hahrt gərth)	circumference of the chest just caudal to the shoulders and cranial to the back

Common Anatomic Terms for Sheep

See Figure 4–6.

poll (pōl)	top of the head; the occiput
crest (krehst)	upper margin of the neck

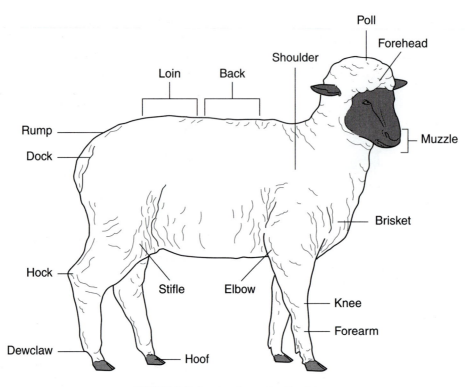

FIGURE 4–6 Anatomic parts of a sheep

muzzle (muh-zuhl)	the two nostrils (including the skin and fascia) and the muscles of the upper and lower lip
shoulder (shōl-dər)	region around the large joint between the humerus and scapula
forearm (fōr-ahrm)	part of the foreleg supported by the radius and ulna, between the elbow and knee
knee (nē)	the carpus in ungulates
pedal (pē-dahl)	pertaining to the foot
cannon (kahn-nohn) **bone**	third and fourth metacarpal (metatarsal) of the ruminant (not commonly used); also called the **shin bone**
fetlock (feht-lohck) **joint**	metacarpophalangeal and metatarsophalangeal joint (joint between the cannon bone and the long pastern bone [phalanx I]) in ungulates
pastern (pahs-tərn) **joint**	proximal interphalangeal joint (joint between the long and short pastern bones [phalanx I and II, respectively]) in ungulates
coffin (kawf-ihn) **joint**	distal interphalangeal joint (joint between the short pastern and coffin bones [phalanx II and III, respectively]) in ungulates
hoof wall (huhf wahl)	the hard, horny, outer layer of the covering of the digit in ungulates

hoof (huhf)	the hard covering of the digit in ungulates
flank (flānk)	side of the body between the ribs and ilium
hock (hohck)	tarsal joint
tail head (tā-uhl hehd)	base of the tail where it connects to the body
loin (loyn)	lumbar region of the back, between the thorax and pelvis
stifle (stī-fuhl) **joint**	femorotibial and femoropatellar joint in quadrupeds
elbow (ehl-bō)	forelimb joint formed by distal humerus, proximal radius, and proximal ulna
brisket (brihs-kiht)	mass of connective tissue, muscle, and fat covering the cranioventral part of the ruminant chest between the forelegs
dewclaw (doo-klaw)	accessory claw of the ruminant foot that projects caudally from the fetlock
heel (hēl)	caudal region of the hoof; also called the **bulb**
toe (tō)	cranial end of the hoof
sole (sōl)	bottom of the hoof
dock (dohck)	amputation of the tail
forehead (fōr-hehd)	region of the head between the eyes and ears

udder (uh-dər)	mammary gland
teat (tēt)	nipple of mammary gland
heart girth (hahrt gərth)	circumference of the chest just caudal to the shoulders and cranial to the back

Common Anatomic Terms for Swine

See Figure 4–7.

shoulder (shōl-dər)	region around the large joint between the humerus and scapula
knee (nē)	the carpus in ungulates
fetlock (feht-lohck) **joint**	metacarpophalangeal and metatarsophalangeal joint in ungulates
pastern (pahs-tərn) **joint**	joint between the long and short pastern bones (phalanx I and II, respectively) in ungulates
coffin (kawf-ihn) **joint**	joint between the short pastern and coffin bones (phalanx II and III, respectively) in ungulates
hoof wall (huhf wahl)	the hard, horny, outer layer of the covering of the digit in ungulates
hoof (huhf)	the hard covering of the digit in ungulates
flank (flānk)	side of the body between the ribs and ilium
hock (hohck)	tarsal joint
loin (loyn)	lumbar region of the back, between the thorax and pelvis
stifle (stī-fuhl) **joint**	femorotibial and femoropatellar joint in quadrupeds
elbow (ehl-bō)	forelimb joint formed by distal humerus, proximal radius, and proximal ulna

dewclaw (doo-klaw)	accessory claw of the porcine foot that projects caudally from the fetlock
rump (ruhmp)	sacral to tailhead region of the back
ham (hahm)	musculature of the upper thigh
tail (tā-uhl)	caudal part of the vertebral column extending beyond the trunk
jowl (jowl)	external throat especially when fat or loose skin is present
snout (snowt)	the upper lip and apex of the nose of swine

Common Anatomic Terms for Dogs and Cats

See Figure 4–8 and 4–9.

muzzle (muh-zuhl)	the two nostrils (including the skin and fascia) and the muscles of the upper and lower lip
shoulder (shōl-dər)	region around the large joint between the humerus and scapula
hock (hohck)	tarsal joint; **tarsus** is used commonly for this joint as well
elbow (ehl-bō)	forelimb joint formed by distal humerus, proximal radius, and proximal ulna
stifle (stī-fuhl) **joint**	femorotibial and femoropatellar joint in quadrupeds
forehead (fōr-hehd)	region of the head between the eyes and ears
pinna (pihn-ah)	projecting part of the ear lying outside the head; the auricle
tail (tā-uhl)	caudal part of the vertebral column extending beyond the trunk

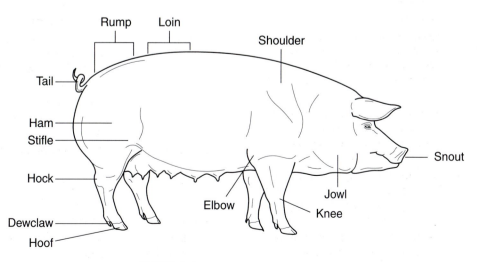

FIGURE 4–7 Anatomic parts of a swine

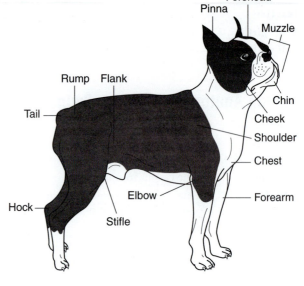

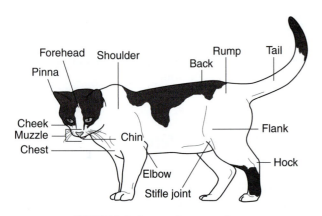

FIGURE 4–8 Anatomic parts of a cat

FIGURE 4–9 Anatomic parts of a dog

cheek (chēk)   fleshy portion of either side of the face; forms the sides of the mouth and continues rostrally to the lips

chin (chihn)   cranioventral protrusion of the mandible

chest (chehst)   part of the body between the neck and abdomen; the thorax

flank (flānk)   side of the body between the ribs and ilium

rump (ruhmp)   sacral to tailhead region of the back; also called the **croup**

dewclaw (doo-klaw)   rudimentary first digit of dogs and cats

REVIEW EXERCISES

Label Diagrams—Figures 4–10, 4–11, 4–12, 4–13, 4–14, and 4–15

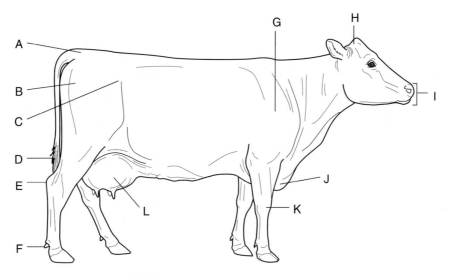

FIGURE 4–10 Brown Swiss. Identify parts.

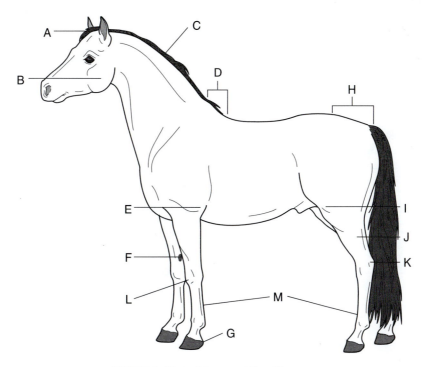

FIGURE 4–11 Welsh pony. Identify parts.

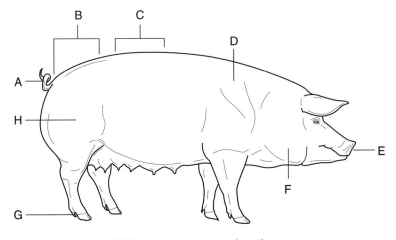

FIGURE 4–12 Duroc hog. Identify parts.

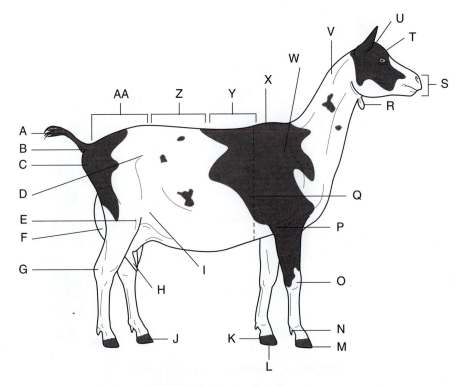

FIGURE 4–13 Goat. Identify parts.

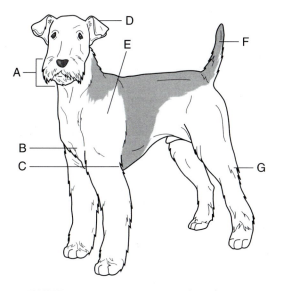

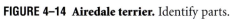

FIGURE 4–14 Airedale terrier. Identify parts.

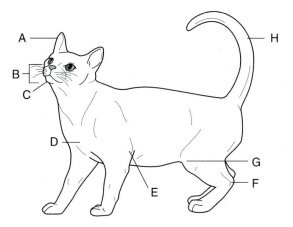

FIGURE 4–15 Tonkinese cat. Identify parts.

What Is in a Name?

Objectives *In this chapter, you should learn to:*

- ► Identify and recognize common terms used for animals
- ► Define common terms used to denote sex and age of animals
- ► Define common terms used to denote birthing and grouping of animals

WHAT IS YOUR NAME?

Lay people as well as professionals use terms to describe in one word the status of an animal. The term may relate to the sexual status of an animal (intact or sexually functional, or altered or sexually nonfunctional) or the age status of an animal. Terms have also been derived to denote the process of giving birth and the grouping of animals. The following lists provide the terms used to describe animals.

canine (kā-nīn) = dogs

dog/stud = intact male dog
bitch = intact female dog
whelp (wehlp) or pup = young dog
whelping (wehl-pihng) = giving birth to whelps/pups
pack = group of dogs
litter = multiple offspring born during same labor

> **Gender versus sex:** In veterinary medicine we only talk about the sex of an animal. Gender is used to denote whether words pertaining to a noun are masculine, feminine, or neutral. Gender can also be used to denote social constructs, for example, the gender or social roles of men and women.

feline (fē-līn) = cats

tom = intact male cat
queen = intact female cat
kitten = young cat
queening = giving birth to kittens

lagomorph (lāg-ō-mōrf) = rabbits

buck = intact male rabbit
doe = intact female rabbit
lapin (lahp-ihn) = neutered male rabbit
kit = young (blind, deaf) rabbit
kindling (kihnd-lihng) = giving birth to rabbits
herd = group of rabbits

ferrets = (fehr-retz)

hob = intact male ferret
jill = intact female ferret
gib (gihb) = neutered male ferret
sprite (sprīt) = spayed female ferret
kit = young ferret
kindling = giving birth to ferrets

psittacine (siht-ah-sēn) = parrots

cock = intact male parrot
hen = intact female parrot
chick = young parrot

murine (moo-rēn) = mice and rats

sire/male (sī-ər) = intact male mouse or rat
dam/female (dahm) = intact female mouse or rat
pup = young mouse or rat

> Dam and sire are terms used frequently to denote female parent or male parent for many species. When animals are bred, these terms may be used instead of the ones in the lists here. Used correctly, these mean that the male and female have bred and hopefully have been produced an embryo or fetus. **Dam** is female parent; **sire** is male parent.

turkey = one kind of poultry

tom = intact male turkey
hen = intact female turkey
poult (pōlt) = young turkey
flock = group of turkeys
clutch (kluhtch) = group of eggs

chickens = one kind of poultry

rooster = sexually mature male chicken; also called cock
hen = intact female chicken
capon (kā-pohn) = young, castrated male chicken or domestic fowl
cockeral (kohck-ər-ahl) = immature male chicken
pullet (puhl-eht) = immature female chicken
poult = young chicken
chick = very young chicken
flock = group of chickens

anserine (ahn-sehr-ihn) = geese

gander = intact male goose
goose = intact female goose
gosling = young goose
gaggle = group of geese

anserine = ducks

drake = intact male duck
duck = intact female duck
duckling = young duck
flock = group of ducks

porcine (poor-sīn) = pigs = swine

boar (bōr) = intact male pig
sow = intact female pig
barrow (bār-ō) = male pig castrated when young
stag = male pig castrated after maturity
gilt (gihlt) = young female pig that has not farrowed
pig = young pig; old term is shoat
farrowing (fār-ō-ihng) = giving birth to pigs
herd = group of pigs

equine (ē-qwīn) = includes horses, ponies, donkeys, and mules

stallion (stahl-yuhn) = intact male equine >4 years old
colt (kōlt) = intact male equine <4 years old
mare (mār) = intact female equine >4 years old
filly (fihl-ē) = intact female equine <4 years old
gelding (gehld-ihng) = castrated male equine
ridgeling (rihdg-lihng) or **rig** = cryptorchid equine (one or both testicles have not descended from the abdomen)
foal = young equine of either sex
weanling = young equine <1 year old
yearling = young equine between 1 and 2 years old
foaling = giving birth to equine
herd = group of equine
brood mare (bruhd mār) = breeding female equine
maiden mare (mā-dehn mār) = female equine never bred
barren mare (bār-ehn mār) = intact female horse that was not bred or did not conceive the previous season = open mare
wet mare = intact female horse that has foaled during the current breeding season
agalactic mare (ā-gahl-ahck-tihck mār) = intact female horse not producing milk
pony = equine between 8.2 and 14.2 hands when mature (not a young horse)

> **What is a mule?**
> Mule is a general term that applies to the hybrid crossing of equines. Mule is also used to denote the offspring of a jack (male donkey) and a mare (female horse). **Hinny** (hihn-ē) is used to denote the offspring of a stallion (male horse) and a jenny (female donkey). Think mule:mare to remember the lineage of this hybrid. Both mules and hinnies are sterile.

donkey = ass = burro

jack or **jack ass** = intact male donkey
jenny = intact female donkey

ovine (ō-vīn) = sheep

ram = intact male sheep
ewe (yoo) = intact female sheep
wether (wheh-thər) = castrated male sheep
lamb = young sheep
lambing = giving birth to sheep
flock = group of sheep

> **What is a bellwether?** From the list, wether is a neutered sheep or goat. Bell is a ringing device. Originally this word was used to describe the practice of putting a bell on the lead wether of a flock/herd.

caprine (kahp-rīn) = goats

buck = intact male goat
doe = intact female goat
wether = castrated male goat
kid = young goat
kidding = giving birth to goats
freshening = giving birth to dairy animals
herd = group of goats

camelid = llamas

bull = intact male llama (also referred to as a stallion)
cow = intact female llama
gelding = castrated male llama
cria (krē-ah) = young llama

cavy (kā-vē) = guinea pigs

boar = intact male guinea pig
sow = intact female guinea pig

bovine = cattle

bull = intact male bovine
jumper bull = intact male bovine that has just reached maturity and is used for breeding
cow = intact female bovine that has given birth
steer = male bovine castrated when young
stag = male bovine castrated after maturity
heifer (hehf-ər) = young female bovine that has not given birth
calf = young bovine
calving = giving birth to cattle
freshening = giving birth to dairy animals
herd = group of cattle
springing heifer = young female pregnant with her first calf = first calf heifer
second calf heifer = female pregnant with her second calf
freemartin (frē-mahr-tihn) = sexually imperfect, usually sterile female calf twinborn with a male
gomer bull = bull used to detect female bovines in heat; bull may have penis surgically deviated to the side, may be treated with androgens, or may be vasectomized so as not to impregnate female; also called **teaser bull**

> The symbols denoting male ♂ and female ♀ originally stood for Mars (the god of war) and Venus (the goddess of love), respectively. Mars and male both begin with "m." Then associate the fact that the symbol for male looks like a shield, which is a protective device used in war.

REVIEW EXERCISES

Multiple Choice—Choose the correct answer.

1. A neutered male sheep or goat is a/an
   a. bull
   b. ovine
   c. wether
   d. caprine

2. A sexually imperfect, usually sterile female calf twinborn with a male calf is a/an
   a. heifer
   b. freemartin
   c. gilt
   d. filly

3. A cross between a stallion and a jenny is a/an
   a. donkey
   b. mule
   c. jenny
   d. hinny

4. Cow is to mare as steer is to
   a. stallion
   b. ridgling
   c. colt
   d. gelding

5. Parrots are in a group of birds called the
   a. amazons
   b. psittacine
   c. lagomorph
   d. murine

6. The act of giving birth in canines is
   a. whelping
   b. pupping
   c. packing
   d. gestation

7. Male and female ferrets are called
   a. jacks and jills
   b. kits and jills
   c. hobs and jills
   d. jacks and kits

8. A young llama is called a
   a. calf
   b. cria
   c. clutch
   d. colt

9. Freshening is a term that means
   a. cleaning an animal to make it smell fresh
   b. giving birth to a dairy animal
   c. the act of mating in cattle
   d. removing the horns of a bovine to enhance mating

10. Giving birth to swine is called
    a. barrowing
    b. queening
    c. farrowing
    d. tupping

Fill in the blanks.

1. A young dog is called a _____ or _____.

2. A young cat is called a _____.

3. A young horse is called a _____.

4. A young bovine is called a _____.

5. A young goat is called a _____.

6. A young sheep is called a _____.

7. A young swine is called a _____ or _____.

8. Young rabbits or ferrets are called _____.

9. Young mice or rats are called _____.

10. A young llama is called a _____.

Matching—Match the species common name with its taxonomic name.

_____canine	cattle	
_____bovine	cat	
_____equine	pig	
_____feline	parrot	
_____caprine	rat	
_____ovine	dog	
_____porcine	sheep	
_____psittacine	mouse	
_____murine	donkey	
	horse	
	goat	

Gut Instincts

Objectives *In this chapter, you should learn to:*

► Identify and describe the major structures and functions of the digestive tract
► Distinguish between monogastric and ruminant digestive system anatomy and physiology
► Describe the processes of digestion, absorption, and metabolism
► Recognize, define, spell, and pronounce terms related to the diagnosis, pathology, and treatment of the digestive system

FUNCTIONS OF THE DIGESTIVE SYSTEM

The **digestive** (dī-jehs-tihv) **system, alimentary** (āl-ih-mehn-tahr-ē) **system, gastrointestinal** (gahs-trō-ihn-tehst-ihn-ahl) **system,** and **GI system** are all terms used to describe the body system that is basically a long, muscular tube that begins at the mouth and ends at the anus. The digestive system is responsible for the intake and digestion of food and water, the absorption of nutrients, and the elimination of solid waste products. The combining form for nourishment is **aliment/o.**

STRUCTURES OF THE DIGESTIVE SYSTEM

The major structures of the digestive system include the oral cavity, pharynx, esophagus, stomach, and small and large intestines. The liver, gallbladder, and pancreas are organs associated with the digestive system.

Down in the Mouth

The mechanical and chemical process of digestion begins in the mouth. The mouth contains the major structures of the oral cavity. The **oral** (ōr-ahl) **cavity** contains the lips and cheeks, hard and soft palates, salivary glands, tongue, teeth, and periodontium. The combining forms **or/o** and **stomat/o** mean mouth.

The maxilla and mandible are bones that are the boundaries of the oral cavity. The combining form for jaw is **gnath/o.** An animal that has **prognathia** (prohg-nah-thē-ah) has an elongated mandible or a mandible that is overshot. Prognathia is sometimes referred to as sow mouth. An animal that has **brachygnathia** (brahk-ē-nah-thē-ah) has a shortened mandible or a mandible that is undershot. Brachygnathia is sometimes referred to as parrot mouth.

The lips form the opening to the oral cavity. The term **labia** (lā-bē-ah) is the medical term for lips. A single lip is a **labium** (lā-bē-uhm). The combining forms for lips are **cheil/o** and **labi/o.** The cheeks form the walls of the oral cavity. The combining form for cheek is **bucc/o.** The term **buccal** (būk-ahl) means pertaining to or directed toward the cheek.

The **palate** (pahl-aht) forms the roof of the mouth. The palate consists of two parts: the hard and soft palates. The **hard palate** forms the bony rostral portion of the palate that is covered with specialized mucous membrane. This specialized mucous membrane contains irregular folds called **rugae** (roo-gā). Rugae are also found in the stomach. **Rug/o** is the combining form for wrinkle or fold. The **soft palate** forms the flexible caudal portion of the palate. The soft palate is involved in closing off the nasal passage during swallowing so that food does not move into the nostrils. The combining form **palat/o** means palate.

The **tongue** (tuhng) is a movable muscular organ in the oral cavity used for tasting and processing food, grooming, and sound articulation. The tongue moves food during chewing and swallowing. The dorsum of the tongue has **papillae** (pah-pihl-ā) or elevations, while the ventral surface of the tongue is highly vascular. The types of papillae located on the dorsum of the tongue may appear threadlike or **filiform** (fihl-ih-fōrm), mushroom like or **fungiform** (fuhn-jih-fōrm), or cup-shaped or **vallate** (vahl-āt). Taste buds are located in the fungiform and vallate papillae. The tongue is connected to the ventral surface of the oral cavity by a band of connective tissue called the **frenulum** (frehn-yoo-luhm). The combining forms for tongue are **gloss/o** and **lingu/o**. The **lingual surface** of the cheek is the side adjacent to the tongue.

The combining forms **dent/o, dent/i,** and **odont/o** are used to refer to teeth. **Dentition** (dehn-tihsh-shuhn) refers to the teeth as a whole, that is, the teeth arranged in the maxillary (upper) and mandibular (lower) arcades. The primary dentition or **deciduous** (deh-sihd-yoo-uhs) **dentition** are the temporary set of teeth that erupt in young animals and are replaced at or near maturity. **Decidu/o** is the combining form for shedding. The **permanent dentition** are the teeth designed to last the lifetime of an animal. Sometimes a deciduous tooth of brachydontic animals is not shed at the appropriate time and both the deciduous and permanent teeth are situated beside each other. The deciduous tooth that has not been shed is referred to as a **retained deciduous tooth** and may be extracted (removed) professionally.

There are four types of teeth that have different functions. The types of teeth are

▶ **incisor** (ihn-sīz-ōr) = front teeth used for cutting; an incision is a cut; abbreviated I
▶ **canine** (kā-nīn) = long, pointed bonelike tooth located between the incisors and premolars; also called **fang** or **cuspid** (cusp means having a tapering projection; cuspid means having one point); abbreviated C
▶ **premolar** (prē-mō-lahr) = cheek teeth found between the canine teeth and molars; also called **bicuspids** because they have two points; abbreviated P
▶ **molar** (mō-lahr) = most caudally located permanent cheek teeth used for grinding; molar comes from the Latin term to grind; abbreviated M

When the number and type of teeth found in an animal are written, a shorthand method is used. This method is called the dental formula. The **dental formula** of an animal represents the type of tooth and the number of each tooth type found in that species. In the dental formula only one side of the mouth is repre-

sented and is preceded by a 2 to indicate that this arrangement is the same on both sides. For example, the dental formula for the adult dog is 2 (I 3/3, C 1/1, P 4/4, M 2/2). This means that an adult dog has 3 upper incisors, 1 upper canine, 4 upper premolars, and 2 upper molars on both left and right sides of its mouth. Similarly, an adult dog has 3 lower incisors, 1 lower canine, 4 lower premolars, and 3 lower molars on both left and right sides of its mouth. Adding all the numbers together gives the total dentition, which in the adult dog is 42.

Teeth are occasionally referred to by terms other than those seen in dental formulas. Examples of lay terms for teeth include the following:

- **cheek teeth**   premolars and molars
- **needle teeth**   deciduous canines and third incisor of pigs
- **wolf teeth**   rudimentary premolar 1 in horses
- **milk teeth**   first set of teeth
- **tusks**   permanent canine teeth of pigs
- **carnassial tooth**   large shearing cheek tooth; upper P4 and
  (kahr-nā-zē-ahl)   lower M1 in dogs; upper P3 and lower M1 in cats
- **fighting teeth**   set of six teeth in llamas that include upper vestigial incisors and upper and lower canines on each side

---

The Donts Animals may be grouped together based on the types of teeth they have. Examples include:

selenodont (sē-lēn-ō-dohnt)	animals with teeth that have crescents on their grinding surfaces, i.e., ruminants
lophodont (lō-fō-dohnt)	animals with teeth that have ridged occlusal surfaces; i.e., equine
bunodont (boon-ō-dohnt)	rounded surfaces, i.e., swine
hypsodont (hihps-ō-dohnt)	animals with teeth that have worn animals with continuously erupting teeth; i.e., cheek teeth of ruminants
pleurodont (pluhr-ō-dohnt)	animals with teeth attached by one side on the inner jaw surface; i.e., lizards
brachydont (brā-kē-dohnt)	animals with permanently rooted teeth; i.e., carnivores

---

The anatomy of the tooth basically consists of the enamel (located in the crown) or cementum (located in the root), dentin, and pulp (Figure 6–1). The **enamel** (ē-nahm-ahl) is the hard, white substance covering the dentin of the crown of the tooth. The enamel is the

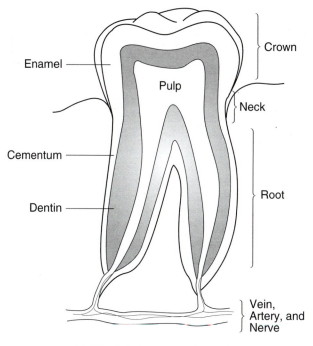

FIGURE 6-1 Anatomy of a tooth

Labels: Enamel, Pulp, Cementum, Dentin, Crown, Neck, Root, Vein, Artery, and Nerve

---

outer surface of the crown. **Cementum** (sē-mehn-tuhm) is the bone-like connective tissue that covers the root of the tooth. Cementum is the outer surface of the root. The **dentin** (dehn-tihn) is the connective tissue surrounding the tooth pulp. The tooth pulp consists of nerves, vessels, and loose connective tissue. The **periodontia** (pehr-ē-ō-dohn-shē-ah) are the structures that support the teeth. Teeth are situated in sockets or sac-like dilations called **alveoli** (ahl-vē-ō-lī). A single socket is called an alveolus (ahl-vē-ō-luhs).

The **gingiva** (jihn-jih-vah) is the mucous membrane that surrounds the teeth and forms the mouth lining. The gingiva is sometimes referred to as the gums. The combining form **gingiv/o** means gums.

Salivary glands (sahl-ih-vahr-ē glahndz) are a group of cells located in the oral cavity that secrete a clear substance containing digestive enzymes (saliva). The **saliva** (sah-lī-vah) moistens food, begins the digestive process by aiding in bolus formation and some digestive enzyme activity, and cleanses the mouth. There are different salivary glands named for the location in which they are found: the **submandibular** (suhb-mahn-dihb-yoo-lahr) **salivary glands** are found under the lower jaw, the **sublingual** (suhb-lihn-gwahl) **salivary glands** are found under the tongue, and the **parotid** (pah-roht-ihd) **salivary glands** are found near the ear. **Para-** is the prefix for near and the combining form **ot/o** means ear. The combining forms for salivary glands are **sialaden/o** and **sial/o** (which also means saliva).

> **Hard to Swallow**
> **Chewing**, also called **mastication** (mahs-tih-kā-shuhn), makes food easier to swallow and can increase the surface area of food particles or ingesta. **Ingesta** (ihn-jehst-ah) is the material taken in orally). This increased surface area increases the contact between digestive enzymes and the food and may speed up the breakdown of food. The first digestive juice that food comes in contact with is saliva. Sometimes animals salivate or drool when they smell food. **Hypersalivation** (hī-pər-sahl-ih-vā-shuhn) is excessive production of saliva. Hypersalivation is also called **ptyalism** (tī-uh-lihz-uhm) and **hypersialosis** (hī-pər-sī-ahl-ō-sihs). The combining forms for saliva are **sial/o** and **ptyal/o**.
>
> The process of swallowing is called **deglutition** (dē-gloo-tih-shuhn). The combining form **phag/o** means eating or ingestion. Swallowed food passes from the mouth to the pharynx and then to the esophagus.

Pharynx

The **pharynx** (fār-ihnks) is the cavity in the caudal oral cavity that joins the respiratory and gastrointestinal systems. The pharynx is also called the **throat**. The combining form for pharynx is **pharyng/o**. The pharynx is covered in Chapter 9 on the respiratory system.

Gullet

The **esophagus** (ē-sohf-ah-guhs) is a collapsible, muscular tube that leads from the oral cavity to the stomach. The esophagus is located dorsal to the trachea. The combining form **esophag/o** means esophagus.

The esophagus enters the stomach through an opening that is surrounded by a sphincter. A **sphincter** (sfingk-tər) is a ringlike muscle that constricts an opening.

Stomach

After leaving the esophagus, the remaining organs of digestion are located in the **peritoneal** (pehr-ih-tō-nē-ahl) **cavity** or abdominal cavity (Figure 6–2). The **abdomen** is the cavity located between the diaphragm and pelvis. The combining forms for abdomen are **abdomin/o** or **celi/o**. **Lapar/o** is the combining form for abdomen or flank. The **peritoneum** (pehr-ih-tō-nē-uhm) is the membrane lining that covers the abdominal and pelvic cavities and some of the organs in this area. The layer of the peritoneum that lines the abdominal and pelvic cavities is termed the **paxetal peritoneum,** while the layer of the peritoneum that covers the abdominal organs is termed the **visceral peritoneum.**

Food enters the stomach from the esophagus where it is stored, and the act of digestion begins. The stomach is connected to other visceral organs by a fold of peritoneum called the **omentum** (ō-mehn-tuhm). The combining form for stomach is **gastr/o.**

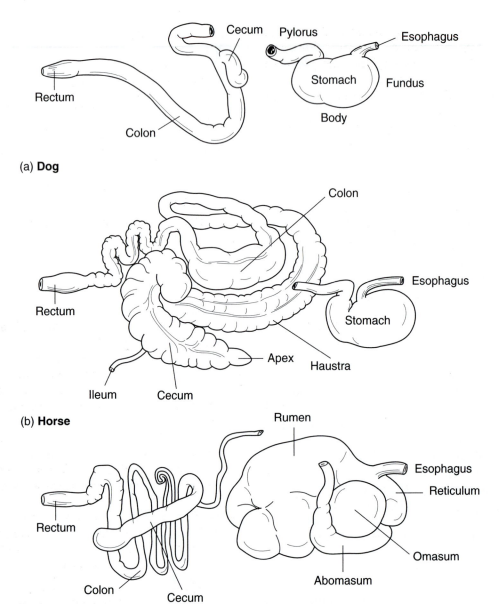

(a) **Dog**

(b) **Horse**

(c) **Ruminant**

FIGURE 6–2 Gastrointestinal tracts. (Small intestinal segments omitted for clarity.) (a) Dog;
(b) Horse; (c) Ruminant

Animals can be classified as **monogastric** (mohn-ō-gahs-trihck) or **ruminant** (roo-mihn-ehnt). Monogastric animals have one true or glandular stomach. The **glandular stomach** is the one that produces secretions for digestion. Ruminants also have one true or glandular stomach (the abomasum), but they also have three forestomachs (the rumen, reticulum, and omasum). The forestomachs of ruminants are actually outpouchings of the esophagus. The parts of the true stomach are (Figure 6–3):

▶ **cardia** (kahr-dē-ah) = entrance area located nearest the esophagus
▶ **fundus** (fuhn-duhs) = base of an organ, which is the cranial, rounded part
▶ **body** (boh-dē) = main portion of an organ, which is the rounded base or bottom; also called the **corpus** (kōr-puhs)
▶ **antrum** (ahn-truhm) = caudal part, which is the constricted part of the stomach that joins the pylorus

Rugae (roo-gā) are the folds present in the mucosa of the stomach. Rugae contain glands that produce gastric juices that aid in digestion, and the mucus forms a protective coating for the stomach lining.

The **pylorus** (pī-lōr-uhs) is the narrow passage between the stomach and the duodenum. The combining form **pylor/o** means gatekeeper and refers to the narrow passage between the stomach and duodenum.

MAKE ROOM FOR THE RUMINANTS

Ruminants are animals that can **regurgitate** (rē-guhr-jih-tāt) and **remasticate** (rē-mahs-tih-kāt) their food. The ruminant stomach is adapted for fermentation of ingested food by bacterial and protozoan microorganisms. Normal microorganisms residing in the gastrointestinal tract are referred to as **intestinal flora** (ihn-tehs-tih-nahl flō-rah). These microbes produce enzymes that can digest plant cells through fermentation. Fermentation is aided by regurgitation (return of undigested material from the stomach to the mouth) and remastication (rechewing). Regurgitation and remastication provide finely chopped material with a greater surface area to the stomach. Regurgitated food particles, fiber, rumen fluid, and rumen microorganisms are called **cud** (kuhd).

The ruminant stomach is divided into four parts (Figure 6–4).

▶ **rumen** (roo-mehn) = largest compartment of the ruminant stomach that serves as a fermentation vat;

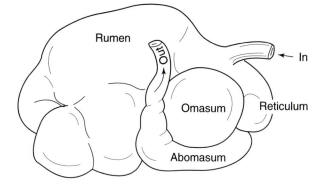

FIGURE 6–4 Parts of the ruminant stomach area

also called the **paunch.** The rumen is divided into a ventral sac and dorsal sac.

▶ **reticulum** (re-tihck-yoo-luhm) = most cranial compartment of the ruminant stomach, also called the honeycomb because it is lined with a mucous membrane that contains numerous intersecting ridges

▶ **omasum** (ō-mā-suhm) = third compartment of the ruminant stomach. The omasum has short blunt papillae that grind food before it enters the abomasum. Omasal contractions also squeeze fluid out of the food bolus.

▶ **abomasum** (ahb-ō-mā-suhm) = fourth compartment of the ruminant stomach. Also called the true stomach. The abomasum is the glandular portion that secretes digestive enzymes.

The layout of the ruminant stomach depends upon the age of the animal. In adult ruminants the rumen is the largest compartment and occupies a prominent portion of the left side of the animal. The abomasum is for the most part on the right side of the animal. In young ruminants the abomasum is the largest compartment. Forestomach development is associated with roughage intake and calves are fed only milk for a period of time after birth.

In adult ruminants food enters the rumen. A contraction transfers the rumen contents into the reticulum. The foodstuff is then regurgitated and either enters the caudal part of the rumen or enters the omasum. The plies of the omasum grind the food, and water is removed. Food enters the abomasum or true stomach, which is similar to the glandular stomach of the monogastric. In young ruminants, a reticular groove shuttles milk from the esophagus to the abomasum.

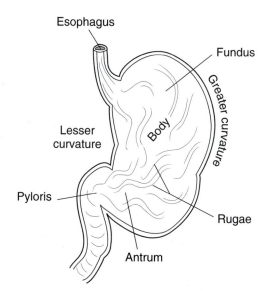

FIGURE 6–3 Parts of the stomach

Rumination

Rumination (roo-mehn-ā-shuhn) is the process of bringing up food material from the stomach to the mouth for further chewing. Rumination is a cycle of four phases: regurgitation, remastication, resalivation, and redeglutition.

Step 1—regurgitation: the animal takes a deep breath (glottis is closed), the thoracic cavity enlarges and intrapleural pressure decreases, the cardia opens, and because of the low pressure in the esophagus the rumen content is aspirated into the esophagus. Reverse peristalsis occurs and the food bolus quickly enters the mouth.

Steps 2 and 3—remastication and resalivation: liquid is squeezed out of bolus, and the liquid is reswallowed. Remastication and resalivation occur together and the animal may chew its cud 100 or more times before swallowing.

Step 4—redeglutition: the bolus is reswallowed, and the next rumination cycle begins.

Rumination usually occurs when the animal is quiet. The time spent ruminating each day varies with species and diet. Smaller particles (like ground feed) require less rumination time.

Vore or Less

An **herbivore** (hərb-ih-vōr) is an animal that is able to sustain life by eating only plants. An **omnivore** (ohm-nih-vōr) is an animal that sustains life by eating plant and animal tissue. A **carnivore** (kahr-nih-vōr) is an animal that is able to sustain life by only eating animal tissue. Remember just because a carnivore may eat a plant does not make it an omnivore. A cat is a carnivore. Cats may occasionally eat grass. This does not make them omnivores.

Small Intestine

The small intestine or small bowel extends from the pylorus to the proximal part of the large intestine. The intestine is attached to the dorsal abdominal wall by a fold of the peritoneum called the **mesentery** (mehs-ehn-tehr-ē). Digestion of food and absorption take place in the small intestine for those animals not requiring extensive fermentation of their ingested food. The combining form **enter/o** means small intestine. **Gastroenterology** (gahs-trō-ehn-tər-ohl-ō-jē) is the study of the stomach and small intestine.

The small intestine has three segments. These are

▶ **duodenum** (doo-ō-də-nuhm or doo-wahd-nuhm) = proximal or first portion of the small intestine. The proximal portion is also known as the most oral portion. The duodenum is that segment of the small intestine located nearest the mouth. The combining forms **duoden/i** and **duoden/o** mean duodenum.

▶ **jejunum** (jeh-joo-nuhm) = middle portion of the small intestine. The combining form **jejun/o** means jejunum.

▶ **ileum** (ihl-ē-uhm) = distal or last portion of the small intestine. The distal portion is also known as the most aboral (ahb-ōr-ahl) portion. The ileum is that segment of the small intestine located furthest from the mouth. The combining form **ile/o** mean ileum. Remember ileum is spelled with an "e" as in eating or entero. Ilium is part of the hip bone.

Once food is digested in the small intestine it forms a milky fluid. This milky fluid is called **chyle** (kī-uhl). Chyle is absorbed through the intestinal wall and travels via the thoracic duct where it is passed into veins (Figure 6–5).

Large Intestine

The large intestine or large bowel extends from the ileum to the anus. The large intestine consists of the **cecum** (sē-kuhm), **colon** (kō-lihn), **rectum** (rehck-tuhm), and **anus** (ā-nuhs). Development of the large intestine varies among species. Fermentation occurs to some extent in the large intestine of all animals, but is a more consuming process in herbivorous animals. In ruminants the forestomachs constitute a fermentation vat (hence they are called foregut fermenters) and in nonruminant herbivores (like rabbits and horses) the cecum and colon provide fermentation (hence they are called hindgut fermenters).

Food enters a pouch called the cecum from the ileum. The cecum may be poorly developed as in the dog or large as in the horse or double pouched as in birds. **Cec/o** is the combining form for cecum.

The colon continues from the cecum to its termination at the anus. The colon consists of ascending, transverse, and descending portions. All animals have a transverse and descending portion. The arrangement from the cecum to the transverse colon varies among species. Dogs and cats have an ascending colon, pigs and ruminants have spiral colons, and horses have large colons. The cecum and colon of pigs and horses are sacculated; these sacculations are called **haustra** (hahw-strah). Haustra act as buckets and prolong retention of material so that the microbes have more time for digestion. Haustra are formed because of the longitudinal smooth muscle bands, or **teniae** (tehn-ē-ā), in the cecal wall.

The Colon

The colon is divided into three parts: ascending, transverse, and descending. **Ascending** is the part that progresses upward or cranially. Ascend means to move up. **Transverse** is the part that travels across. **Trans-** is the prefix for across. **Descending** is the part that progresses downward or caudally. To descend a flight of stairs is to move down. Bends or curves are called **flexures** (flehck-shərz); therefore, the **pelvic flexure** is a bend in the colon near the pelvis, and the **diaphragmatic flexure** is a bend in the colon near the diaphragm.

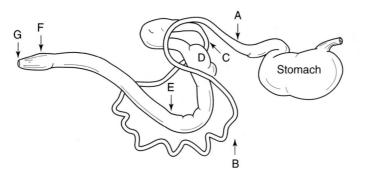

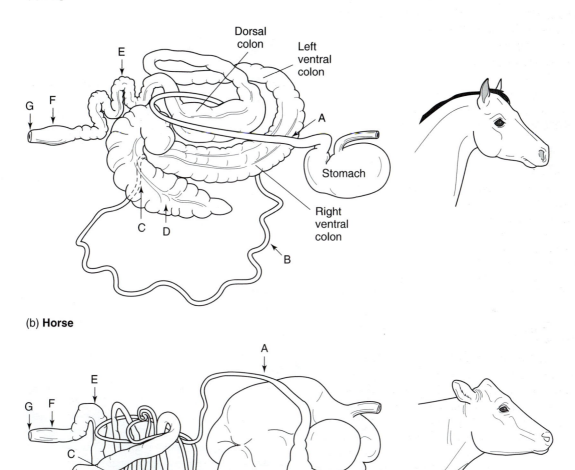

(a) **Dog**

(b) **Horse**

(c) **Ruminant**

Dorsal colon

Left ventral colon

Right ventral colon

Stomach

Spiral loops

A = Duodenum
B = Jejunum
C = Ileum
D = Cecum
E = Colon
F = Rectum
G = Anus

FIGURE 6–5 Intestines. (a) Dog; (b) Horse; (c) Ruminant

The rectum is the caudal portion of the large intestine. The combining form **rect/o** means rectum. The anus is the caudal opening of the gastrointestinal tract. The anus is controlled by two anal sphincter muscles that tighten or relax to allow or control defecation. The combining form for anus is **an/o. Anorectal** (ā-nō-rehck-tahl) is a term that means pertaining to the anus and rectum. The combining form **proct/o** refers to the anus and rectum collectively.

The anal canal is a short terminal portion of the digestive tract. In dogs and cats there are a pair of pouches in the skin between the internal and external anal sphincters. These pouches are called **anal sacs.** Anal sacs are lined with microscopic anal glands that secrete a foul-smelling fluid. Normally, the anal sacs are compressed during defecation. The fluid may be related to social recognition in dogs and cats.

Accessories

Accessory organs aid the digestive tract in different ways. The accessory organs of the digestive tract include the salivary glands (covered previously), the liver, the gallbladder, and the pancreas (Figure 6–6).

LIVER

The liver is located caudal to the diaphragm and has several important functions. The combining form for liver is **hepat/o.** The liver removes excess **glucose** (gloo-kohs) from the bloodstream and stores it as **glycogen** (glī-koh-jehn). When blood sugar is low, a condition called **hypoglycemia** (hī-pō-glī-sē-mē-ah) results, the liver converts glycogen back into glucose and releases it. The liver also destroys old erythrocytes, removes toxins from the blood, stores iron, vitamins A, B$_{12}$, and D, and produces some blood proteins. Liver cells are called **hepatocytes** (heh-paht-ō-sītz). The liver also contains **sinusoids** (sīn-yoo-soidz) or channels. The hepatocytes make up the liver **parenchyma** (pahr-ehnk-ih-mah) or the functional elements of a tissue or organ.

The digestive function of the liver involves the production of **bile** (bīl). The term **biliary** (bihl-ē-ār-ē) means pertaining to bile. Bile travels down the hepatic duct to the cystic duct that leads to the gallbladder (in those species that have a gallbladder). Bile alkalinizes the small intestine and bile salts play a part in fat digestion. Fat digestion is called **emulsification** (ē-muhl-sih-fih-kā-shuhn). **Bilirubin** (bihl-ē-roo-bihn) is a

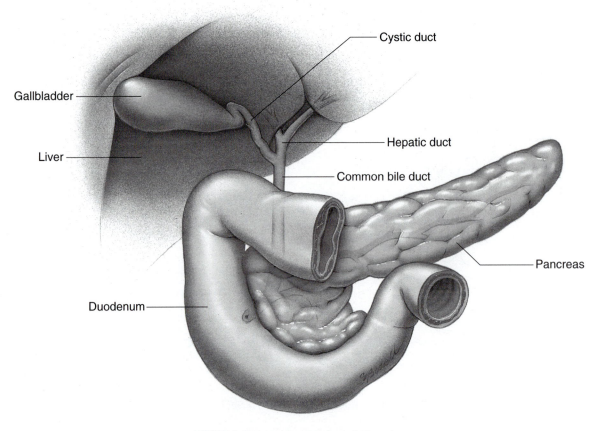

Cystic duct

Gallbladder

Hepatic duct

Liver

Common bile duct

Pancreas

Duodenum

FIGURE 6–6 Accessory organs of digestion

pigment produced from the destruction of hemoglobin that is released by the liver in bile.

GALLBLADDER

The gallbladder is a sac located under the liver that stores bile for later use. When bile is needed the gallbladder contracts, which forces bile out of the cystic duct into the common bile duct. The common bile duct carries bile into the duodenum. The rat and horse do not have gallbladders, but have a continuous flow of bile into the duodenum. The combining form **cyst/o** means cyst, sac of fluid, or urinary bladder. The combining form **chol/e** means bile or gall. The combining form **doch/o** means receptacle. Together these combining forms are used to refer to the bile and its associated structures. **Cholecystic** (kō-lē-sihst-ihck) means pertaining to the gallbladder. **Choledochus** (kō-lehd-uh-kuhs) means common bile duct.

PANCREAS

The **pancreas** (pahn-krē-ahs) is an elongated gland located near the cranial portion of the duodenum. The main pancreatic duct enters the duodenum close to the common bile duct (in some species like sheep and goats it empties directly into the common bile duct). **Pancreat/o** is the combining form for pancreas. The pancreas is an organ that has both exocrine and endocrine functions. The endocrine functions are covered in Chapter 11 on the endocrine system. The exocrine function of the pancreas involves the production of pancreatic juices, which are filled with digestive enzymes. **Trypsin** (trihp-sihn) is an enzyme that digests protein, **lipase** (lī-pās) is an enzyme that digests fat, and **amylase** (ahm-ih-lās) is an enzyme that digests starch. All are produced in the pancreas.

Digestion

Digestion (dī-gehst-shuhn) is the process of breaking down foods into nutrients that the body can use. **Enzymes** (ĕhn-zīmz) are substances that chemically change another substance. Digestive enzymes are responsible for the chemical changes that break foods into smaller forms for the body to use. Enzymes typically end with the **-ase** ending. For example, proteases are enzymes that work on proteins, and lipases are enzymes that work on fats. One enzyme that does not end in -ase is pepsin, which is an enzyme that digests protein.

Metabolism (meh-tahb-ō-lihzm) is the processes involved in the body's use of nutrients. The prefix **meta-** means change or beyond. **Anabolism** (ahn-nahb-ō-lihzm) is the building of body cells and substances. **Catabolism** (kah-tahb-ō-lihzm) is the breaking down of body cells and substances.

Absorption (ahb-sōrp-shuhn) is the process of taking digested nutrients into the circulatory system. A nutrient (nū-trē-ehnt) is a substance that is necessary for normal functioning of the body. Absorption occurs mainly in the small intestine. The small intestine contains tiny hairlike projections called **villi** (vihl-ī). A single projection is called a **villus** (vihl-uhs). The combining form **vill/i** means tuft of hair. Because the small intestine has villi or projections, it also has blind sacs or valleys. These blind sacs are called **crypts** (krihptz). Think of a crypt as something hidden. An intestinal crypt is a valley of the intestinal mucous membrane lining the small intestine.

PATH OF DIGESTION

- ▶ **prehension** (prē-hehn-shuhn) or grasping of food involves collecting food into the oral cavity
- ▶ **mastication** breaks food into smaller pieces and mixes the ingesta with saliva
- ▶ **deglutition** moves chewed ingesta into the pharynx and on into the esophagus (the epiglottis closes off the entrance to the trachea and allows food to move into the esophagus)
- ▶ food moves down the esophagus by gravity and peristalsis
- ▶ **peristalsis** (pehr-ih-stahl-sihs) is the series of wavelike contractions of smooth muscles (the suffix **-stalsis** means contraction)

Ingesta moves into the stomach area. In ruminants it enters the rumen, reticulum, and omasum before entering the true stomach or abomasum. Ruminant function is described above. The true glandular stomach contains **hydrochloric** (hī-drō-klōr-ihck) **acid** and the enzymes **protease** (prō-tē-ās), **pepsin** (pehp-sihn), and **lipase** (lī-pās). The muscular action of the stomach mixes the ingesta with the gastric juices to convert the food to chyme. **Chyme** (kīm) is the semifluid mass of partly digested food that passes from the stomach.

Chyme passes from the stomach into the duodenum. Food moves through the small intestines by peristaltic action and **segmentation** (sehg-mehn-tā-shuhn). Peristalsis represents contractile waves that propel ingesta caudally towards the anus; segmentation represents mixing and thus delays movement of ingesta (Figure 6–7). Digestion is completed in the duodenum after chyme has mixed with bile and pancreatic secretions.

Digested food is absorbed in the small intestines. Another term for absorption is **assimilation** (ah-sihm-ih-lā-shuhn)

The large intestine receives waste products of digestion and in some species is responsible for fermentation

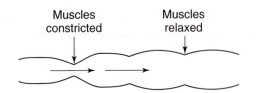

(a) Peristaltic wave of contraction
Peristalsis propels food through
the digestive system.

(b) Segmentation
Segmentation helps break down
and mix food through cement-mixer
type action.

FIGURE 6–7 Peristalsis versus segmentation. (a) Peristaltic wave of contraction. Peristalsis propels food through the digestive system. (b) Segmentation helps break down and mix food through cement-mixer type action.

of plant material and vitamin synthesis. Excess water is absorbed from the waste, and solid feces are formed. **Defecation** (dehf-eh-kā-shuhn) is the emptying of the bowels.

The scoop on poop: Terms used to describe solid waste include

▶ feces
▶ dung
▶ manure
▶ stool
▶ bowel movement
▶ excrement
▶ poop

▶ **Test Me: Digestive System** ◀

Diagnostic tests performed on the digestive system include

- **radiography** (rā-dē-ohg-rah-fē) = imaging of internal structures is created by the exposure of sensitized film to X-rays. Radiographs of the gastrointestinal system demonstrate foreign bodies, torsions, organ distension or enlargement, and some masses.
- **barium** (bār-ē-uhm) = contrast material used for radiographic studies. To evaluate the gastrointestinal tract barium sulfate may be given orally (and the

resulting test is called a barium swallow or upper GI) or barium sulfate may be given rectally (and the resulting test is called a barium enema or lower GI). An **enema** (ehn-ah-mah) is introduction of fluid into the rectum (Figure 6–8).

- **hemoccult** (hēm-ō-kuhlt) = test for hidden blood in the stool. **Occult** means hidden.
- **esophagoscopy** (ē-sohf-ah-gohs-kō-pē) = endoscopic visual examination of esophagus; the scope is passed from the oral cavity through the esophagus (Figure 6–9).
- **gastroscopy** (gahs-trohs-kō-pē) = endoscopic visual examination of the inner surface of the stomach; the scope is passed from the oral cavity through the stomach. An **endoscope** (ehn-dō-skōp) is a tube-like instrument with lights and refracting mirrors that is used to internally examine the body or organs.

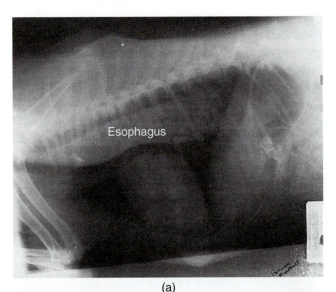

Esophagus

(a)

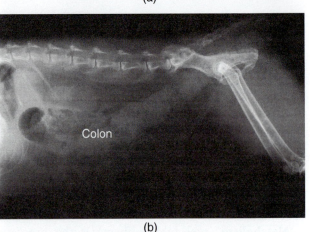

Colon

(b)

FIGURE 6–8 Contrast radiography. (a) Megaesophagus is diagnosed in the dog using contrast radiography. (b) Megacolon is diagnosed in the dog using contrast radiography. *Source:* (b) Lodi Veterinary Hospital, S.C.

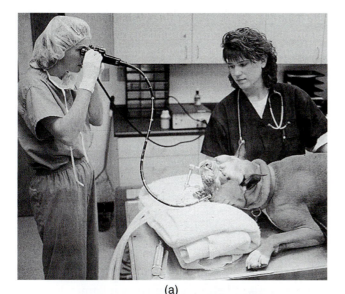

(a)

(b)

FIGURE 6–9 Endoscopy. (a) An endoscope is passed through the esophagus (esophagoscopy) to view parts of the gastrointestinal tract. (b) View of the intestine from an endoscope. A string foreign body is identified. *Sources:* (a) Lodi Veterinary Hospital, S.C.; (b) Mark Jackson, DVM, PhD.

- **colonoscopy** (kō-luhn-ohs-kō-pē) = endoscopic visual examination of the inner surface of the colon; the scope is passed from rectum through colon
- **ultrasound** (uhl-trah-sownd) = imaging of internal body structures by recording echoes of pulse of sound waves
- **biopsy** (bī-ohp-sē) = removing tissue to examine. Biopsies can be **incisional** (ihn-sih-shuhn-ahl) in which part of the tissue is removed and examined or they can be **excisional** (ehcks-sih-shuhn-ahl) in which the entire tissue is removed.
- **blood tests** can be used to determine some disease of the gastrointestinal tract. For example bile acids are used to assess liver disease and elevated amylase levels may indicate pancreatitis. Sometimes it is important to know if the blood sample was taken as a preprandial or postprandial sample. **Preprandial**

(prē-prahn-dē-ahl) is before eating and **postprandial** (pōst-prahn-dē-ahl) is after eating.
- **fecal examinations** can be used to determine parasitic diseases of animals. Specialized fecal tests can also identify bacteria or isolate viruses or demonstrate abnormal substances present in the stool.
- **ballottement** (bahl-oht-mehnt) = diagnostic technique of hitting or tapping the wall of a fluid-filled structure to bounce a solid structure against a wall; used for pregnancy diagnosis and determination of abdominal contents

▶ Path of Destruction: Digestive System ◀

Pathologic conditions of the digestive system include
- **palatoschisis** (pahl-ah-tohs-kih-sihs) = congenital fissure of the roof of the mouth that may involve the upper lip, hard palate, and soft palate; more commonly called a **cleft palate**
- **plaque** (plahck) = small differentiated area on a body surface. In the gastrointestinal system it is used to refer to the mixed colony of bacteria, leukocytes, and salivary products that adhere to the tooth enamel; also referred to as dental plaque.
- **dental calculus** (dehn-tahl kahl-kyoo-luhs) = abnormal mineralized deposit that forms on teeth. Calculus is mineral deposit. Dental calculus is also called **tartar** (tahr-tahr).
- **periodontitis** (pehr-ē-ō-dohn-tī-tihs) = inflammation of the tissue surrounding and supporting the teeth; also called **periodontal disease. Inflammation** is a localized protective response elicited by injury or destruction of tissue. The signs of inflammation are heat, redness, pain, swelling, and loss of function.
- **gingivitis** (jihn-jih-vī-tihs) = inflammation of the gums
- **bruxism** (bruhck-sihzm) = involuntary grinding of the teeth
- **malocclusion** (mahl-ō-kloo-shuhn) = abnormal contact between the teeth. The prefix **mal-** means bad and **occlusion** means any contact between the chewing surface of the teeth.
- **esophageal reflux** (ē-sohf-ah-jē-ahl rē-fluhcks) = return of stomach contents into the esophagus; also called gastroesophageal reflux
- **hiatal hernia** (hī-ā-tahl hər-nē-ah) = protrusion of part of the stomach through the esophageal opening in the diaphragm
- **dysphagia** (dihs-fā-jē-ah) = difficulty swallowing or eating. The prefix **dys-** means difficult and the suffix **-phagia** means eating or swallowing.
- **polyphagia** (pohl-ē-fā-jē-ah) = increased eating or swallowing. The prefix **poly-** means many or much.
- **stomatitis** (stō-mah-tī-tihs) = inflammation of the mouth

- **gastritis** (gahs-trī-tihs) = inflammation of the stomach
- **ulcer** (uhl-sihr) = erosion of tissue
- **perforating ulcer** (pər-fohr-āt-ihng uhl-sihr) = erosion through the entire thickness of a surface
- **anorexia** (ahn-ō-rehck-sē-ah) = lack or loss of appetite
- **inappetence** (ihn-ahp-eh-tehns) = lacking the desire to eat. Anorexia and inappetence are not the same.
- **regurgitation** (rē-gərg-ih-tā-shuhn) = return of swallowed food into the oral cavity; a passive event versus the force involved with vomiting
- **nausea** (naw-sē-ah) = stomach upset or sensation of urge to vomit; difficult to use descriptively in animals
- **lethargy** (lehth-ahr-jē) = condition of drowsiness or indifference
- **flatulence** (flaht-yoo-lehns) = excessive gas formation in the gastrointestinal tract
- **dehydration** (dē-hī-drā-shuhn) = condition of excessive loss of body water or fluid
- **dental caries** (dehn-tahl kār-ēz) = decay and decalcification of teeth; producing a hole in the tooth
- **emesis** (ehm-eh-sihs) = vomiting. When an animal is vomiting, the recommendation is not to give it anything orally. The term for orally is **per os,** which is abbreviated **PO.** If the desire is to give nothing orally, the abbreviation is **NPO.**
- **hematemesis** (hēm-ah-tehm-eh-sihs) = vomiting blood
- **aerophagia** (ār-ō-fā-jē-ah) = swallowing of air
- **eructation** (ē-ruhk-ta-shuhn) = belching or raising gas orally from the stomach
- **borborygmus** (bohr-bō-rihg-muhs) = gas movement in the gastrointestinal tract that produces a rumbling noise
- **gastroenteritis** (gahs-trō-ehn-tehr-ī-tihs) = inflammation of the stomach and small intestine. Note that the stomach anatomically occurs first followed by the small intestine. This order is reflected in the order of the medical terms as well.
- **enteritis** (ehn-tər-ī-tihs) = inflammation of the small intestines
- **ileitis** (ihl-ē-ī-tihs) = inflammation of the ileum
- **ileus** (ihl-ē-uhs) = stoppage of intestinal peristalsis
- **diverticulitis** (dī-vər-tihck-yoo-lī-tihs) = inflammation of a pouch or pouches occurring in the wall of a tubular organ. A **diverticulum** (dī-vər-tihck-yoo-luhm) is a pouch occurring on the wall of a tubular organ; diverticula are pouches occurring on the wall of a tubular organ.
- **colitis** (kō-lī-tihs) = inflammation of the colon

- **melena** (meh-lē-nah) = black stools containing digested blood. Melena suggests an upper gastrointestinal tract bleeding problem.
- **obstruction** (ohb-struhck-shuhn) = complete stoppage or impairment to passage. Obstructions are usually preceded by a term that describes its location, as in **intestinal obstruction.** If the obstruction is not complete it is termed a **partial obstruction.**
- **volvulus** (vohl-vū-luhs) = twisting on itself
- **intussusception** (ihn-tuhs-suhs-sehp-shuhn) = telescoping of one part of the intestine into an adjacent part (Figure 6–10)
- **inguinal hernia** (ihng-gwih-nahl hər-nē-ah) = protrusion of bowel through a weakened place in the groin
- **dysentery** (dihs-ehn-tər-ē) = number of disorders marked by inflammation of the intestine
- **hepatitis** (hehp-ah-tī-tihs) = inflammation of the liver
- **cirrhosis** (sihr-rō-sihs) = degenerative disease that disturbs the structure and function of the liver
- **jaundice** (jawn-dihs) = yellow discoloration of the skin and mucous membranes due to greater than normal levels of bilirubin; also called **icterus** (ihck-tər-uhs)
- **ascites** (ah-sī-tēz) = abnormal accumulation of fluid in the abdomen
- **hepatomegaly** (hehp-ah-tō-mehg-ah-lē) = abnormal enlargement of the liver
- **cholecystitis** (kō-lē-sihs-tī-tihs) = inflammation of the gallbladder
- **bloat** (blōt) = accumulation of gas in the digestive tract. In monogastric animals, bloat is accumulation of gas in the stomach. In ruminants bloat is accumulation of gas in the rumen, abomasum, stomach, or cecum. In ruminants gas accumulation in the rumen is also called **ruminal tympany** (tihm-pahn-ē).

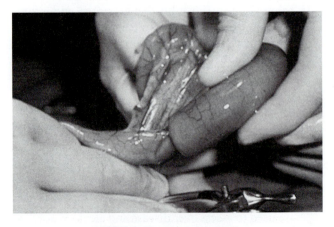

FIGURE 6–10 Intussusception

Gastric Dilatation Volvulus

Pyloric antrum is displaced aborally	Pylorus crosses midline, passes underneath distended oral part of stomach	Fundus moves ventrally and becomes located in ventral abdomen	Gastric dilatation displaces greater curvature ventrally

Necrotic Tissue

FIGURE 6–11 Gastric dilation volvulus formation

- **pica** (pī-kah) = eating and licking abnormal substances or a depraved appetite
- **coprophagia** (kō-prō-fā-jē-ah) = ingestion of fecal material. The combining form **copr/o** means feces.
- **ascariasis** (ahs-kah-rī-ah-sihs) = parasitic infestation with roundworms of the genus *Ascaris*
- **colic** (kohl-ihck) = severe abdominal pain
- **displaced abomasum** (dihs-plāsd ahb-ō-mā-suhm) = disease of ruminants in which the fourth stomach compartment becomes trapped under the rumen; also called DA. It is denoted LDA (left displaced abomasum) or RDA (right displaced abomasum) depending on its location. LDAs are more common.
- **enterocolitis** (ehn-tehr-ō-kō-lī-tihs) = inflammation of the small intestine and large intestine
- **eviscerate** (ē-vihs-ər-āt) = removal or exposure of internal organs. Evisceration is used to describe the exposure of internal organs after unsuccessful surgical closure of the abdomen (or other area containing organs).
- **gastric dilatation** (gahs-trihck dihl-ah-tā-shuhn) = condition usually seen in deep chested canines in which the stomach fills with air and expands
- **gastric dilation volvulus** (gahs-trihck dihl-ah-shuhn vohl-vū -luhs) = condition usually seen in deep-chested canines in which the stomach fills with air, expands, and twists on itself (Figure 6–11)
- **shunt** (shuhnt) = to bypass or divert. A **portosystemic** (poor-tō-sihs-tehm-ihck) **shunt** is where blood vessels bypass the liver and the blood does not get detoxified properly.
- **glossitis** (glohs-ī-tihs) = inflammation of the tongue

- **diarrhea** (dī-ah-rē-ah) = abnormal frequency and liquidity of fecal material
- **scours** (skowrz) = diarrhea in livestock
- **atresia** (ah-trēz-ah) = occlusion or absence of normal body opening or tubular organ
- **megacolon** (mehg-ah-kō-lihn) = abnormally large colon
- **megaesophagus** (mehg-ah-ē-sohf-ah-guhs) = abnormally large esophagus
- **achalasia** (ahk-ah-lahz-ah) = inability to relax the smooth muscle of the gastrointestinal tract
- **constipation** (kohn-stah-pā-shuhn) = condition of prolonged gastrointestinal transit time, making the stool hard, dry, and difficult to pass
- **cribbing** (krihb-ihng) = vice of equine in which an object is grasped between the teeth and pressure is applied
- **fecalith** (fēk-ah-lihth) = stonelike fecal mass. The suffix **-lithiasis** (lih-thī-ah-sihs) means presence of stones. **Coprolith** (kō-prō-lihth) is another name for a fecalith.
- **hemoperitoneum** (hēm-ō-pehr-ih-tō-nē-uhm) = blood in the peritoneum. The peritoneum is the membrane lining the wall of the abdominal and pelvic cavities and covers some of the organs in this area
- **hydrops** (hī-drohps) = abnormal accumulation of fluid in tissues or a body cavity; also called **dropsy** (drohp-sē)
- **hyperglycemia** (hī-pər-glī-sē-mē-ah) = elevated blood sugar levels
- **hypoglycemia** (hī-pō-glī-sē-mē-ah) = lower than normal blood sugar levels
- **incontinence** (ihn-kohn-tihn-ehns) = inability to control. When using the term incontinence a

descriptive term is usually applied in front of it. For example, fecal incontinence is the inability to control bowel movements.

- **polydipsia** (pohl-ē-dihp-sē-ah) = increased thirst or drinking
- **stenosis** (steh-nō-sihs) = narrowing of an opening. The term stenosis is usually used with a descriptive term in front of it. For example a pyloric stenosis is narrowing of the pylorus as it leads into the duodenum.
- **tenesmus** (teh-nehz-muhs) = painful, ineffective defecation. Tenesmus also means painful, ineffective urination but is rarely used in this context.
- **dyschezia** (dihs-kē-zē-ah) = difficulty defecating. Chezein is Greek for stool.
- **hematochezia** (hēm-aht-ō-kē-zē-ah) = passage of bloody stool
- **prolapse** (prō-lahpz) = protrusion of viscera, "to fall forward." A descriptive term usually precedes the term prolapse. For example a rectal prolapse is protrusion of the rectum through the anus.
- **anal sacculitis** (ā-nahl sahck-yoo-lī-tihs) = inflammation of the pouch(es) located around the anus. The term **inspissation** (ihn-spihs-sā-shuhn) is the process of rendering dry or thick by evaporation and is used to describe the anal sac fluid in animals with anal sacculitis.
- **epulis** (ehp-uhl-uhs) = benign tumor arising from periodontal mucous membranes
- **hepatoma** (heh-pah-tō-mah) = tumor of liver (malignant, but cannot determine by medical term)
- **polyp** (pohl-uhp) = small growth on mucous membrane

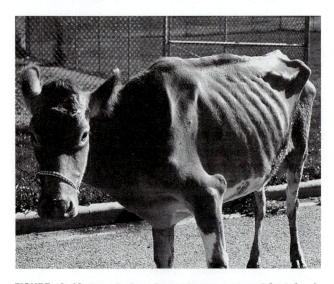

FIGURE 6–12 Emaciation in a Jersey cow with Johne's disease. *Source:* Photo by Michael T. Collins, DVM, PhD, University of Wisconsin School of Veterinary Medicine.

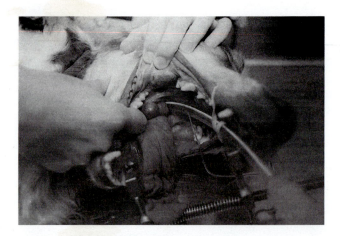

FIGURE 6–13 Salivary mucocele in a dog. *Source:* Lodi Veterinary Clinic S.C.

- **cachexia** (kah-kehcks-ē-ah) = general ill health and malnutrition; used in describing the condition of cancer patients
- **emaciation** (ē-mā-sē-ā-shuhn) = marked wasting or excessive leanness (Figure 6–12).
- **malabsorption** (mahl-ahb-sōrp-shuhn) = impaired uptake of nutrients from the intestine
- **exocrine pancreatic insufficiency** (ehck-sō-krihn pahn-krē-ah-tihck ihn-suh-fih-shehn-sē) = metabolic disease in which the pancreas does not secrete adequate amounts of digestive enzymes and is associated with weight loss, fatty stools, and borborygmus; abbreviated EPI
- **trichobezoar** (trī-kō-bē-zōr) = hairball. The combining form **trich/o** means hair.
- **quidding** (kwihd-ihng) = condition in which food is taken into the mouth and chewed but falls from the mouth.
- **salivary mucocele** (sahl-ih-vahr-ē myoo-kō-sēl) = collection of saliva that has leaked from a damaged salivary gland or duct and is surrounded by granulation tissue (Figure 6–13).

▶ Procede with Caution: Digestive System ◀

Procedures performed on the digestive system include

- **palatoplasty** (pahl-ah-tō-plahs-tē) = surgical repair of a cleft palate
- **extraction** (ehcks-trahck-shuhn) = to remove; used to describe surgical removal of a tooth
- **gingivectomy** (jihn-jih-vehck-tō-mē) = surgical removal of the gum tissue
- **esophagoplasty** (ē-sohf-ah-gō-plahs-tē) = surgical repair of the esophagus
- **emetic** (ē-meh-tihck) = producing vomiting. An **antiemetic** (ahn-tih-ē-meh-tihck) prevents vomiting.

- **antidiarrheal** (ahn-tih-dī-ər-rē-ahl) = substance that prevents frequent and extremely liquid stool
- **orogastric intubation** (ōr-ō-gahs-trihck ihn-too-bā-shuhn) = passage of a tube from the mouth to the stomach; also called a stomach tube. Orogastric means pertaining to the mouth and stomach.
- **nasogastric intubation** (nā-zō-gahs-trihck ihn-too-bā-shuhn) = placement of a tube through the nose into the stomach (Figure 6–14).
- **gastrectomy** (gahs-trehck-tō-mē) = surgical removal of all or part of the stomach. To clarify the extent of the excision, the term **partial gastrectomy** is used to denote surgical removal of part of the stomach
- **gastrotomy** (gahs-troht-ō-mē) = surgical incision into the stomach (Figure 6–15).
- **laparotomy** (lahp-ah-roht-ō-mē) = surgical incision into the abdomen; **lapar/o** is the combining form for abdomen or flank
- **colotomy** (kō-loht-ō-mē) = surgical incision into the colon
- **colectomy** (kō-lehck-tō-mē) = surgical removal of the colon
- **fistula** (fihs-tyoo-lah) = abnormal passage from an internal organ to the body surface or between two internal organs. A cow that has an artificial opening created between the rumen and the body surface has a rumen fistula. This is also called a **rumenostomy** (roo-mehn-ah-stō-mē). A **perianal fistula** (pehr-ih-ā-nahl fihsh-too-lah) is an abnormal passage around the caudal opening

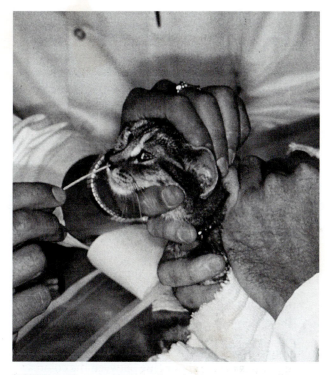

FIGURE 6–14 Nasogastric tube placement in a cat. *Source:* Mark Jackson, DVM, PhD

of the gastrointestinal tract. Perianal means around the anus.
- **gastroduodenostomy** (gahs-trō-doo-ō-deh-nohs-tō-mē) = removal of part of the stomach and duodenum and making a connection between them

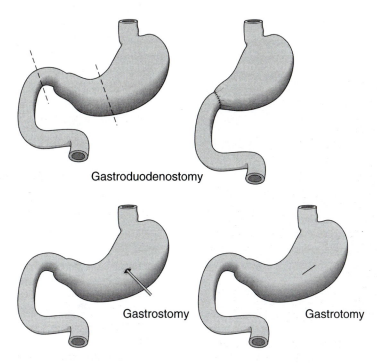

Gastroduodenostomy

Gastrostomy

Gastrotomy

FIGURE 6–15 Gastrotomy, gastrostomy, and gastroduodenostomy

- **anastomosis** (ah-nahs-tō-mō-sihs) = surgical connection between two tubular or hollow structures
- **ileectomy** (ihl-ē-ehck-tō-mē) = surgical removal of the ileum
- **anoplasty** (ah-nō-plahs-tē) = surgical repair of the anus
- **gastrostomy** (gahs-trohs-tō-mē) = surgical production of an artificial opening between the stomach and abdominal wall. The suffix -**stomy** means to surgically produce an artificial opening between an organ and a body surface. The opening created during this procedure is a **stoma** (stō-mah). **Effluent** (ehf-floo-ehnt) means discharge and an effluent flow from the stoma created by astomy surgery.
- **enterostomy** (ehn-tər-ohs-tō-mē) = surgical production of an artificial opening between the small intestine and the abdominal wall
- **ileostomy** (ihl-ē-ohs-tō-mē) = surgical production of an artificial opening between the ileum and abdominal wall
- **colostomy** (kō-lahs-tō-mē) = surgical production of an artificial opening between the colon and the body surface
- **hepatotomy** (hehp-ah-toht-ō-mē) = surgical incision into the liver
- **abdominocentesis** (ahb-dohm-ihn-ō-sehn-tē-sihs) = surgical puncture to remove fluid from the abdomen

- **abomasopexy** (ahb-ō-mahs-ō-pehcks-ē) = surgical fixation of the abomasum of ruminants to the abdominal wall. The suffix -**pexy** means to surgically fix something to a body surface.
- **gastropexy** (gahs-trō-pehcks-ē) = surgical fixation of the stomach to the abdominal wall
- **cholecystectomy** (kō-lē-sihs-ehck-tō-mē) = surgical removal of the gallbladder
- **float** (flōt) = instrument or procedure used to file or rasp an equine's premolar and/or molar teeth
- **gavage** (gah-vahzh) = forced feeding or irrigation through a tube passed into the stomach
- **drench** (drehnch) = to give medication in liquid form by mouth and forcing the animal to drink
- **bolus** (bō-luhs) = rounded mass of food on large pharmaceutical preparation or to give something rapidly
- **trocarization** (trō-kahr-ih-zā-shuhn) = insertion of a sharp pointed instrument (trocar) into a body cavity or organ. The trocar usually is inside a cannula so that once the trocar penetrates the membrane it can be withdrawn and the cannula remains in place. Trocarization is usually performed for acute cases of bloat to relieve pressure. When trocarization is performed for treatment of ruminal bloat it may be referred to as ruminal **paracentesis** (pahr-ah-sehn-tē-sihs).

REVIEW EXERCISES

Multiple Choice—Choose the correct answer.

1. Mixing of ingesta in the intestine is called
   a. propulsion
   b. peristalsis
   c. segmentation
   d. separation

2. Abnormal accumulation of fluid in the abdomen is called
   a. ascites
   b. effusion
   c. icterus
   d. bloat

3. Telescoping of one part of the intestine into an adjacent part is termed
   a. volvulus
   b. diverticulum
   c. parenchyma
   d. intussusception

4. The intestine is attached to the dorsal abdominal wall via the
   a. peritoneum
   b. emesis
   c. mesentery
   d. omentum

5. Eating and licking of abnormal substances is called
   a. coprophagy
   b. pica
   c. dysphagia
   d. polyphagia

6. Inflammation of the mouth is
   a. stomatitis
   b. orititis
   c. dentitis
   d. osititis

7. Straining, painful defecation is termed
   a. strangstolia
   b. colostrangia
   c. tenesmus
   d. epulis

8. Tumor of the liver is a/an
   a. hematoma
   b. hemoma
   c. hepatoma
   d. hemotoma

9. Marked wasting or excessive leanness is
   a. evaluation
   b. elimination
   c. emesis
   d. emaciation

10. Forced feeding or irrigation through a tube passed into the stomach is called
    a. gavage
    b. drench
    c. bolus
    d. cachexia

Label Diagram in Figure 6–16.

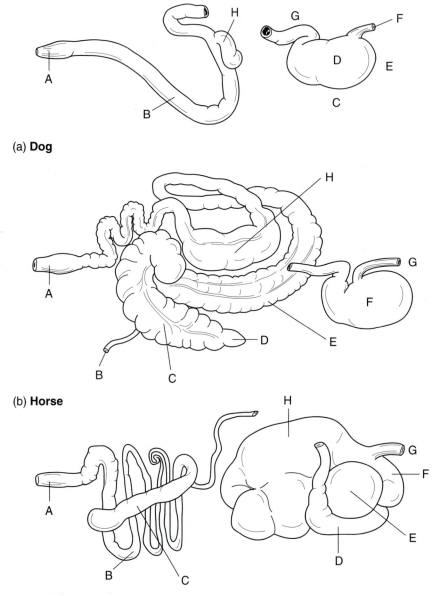

(a) **Dog**

(b) **Horse**

(c) **Ruminant**

FIGURE 6–16 Gastrointestinal tracts. Label parts of the gastrointestinal tracts.

Case Studies—Using terms learned in this chapter, the appendices, and a dictionary, define the underlined terms in each case study.

A 3 yr old <u>F/S</u> New Zealand White rabbit is presented for a few months <u>hx</u> of <u>quidding</u> and <u>ptyalism.</u> The problem has become progressively worse, and although the rabbit has a good appetite, it has difficulty <u>prehending</u> food. On <u>PE</u> it is noted that there is severe overgrowth of the upper and lower <u>incisors</u>. This problem in rabbits is most often due to <u>mal-occlusion</u> of the <u>dental arcades</u>. The <u>premolars</u> and <u>molars</u> are found to be normal. Conservative treatment includes periodic trimming of the overgrown teeth. A more permanent solution involves the extraction of the incisors. Under general anesthesia, all incisors are removed. Analgesics are provided for the first few days postoperatively as rabbits are easily stressed by pain. Once healed, the rabbit returned to its normal eating habits.

yr _____

F/S _____

hx _____

quidding _____

ptyalism _____

prehending _____

incisors _____

malocclusion _____

dental arcades _____

premolars _____

molars _____

extraction _____

anesthesia _____

analgesics _____

postoperatively _____

An 8 wk old M coonhound is presented to the clinic with an acute history of emesis, hemorrhagic diarrhea, lethargy, and anorexia. The pup was not vaccinated and previously was healthy until yesterday. Upon PE it was noted that the pup was pyrexic, dehydrated, and lethargic. Heart and lungs ausculted normally. Stool was collected for parasitic examination, and blood was collected for a CBC and chemistry panel. The stool was negative for parasites, the blood count revealed lymphopenia, and the chemistry panel was normal except for indications of dehydration. A dx of canine parvoviral enteritis was suspected due to the lymphopenia and clinical signs, so virus isolation was performed on a stool sample. Pending virus isolation results the pup was hospitalized and isolated, IV fluids even administered, and antibiotics were given to prevent a secondary septicemia. Twelve hours after hospitalization the pup expired. A necropsy was done and the intestines demonstrated loss of intestinal villi and crypt necrosis. The virus isolation test was positive for canine parvovirus infection. The facility was thoroughly disinfected and the owners were advised to disinfect their facility and vaccinate any future pups.

wk _____

acute _____

emesis _____

hemorrhagic _____

diarrhea _____

lethargy _____

anorexia _____

pyrexic _____

dehydrated _____

ausculted _____

stool _____

lymphopenia_____

dx _____

canine _____

enteritis_____

IV _____

secondary _____

septicemia_____

expired _____

necropsy _____

intestinal villi _____

intestinal crypt _____

disinfected_____

7 Null and Void

Objectives In this chapter, you should learn to:

► Identify the major organs/tissues of the urinary system
► Describe the major functions of the urinary system
► Recognize, define, spell, and pronounce terms relating to diagnosis, pathology, and treatment of the urinary system

FUNCTIONS OF THE URINARY SYSTEM

The urinary system's main responsibility is the removal of wastes from the body. The *urinary* (yoo-rih-nār-ē) *system* removes wastes from the body by constantly filtering blood. The major waste product of protein metabolism is **urea** (yoo-rē-ah) which is filtered by the kidney and used in some diagnostic tests to determine the health status of the kidney.

In addition to filtering of wastes the urinary system also maintains proper balance of water, electrolytes, and acids in body fluids and removes excess fluids from the body. Maintenance of a proper balance of water, electrolytes, and acids in the body allows the body to have a constant internal environment. Maintaining a constant internal environment is termed **homeostasis** (hō-mē-ō-stā-sihs).

One part of the urinary system, the kidney, also produces hormones and affects the secretory rate of other hormones.

Urin/o and **ur/o** are both combining forms meaning urine or pertaining to the urinary organs.

GO WITH THE FLOW

The structures of the normal urinary system include a pair of kidneys, a pair of ureters, a single urinary bladder, and a single urethra (Figure 7–1).

Urine is formed in the kidneys, flows through the ureters to the urinary bladder, is stored in the urinary bladder, and flows through the urethra to outside the body.

Kidney

Kidneys are located **retroperitoneally** (reh-trō-pehr-ih-tō-nē-ahl-lē), which means they are located behind the lining of the abdominal cavity or outside the peritoneal cavity. One kidney sits on each side of the vertebral column below the diaphragm. **Ren/o** (Latin for kidney) and **nephr/o** (Greek for kidney) are both combining forms for kidney. Ren/o (the Latin form) is used as an adjective as in renal pelvis or renal disease. Nephr/o (the Greek term) tends to be used to describe pathological and surgical conditions as in nephritis and nephrectomy.

Blood flows into each kidney through the renal artery and leaves the kidney via the renal vein. Filtration of waste products by the kidney is dependent on this blood flow; hence, blood pressure can affect the rate at which filtration takes place.

Each kidney has two layers that surround the renal pelvis. The outer layer/region of the kidney is known as the **cortex** (kōr-tehckz). **Cortic/o** means outer region and is used to describe the outer region of many organs. The **medulla** (meh-doo-lah) of the kidney is

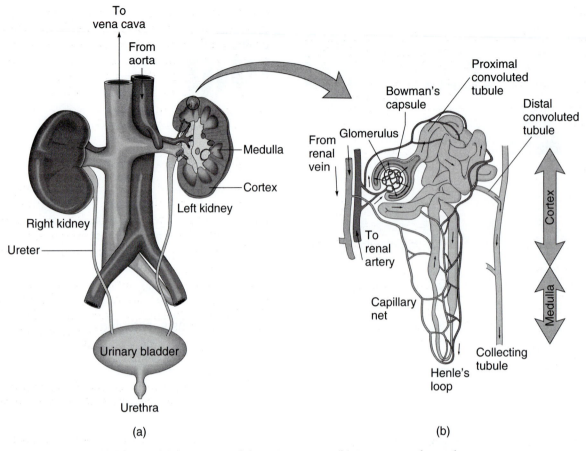

FIGURE 7–1 (a) Structures of the urinary tract; (b) Structures of a nephron

the inner layer/region. **Medull/o** means middle or inner portion and is used to describe the middle or inner region of many organs. The cortex contains the majority of the nephron, and the medulla contains most of the collecting tubules.

The **nephron** (nehf-rohn) is the functional unit of the kidney. The nephron consists of the glomerulus, Bowman's capsule, a proximal convoluted tubule, a loop of Henle, a distal convoluted tubule, and a collecting duct. The nephrons form urine by the processes of filtration, reabsorption, and secretion. Filtration occurs in the glomerulus, reabsorption occurs in the proximal convoluted tubule, loop of Henle, and collecting tubule, and secretion occurs in the distal convoluted tubule (Table 7–1).

The **glomerulus** (glō-mər-yoo-luhs) is a cluster of capillaries surrounded by Bowman's capsule. The combining form for glomerulus is **glomerul/o** which means to wind into a ball, which is what the glomerulus looks like microscopically. The plural form of glomerulus is **glomeruli** (glō-mər-yoo-lī).

The renal pelvis is the area of the kidney where the nephrons collect before entering the ureters. The com-

bining form for renal pelvis is **pyel/o.** The renal pelvis is made up of **calyces** (kā-lah-sēz), which are irregular cuplike spaces that collect urine from the kidney. **Calyx** (kā-lihcks) is the singular form of calyces.

Each kidney has a concave depression called the **hilus** (hī-luhs) that serves as the point of attachment of the renal blood vessels, nerves, and ureter. The hilus is located on the medial surface of the kidney and gives some kidneys that "kidney bean" look. Not all species have kidneys that look "kidney bean shaped." Cattle have lobulated kidneys (and no renal pelvis), and the right kidney of horses is heart-shaped (Figure 7–2).

Descriptive Structural Terms of the Kidney

▶ **distal** = farthest from midline or center
▶ **proximal** = closest to midline or center
▶ **convoluted** (kohn-vō-lūt-ehd) = rolled or coiled
▶ **cortex** = outer
▶ **medulla** = middle or inner
▶ **hilus** = point of attachment or depression
▶ **calyx** or **calix** = cuplike organ
▶ **ascending** (ā-sehnd-ihng) = moving upward
▶ **descending** (dē-sehnd-ihng) = moving downward

TABLE 7–1
Parts of the Nephron

glomeruli	cluster of capillaries that filter blood
Bowman's capsule	cup-shaped structure that contains the glomerulus
proximal convoluted tubules	hollow tubes located between Bowman's capsule and loops of Henle that are involved in reabsorption
loop of Henle	U-shaped turn in the convoluted tubule of the kidney located between the proximal and distal convoluted tubules that is involved in reabsorption; has ascending and descending loop
distal convoluted tubules	hollow tubes located between the loops of Henle and the collecting tubules that are involved in secretion
collecting tubules	hollow tubes that carry urine from the cortex to the renal pelvis

Ureters

The **ureters** (yoo-rē-tərz) are a pair of narrow tubes that carry urine from the kidneys to the urinary bladder. The combining form for ureter is **ureter/o**. The ureters enter the urinary bladder at the **trigone** (trī-gōn). Trigone comes from the Greek term meaning triangle. The trigone of the urinary bladder is a triangular portion at the base of that organ, where the three angles are marked by the two ureteral openings and one urethral opening.

Urinary Bladder

The **urinary bladder** (yoo-rihn-ār-ē blah-dər) is a singular hollow muscular organ that holds urine. **Cyst/o** is the combining form for urinary bladder. The urinary

💡 ***Five Words, One Meaning?***
The terms **urination** (yoo-rih-nā-shuhn), **excretion** (ehcks-krē-shuhn), **voiding** (voy-dihng), **elimination** (i-li-mə-nā-shən) and **micturition** (mihck-too-rihsh-uhn) have all been used to describe the act of excreting urine. Actually the terms excretion, elimination, and voiding are words that mean elimination of a substance and can be used in other body systems. The term urination means the elimination of urine from the body. The term micturition means the elimination of urine from the body; however, it implies voluntary control of the sphincter muscles of the urinary tract. This voluntary control of voiding urine is learned and implies a more intelligent form of animal life.

(There are more than five words to describe urine, but sticking to these terms rather than the slang terms is beneficial.)

bladder is very elastic and its shape and size depend upon the amount of urine it is holding. The flow of urine into the urinary bladder enters from the ureters at such an angle that it serves as a natural valve to control backflow. The flow of urine out of the urinary bladder to the urethra is controlled by **sphincters** (sfhingk-tərz). Sphincters are ring-like muscles that close a passageway (Figure 7–3).

Urethra

The **urethra** (yoo-rē-thrah) is a tube extending from the urinary bladder to the outside of the body. **Urethr/o** is the combining form for urethra. The external opening of the urethra is the **urethral meatus** (yoo-rē-thrahl mē-ā-tuhs) or urinary meatus. The combining form **meat/o** means opening.

In females the only function of the urethra is to transport urine from the urinary bladder to the outside. In males the urethra transports urine from the urinary bladder and reproductive fluids from the reproductive organs out of the body.

💡 ***Urinary System Hormones:*** *Hormones that affect or are produced by the urinary system include*

▶ **erythropoietin** (ē-rihth-rō-poy-ē-tihn) = hormone produced by the kidney that stimulates red blood cell production in the bone marrow. Also pronounced ē-rihth-rō-pō-ih-tihn.
▶ **antidiuretic** (ahn-tih-dī-yoo-reht-ihck) **hormone** = hormone released by the posterior pituitary gland that suppresses urine formation by reabsorbing more water; also referred to and abbreviated as **ADH**
▶ **aldosterone** (ahl-dah-stər-own) = hormone secreted by the adrenal cortex that regulates electrolyte balance via the reabsorption of sodium.

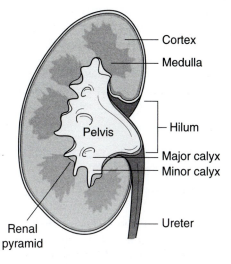

FIGURE 7–2 Structures of the kidney

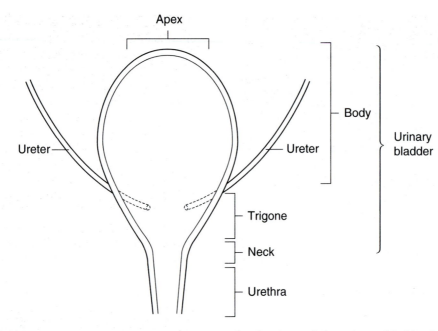

Apex

Ureter

Ureter

Body

Urinary
bladder

Trigone

Neck

Urethra

FIGURE 7–3 Divisions of the urinary bladder. The divisions of the urinary bladder include the apex (cranial free end), the body (central, main part), the trigone (triangular portion where ureters enter the urinary bladder), and the neck (constricted portion that joins the urethra).

IT IS CLEAR AS DAY—MAYBE

The end product of renal filtration of wastes is urine. The process of urine production is **uropoiesis** (yoo-rō-pō-ē-sihs). The suffix **-poiesis** means formation. The normal urine of most species is clear and pale yellow in color. The suffix for color is **-chrome.** However, some species may have normal urine that is **turbid** (tər-bihd) (turbid means cloudy) and may be brown, white, or another color. Sometimes the color of urine is dependent on the diet (as in rabbits) or reproductive cycle. Urine color can also be a reflection of hydration status. In dehydrated animals, urine is more concentrated and, therefore, a deeper shade of yellow. Urine also has a pH, which is dependent on species and diet. Herbivores tend to have basic urine (a higher pH), while carnivores tend to have acidic urine (a lower pH). Through dietary management urine pH can also be manipulated in the treatment or prevention of disease.

▶ Test Me: Urinary System ◀

Diagnostic procedures performed on the urinary system include

- **urinary catheterization** (kahth-eh-tər-ih-zā-shuhn) = insertion of a tube through the urethra and into the urinary bladder (usually to collect urine). A **catheter** (kahth-eh-tər) is the hollow tube that is inserted into a body cavity to inject or remove fluid (Figure 7–4).
- **cystocentesis** (sihs-tō-sehn-tē-sihs) = the surgical puncture of the urinary bladder usually to collect urine. A cystocentesis is usually performed with a needle and syringe (Figure 7–5).
- **cystoscopy** (sihs-tohs-kō-pē) = visual examination of the urinary bladder using a fiberoptic instrument. A **cystoscope** (sihs-toh-skōp) is the fiberoptic instrument used to access the interior of the urinary bladder.
- **radiography** (rā-dē-ohg-rah-fē) = imaging of internal structures that is created by exposing specialized film to X-rays. A **scout film** is a term used to describe a plain X-ray without the use of contrast material (Figure 7–6).

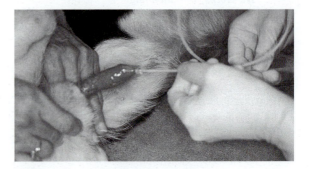

FIGURE 7–4 Urinary catheterization in a male dog. *Source:* Terri Raffel, CVT.

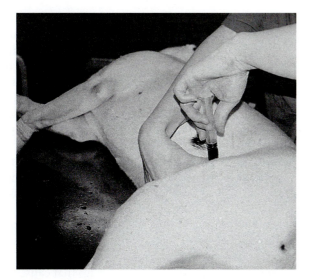

FIGURE 7–5 Cystocentesis in a dog. *Source:* Lodi Veterinary Hospital, S.C.

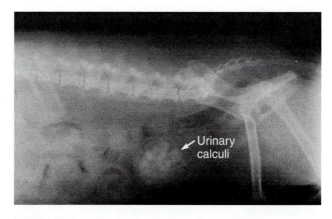

FIGURE 7–6 Radiograph of the urinary bladder. Urinary calculi in the canine urinary bladder are seen on this lateral scout radiograph. *Source:* Lodi Veterinary Hospital, S.C.

- **cystography** (sihs-tohg-rah-fē) = radiographic study of the urinary bladder after contrast material has been placed into the urinary bladder via a urethral catheter. The contrast material that is used in the urinary bladder is water soluble. Cystography can be single contrast, with one contrast material used, or double contrast, when more than one contrast material is used. Double contrast cystography involves the use of contrast material and air. A **cystogram** (sihs-tō-grahm) is the radiographic film of the urinary bladder after contrast material has been placed into the urinary bladder via a urethral catheter. **Retrograde** (reh-trō-grād) means going backwards and can be used to describe the path that contrast material takes. If the contrast material goes in reverse order of how urine normally flows in the body, it is referred to as retrograde.
- **pneumocystography** (nū-mō-sihs-tohg-rah-fē) = radiographic study of the urinary bladder after air has been placed into it via a urethral catheter. Double-contrast cystography is a radiographic study of the urinary bladder after air and contrast material have been placed into it via a urethral catheter.
- **intravenous pyelogram** (ihn-trah-vē-nuhs pī-eh-lō-grahm) = radiographic study of the kidney (especially the renal pelvis) and ureters in which a dye is injected into a vein to define structures more clearly. The urinary bladder may also be visualized better with an intravenous pyelogram. It is also referred to and abbreviated as IVP (Figure 7–7).
- **retrograde pyelogram** (reh-trō-grād pī-eh-lō-grahm) = radiographic study of the kidney and ureters in which a contrast material is placed directly into the urinary bladder

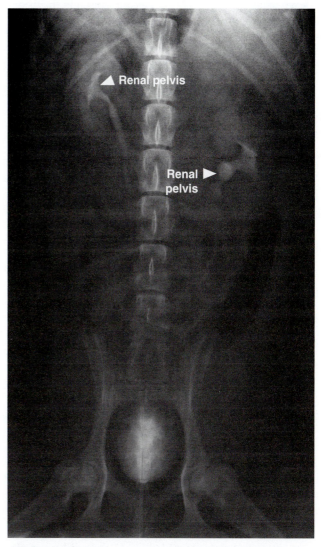

FIGURE 7–7 IVP in a dog. Contrast material outlines the renal pelvis in this radiograph. Some contrast material is also present in the urinary bladder. *Source:* University of Wisconsin Hospital—Radiology.

TABLE 7–2
Descriptive Terms for Urine

▶ **anuria** (a-nū-rē-ah)	complete suppression of urine production
▶ **dysuria** (dihs-yoo-rē-ah)	difficult or painful urination
▶ **oliguria** (ohl-ih-goo-rē-ah)	scanty or little urine
▶ **nocturia** (nohck-too-rē-ah)	excessive urination at night
▶ **stranguria** (strahng-yoo-rē-ah)	slow or painful urination
▶ **hematuria** (hēm-ah-toor-ē-ah)	blood in urine
▶ **pollakiuria** (pōl-lahck-ē-yoo-rē-ah)	frequent urination
▶ **bacteriuria** (bahck-tē-rē-yoo-rē-ah)	presence of bacteria in urine
▶ **glycosuria** (glī-kohs-yoo-rē-ah)	glucose (sugar) in the urine
▶ **glucosuria** (gloo-kohs-yoo-rē-ah)	glucose (sugar) in the urine
▶ **ketonuria** (kē-tō-nū-rē-ah)	presence of ketones in the urine (ketones are produced during increased fat metabolism)
▶ **proteinuria** (prō-tēn-yoo-rē-ah)	presence of proteins in urine
▶ **albuminuria** (ahl-bū-mihn-yoo-rē-ah)	presence of the major blood protein in urine
▶ **pyuria** (pī-yoo-rē-ah)	pus in urine
▶ **polyuria** (pohl-ē-yoo-rē-ah)	increased urination
▶ **crystalluria** (krihs-tahl-yoo-rē-ah)	urine with naturally produced angular solid of definitive form (crystals)

- **urinalysis** (yoo-rih-nahl-ih-sihs) = examination of urine components. It is also referred to and abbreviated UA. Urinalyses can tell us about pH (hydrogen ion concentration that indicates acidity or alkalinity), leukocytes, erythrocytes, protein, glucose, specific gravity (measurement that reflects the amount of wastes, minerals, and solids in urine), and other factors (Table 7–2 and Figure 7–8).

▶ Path of Destruction: Urinary System ◀

Pathologic conditions of the urinary system include
- **renal failure** (rē-nahl fāl-yər) = inability of the kidney(s) to function. Renal failure may be **acute**

(ā-kūt) or **chronic** (krohn-ihck). Acute means occurring suddenly or over a short period. Acute renal failure (ARF) is the sudden onset of the inability of the kidney(s) to function. ARF may be caused by a **nephrotoxin** (nehf-rō-tohcks-ihn), which is a poison having destructive affects on the kidney(s). Chronic means having a longer onset. Chronic renal failure (CRF) is the progressive onset of the inability of the kidney(s) to function. Signs of renal failure may include **polyuria** and **polydipsia** (pohl-ē-dihp-sē-ah), which is abbreviated as PU/PD. Polyuria is increased urine production and polydipsia is increased thirst or drinking. Animals with renal failure may undergo **diuresis** (dī-yoo-rē-sihs). Diuresis is the increased excretion of urine. Diuresis may be produced by fluid therapy or drug therapy. Drugs that cause increased urine production are **diuretics.**
- **uremia** (yoo-rē-mē-ah) = waste products in the blood. Uremia may be seen with many types of kidney disease.
- **nephrosis** (neh-frō-sihs) = abnormal condition of the kidney(s)
- **nephropathy** (neh-frohp-ah-thē) = disease of the kidney(s)
- **nephritis** (neh-frī-tihs) = inflammation of the kidneys
- **nephrectasis** (neh-frehck-tah-sihs) = distention of the kidneys. Distention means enlargement and the suffix **-ectasis** means distention or stretching.
- **nephrosclerosis** (nehf-rō-skleh-rō-sihs) = abnormal hardening of the kidney
- **nephromalacia** (nehf-rō-mah-lā-shē-ah) = abnormal softening of the kidney

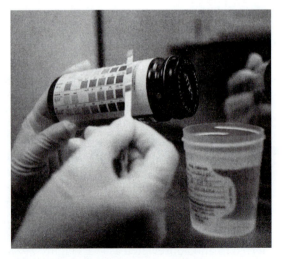

FIGURE 7–8 Urinalysis. Chemical properties of urine, such as pH, glucose, ketones, and bilirubin, are tested via a dipstick.

- **prolapse** (prō-lahps) = downward displacement of a body organ. The suffix **-ptosis** means drooping or dropping down and **nephroptosis** (nef-rohp-tō-sihs) may be used to describe a prolapsed kidney.
- **hydronephrosis** (hī-drō-neh-frō-sihs) = dilation of the renal pelvis as a result of an obstruction to urine flow (Figure 7–9).
- **glomerulonephritis** (glō-mər-yoo-lō-neh-frī-tihs) = inflammation of the kidney involving the glomeruli
- **pyelitis** (pī-eh-lī-tihs) = inflammation of the renal pelvis
- **pyelonephritis** (pī-eh-lō-neh-frī-tihs) = inflammation of the renal pelvis and kidney
- **crystals** (krihs-tahlz) = naturally produced angular solid of definitive form (Figure 7–10)
- **calculus** (kahl-kyoo-luhs) = abnormal mineral deposit. Urine may contain crystals that remain in solution throughout urine. Stones, calculi, or liths are formed when crystals precipitate and form solids. **Lith/o** is the combining form for stone or calculus, **-lith** is the suffix for stone or calculus. When used in relationship to the urinary system, calculus needs to be modified to urinary calculus (versus dental calculus), liths need to be modified to urolith (versus enterolith), and stones need to be modified to urinary stones or kidney stones to clarify which system is involved.

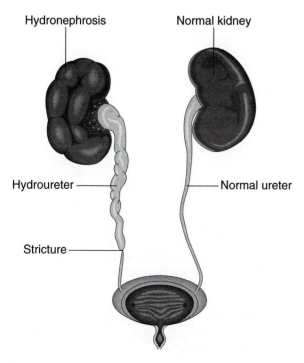

FIGURE 7–9 Hydroureter and hydronephrosis due to urethral stricture

- **casts** (kahstz) = fibrous or protein materials found in the urine with renal disease or other abnormality (Figure 7–10).
- **urolith** (yoo-rō-lihth) = urinary bladder stone; also called **cystolith** (sihs-tō-lihth)
- **nephrolith** (nehf-rō-lihth) = kidney stone or renal calculus
- **ureterolith** (yoo-rē-tər-ō-lihth) = stone in the urethra
- **-lithiasis** (lih-thī-ah-sihs) = suffix meaning the presence of stones or calculi; nephrolithiasis (disorder characterized by the presence of kidney stones), urolithiasis (disorder characterized by the presence of urinary bladder stones), and ureterolithiasis (disorder characterized by the presence of stones in the hollow tube that connects the kidney and urinary bladder)
- **ureterectasis** (yoo-rē-tər-ehck-tah-sihs) = distention of the ureter
- **hydroureter** (hī-drō-yoo-rē-tər) = distention of the ureter with urine due to any blockage
- **cystalgia** (sihs-tahl-jē-ah) = urinary bladder pain = **cystodynia** (sihs-tō-dihn-ē-ah)
- **cystitis** (sihs-tī-tihs) = inflammation of the urinary bladder
- **interstitial cystitis** (ihn-tər-stihsh-ahl sihs-tī-tihs) = inflammation within the wall of the urinary bladder
- **urinary tract infection** (yoo-rihn-ār-ē trahck ihn-fehck-shuhn) = invasion of microorganisms into the urinary system which results in local cellular injury; abbreviated UTI
- **cystocele** (sihs-tō-sēl) = displacement of the urinary bladder through the vaginal wall
- **urethritis** (yoo-rē-thrī-tihs) = inflammation of the urethra
- **urethrostenosis** (yoo-rē-thrō-steh-nō-sihs) = stricture of the urethra. A **stricture** (strihck-shər) is an abnormal band of tissue narrowing a passage.
- **incontinence** (ihn-kohn-tih-nehnts) = inability to control excretory functions. The term urinary needs to be applied in front of this term to imply the inability to control urine.
- **inappropriate urination** (ihn-ah-prō-prē-ət yoo-rih-nā-shuhn) = eliminating urine either at the wrong time or in the wrong place
- **urinary retention** (yoo-rih-nār-ē rē-tehn-shuhn) = inability to completely empty the urinary bladder
- **hypospadias** (hī-pō-spā-dē-uhs) = abnormal condition where the urethra opens on the ventral surface of the penis
- **epispadias** (ehp-ih-spā-dē-uhs) = abnormal condition where the urethra opens on the dorsum of the penis

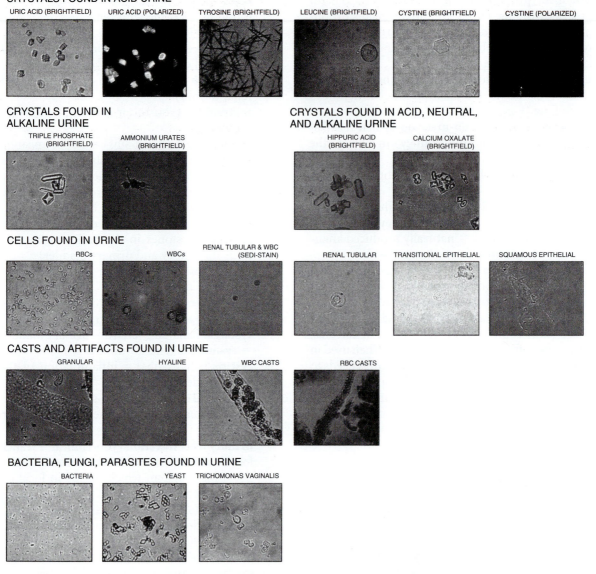

FIGURE 7–10 Crystals, cells, and casts found in urine

- **azotemia** (ā-zō-tē-mē-ah) = presence of urea or other nitrogenous elements in the blood
- **renal infarction** (rē-nahl ihn-fahrck-shuhn) = decreased blood flow to the kidney(s)
- **prerenal** (prē-rē-nahl) and postrenal (pōst-rē-nahl) = before and after the kidney, respectively. These terms are used to describe other pathologic conditions, such as prerenal azotemia and postrenal obstruction.
- **feline lower urinary tract infection** = a common disease of cats in which cystitis, urethritis, and crystalluria are found; formerly called feline urologic syndrome (FUS). In male cats urethral obstruction is commonly associated with this disease.

▶ Procede With Caution: Urinary System ◀

Procedures performed on the urinary system include
- **dialysis** (dī-ahl-ih-sihs) = procedure to remove blood waste products when the kidney (s) are no longer functioning. **Peritoneal dialysis** is the removal of blood waste products by fluid exchange through the peritoneal cavity; **hemodialysis** (hē-mō-dī-ahl-ih-sihs) is the removal of blood waste products by filtering blood through a machine.
- **nephrectomy** (neh-frehck-tō-mē) = surgical removal of a kidney
- **lithotripsy** (lihth-ō-trihp-sē) = destruction of stone using ultrasonic waves traveling through water (the suffix **-tripsy** means to crush)

- **nephropexy** (nehf-rō-pehcks-sē) = surgical fixation of a kidney to the abdominal wall
- **ureterectomy** (yoo-rē-tər-ehck-tō-mē) = surgical removal of the ureter
- **ureteroplasty** (yoo-rē-tər-ō-plahs-tē) = surgical repair of the ureter
- **cystectomy** (sihs-tehck-tō-mē) = surgical removal of all or part of the urinary bladder
- **cystopexy** (sihs-tō-pehck-sē) = surgical fixation of the urinary bladder to the abdominal wall
- **cystotomy** (sihs-tah-tō-mē) = surgical incision into the urinary bladder
- **cystoplasty** (sihs-tō-plahs-tē) = surgical repair of the urinary bladder
- **urethrostomy** (yoo-re-throhs-tō-mē) = surgical creation of a permanent opening between the urethra and the skin. **Perineal urethrostomy** (pər-ih-nē-ahl yoo-rē-throhs-tō-mē) is the surgical creation of a permanent opening between the urethra and the skin between the anus and scrotum.
- **cystostomy** (sihs-tohs-tō-mē) = surgical creation of a new opening between the skin and urinary bladder

REVIEW EXERCISES

Multiple Choice—Choose the correct answer.

1. The combining forms for kidney are
   a. ren/o and ureter/o
   b. ren/o and nephr/o
   c. ren/o and cyst/o
   d. ren/o and periren/o

2. Inflammation of the kidney is
   a. nephrosis
   b. nephroptosis
   c. nephritis
   d. nephropathy

3. Insertion of a hollow tube through the urethra and into the urinary bladder is termed
   a. cystocentesis
   b. cystogram
   c. urinary injection
   d. urinary catheterization

4. The hormone produced by the kidney that stimulates red blood cell production in the bone marrow is
   a. ADH
   b. erythropoietin
   c. aldosterone
   d. renerythogenin

5. Retrograde means
   a. going backward
   b. going forward
   c. going sideways
   d. repeating

6. Examination of the components of urine is a/an
   a. urinoscopy
   b. cystoscopy
   c. urinalysis
   d. cystolysis

7. Inflammation of the urinary bladder is
   a. cystitis
   b. urolithiasis
   c. urology
   d. uritis

8. UTI is the abbreviation for
   a. urinary treatment for infection
   b. urinary tract infection
   c. urinary tract inflammation
   d. urinary trigone infarct

9. Constant internal environment is
   a. stricture
   b. status
   c. homeostasis
   d. isostatic

10. Diuretics are chemical substances that
    a. cause painful urination
    b. cause complete cessation of urine
    c. cause nighttime urination
    d. cause increased urine production

Diagrams—Label the parts of the urinary system on Figure 7–11. Provide the combining forms for the parts labeled.

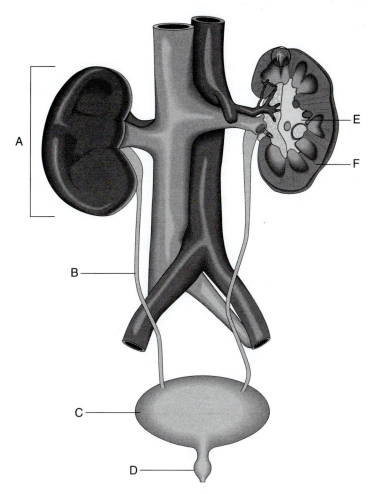

A

E

F

B

C

D

FIGURE 7–11 Urinary system structures

Matching—Match the urinary term in Column I with the definition in Column II.

Column I
A. bacteriuria

B. glycosuria or
   glucosuria

C. nocturia

D. proteinuria

E. anuria

F. oliguria

G. albuminuria

H. stranguria

I. polyuria

J. pyuria

K. pollakiuria

L. ketonuria

M. crystalluria

N. dysuria

O. hematuria

Column II
_____complete suppression of urine production

_____difficult or painful urination

_____scanty or little urine

_____excessive urination at night

_____slow or painful urination

_____blood in urine

_____frequent urination

_____presence of bacteria in urine

_____glucose (sugar) in the urine

_____presence of ketones in the urine

_____presence of proteins in urine

_____presence of the major blood protein in urine

_____pus in urine

_____increased urination

_____crystals in the urine

Case Study—Supply the proper abbreviation or medical term for the lay terms/definitions underlined.

A 5 <u>year</u> old <u>male neutered</u> springer spaniel was presented with <u>difficulty urinating</u> and <u>blood in the urine</u>. An <u>examination of the urine by breaking it into its components</u> was obtained by <u>inserting a needle into the urinary bladder and withdrawing urine</u>. The test revealed large numbers of <u>erythrocytes</u> and struvite crystals. Suspecting more than <u>inflammation of the urinary bladder</u>, the veterinarian took an <u>X-ray</u> of the dog's urinary bladder and <u>urinary bladder stones</u> were detected. An <u>incision into the urinary bladder</u> was performed to remove the urinary stones and the dog recovered uneventfully.

year _____

male neutered _____

difficulty urinating _____

blood in the urine_____

examination of the urine by breaking it into its components _____

insertion of a needle into the urinary bladder and withdrawing urine _____

erythrocytes _____

inflammation of the urinary bladder _____

X-ray_____

urinary bladder stones _____

incision into the urinary bladder _____

Case Studies—Define the underlined terms in each case study.

A 2 yr old <u>M/N</u> <u>DLH</u> cat was presented to the clinic for crying when urinating in the litterbox. The owner states that the cat had been <u>inappropriately urinating</u> on the carpet for the past few days, but otherwise seems normal. Upon examination, the cat was about 7% <u>dehydrated</u>, T = 102° F , HR = 180 bpm, and RR = panting. Heart and lungs <u>ausculted</u> normally. Abdominal palpation revealed <u>cystomegaly</u>. The <u>urethra</u> was red and swollen. When the cat was put in a litterbox, he was straining to urinate but was unable to <u>urinate</u>. The cat was <u>dx'd</u> with <u>urethral obstruction</u>. The owner was informed of the emergency status of relieving the obstruction, and the cat was hospitalized. The cat was <u>anesthetized</u> using <u>inhalant</u> <u>anesthesia</u> via a mask, and a <u>urinary catheter</u> was passed and sutured to the <u>perineum</u>. <u>IV</u> fluids were administered to reverse the cat's dehydration. Urine was collected via the urinary catheter, and a <u>UA</u> was performed. The UA revealed a <u>pH</u> = 7.0, large amounts of blood in the urine, a large number of <u>leukocytes</u> in the urine, a <u>specific gravity</u> of 1.040, and large amounts of <u>struvite crystals</u> and bacteria in the urine. In addition to the urethral obstruction, the cat also had <u>cystitis</u> and was treated with antibiotics and a urinary acidifying diet. The cat was hospitalized until the urinary catheter was removed. The cat was discharged with its medication, and the owner was advised to have a recheck UA when the medication was completed.

M/N _____

DLH _____

inappropriate urination _____

dehydrated _____

ausculted_____

cystomegaly_____

urethra _____

urinate _____

dx'd _____

urethral obstruction _____

anesthetized _____

inhalant anesthesia _____

urinary catheter _____

perineum _____

IV _____

UA _____

pH _____

leukocytes _____

specific gravity _____

struvite crystals _____

cystitis _____

A 2 yr old milking <u>cow</u> has been showing a decrease in milk production and weight loss over the past few weeks. She is now <u>off feed</u> and restless per the farmer. Upon arrival to the farm, the veterinarian noted that the cow was switching its tail, vital signs were normal, and the cow was <u>polyuric.</u> <u>Hematuria</u> was noted, and urine was collected for analysis. <u>Rectal palpation</u> revealed <u>nephromegaly</u>. A tentative <u>diagnosis</u> of <u>pyelonephritis</u> was made. Antibiotics were given to this cow, the cow's milk was pulled due to antibiotic withdrawal times, and the farmer was advised to monitor the other cows for abnormalities similar to this cow. The veterinarian returned to the clinic and performed the <u>UA</u> that showed a large amount of blood in the urine and rod-shaped bacteria on microscopic examination. The sample was sent in for bacterial <u>culture and sensitivity</u> to determine if this cow had a *Corynebacterium renale* infection. *C. renale* infections may indicate a herd with subclinical infection, which would affect the treatment and hygiene practices on this farm.

cow _____

off feed _____

polyuric _____

hematuria _____

rectal palpation _____

nephromegaly _____

diagnosis _____

pyelonephritis _____

UA _____

culture and sensitivity _____

8 Have a Heart

Objectives *In this chapter, you should learn to:*

▶ Identify the structures of the cardiovascular system
▶ Differentiate between the types of blood vessels and describe their functions
▶ Describe the functions of the cardiovascular system
▶ Describe the flow of blood throughout the body
▶ Recognize, define, spell, and pronounce the terms related to the diagnosis, pathology, and treatment of the cardiovascular system

FUNCTIONS OF THE CARDIOVASCULAR SYSTEM

The functions of the cardiovascular system are delivery of oxygen, nutrients, and hormones to the various body tissues and transport of waste products to the appropriate waste removal system. The cardiovascular system is occasionally called the circulatory system; however, the circulatory system is divided into systemic circulation (blood flow to all parts of the body except the lungs) and pulmonary circulation (blood flow out of the heart through the lungs and back to the heart).

STRUCTURES OF THE CARDIOVASCULAR SYSTEM

Cardiovascular (kahr-dē-ō-vahs-qoo-lər) means pertaining to the heart and vessels (blood vessels in this context). There are three major parts of this system: the heart, the blood vessels, and the blood. Blood will be discussed in Chapter 15 on the hematologic, lymphatic, and immunologic systems.

The Heart of the Matter

The heart is a hollow muscular organ that provides the power to move blood through the body. The combining form for heart is **cardi/o.** The heart is located inside the **thoracic** (thō-rahs-ihck) **cavity** or **chest cavity.** The heart lies between the lungs in a cavity called the **mediastinum** (mē-dē-ahs-tī-nuhm). The mediastinum also contains the large blood vessels, trachea, esophagus, lymph nodes, and other structures.

AROUND AND AROUND

Surrounding the heart is a double-walled membrane called the **pericardium** (pehr-ih-kahr-dē-uhm). The two layers of the pericardium are the fibrous and serous. The **fibrous pericardium** is the tough external layer. The **serous layer** is the inner layer and it is divided into two parts: the **parietal** (pahr-ī-ih-tahl) **layer** and **visceral** (vihs-ər-ahl) **layer.** The parietal layer (parietal means belonging to the wall in Latin) is the serous layer that lines the fibrous pericardium. The visceral layer (viscus is Latin for an organ) is the serous layer that lines the heart. The visceral layer is also called the

epicardium (ehp-ih-kahr-dē-uhm). Between the two serous layers of the pericardium is a space called the **pericardial** (pehr-ih-kahr-dē-ahl) **space**. **Pericardial fluid** is a liquid in the pericardial space. Pericardial fluid prevents friction between the heart and the pericardium when the heart beats.

UP AGAINST A WALL

The heart is made up of three walls as follows:

epicardium (ehp-ih-kahr-dē-uhm) = external layer of the heart; also part of the serous layer of the pericardium. The prefix epi- means upper.

myocardium (mī-ō-kahr-dē-uhm) = middle and thickest layer of the heart; the actual "heart muscle." The combining form **my/o** means muscle.

endocardium (ehn-dō-kahr-dē-uhm) = inner layer of the heart; lines the heart chambers and valves. The prefix **endo-** means within.

SELF-SERVING

Heart tissue beats constantly and must have a continuous supply of oxygen and nutrients and the prompt removal of waste. The problem is that it can-not fulfill these needs from the blood that it is pumping. These needs are supplied by its own arteries, and its wastes are removed by its own veins. The arteries that serve the heart are known as the **coronary arteries** (kōr-oh-nār-ē ahr-tər-ēz), because they resemble a crown. **Coron/o** is the combining form meaning crown. The **coronary veins** remove waste products from the myocardium.

If the blood supply to the heart is disrupted, the myocardium cannot function. Disruption of blood to the myocardium may be caused by **coronary occlusion** (ō-kloo-shuhn). Occlusion means blocked. Coronary occlusion may lead to **ischemia** (ihs-kē-mē-ah), which is deficiency in the blood supply to an area. Ischemia can lead to **necrosis** (neh-krō-sihs), which is tissue death. The area of necrosis is called an **infarct** (ihn-fahrck) or **infarction** (ihn-fahrck-shuhn). An infarct is a localized area of necrosis caused by an interrupted blood supply.

MAKE ROOM

The heart is divided into right and left sides. The right and left sides are further subdivided into chambers. Mammalian and avian hearts are four chambered; reptile hearts have three chambers. The superior chamber

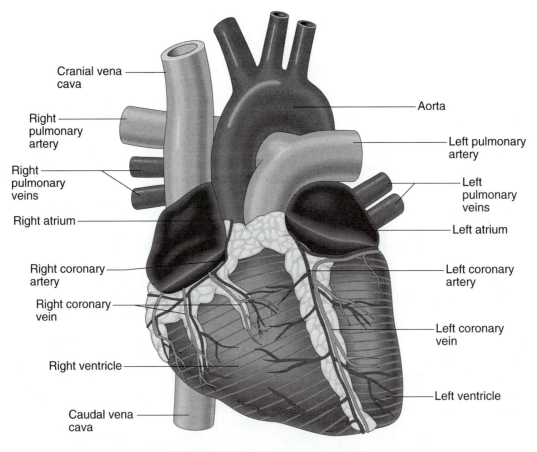

FIGURE 8–1 External heart structures

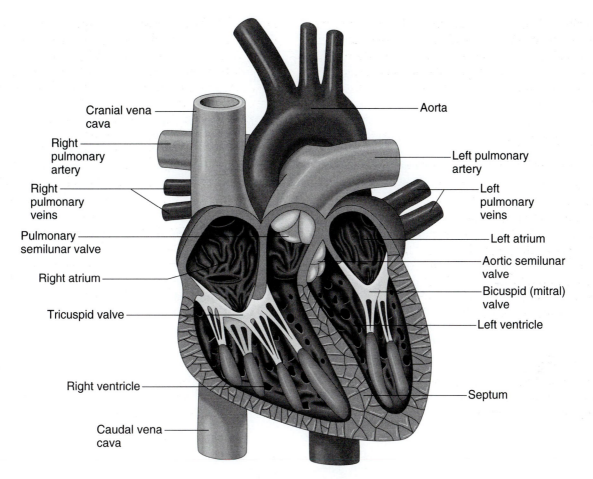

Cranial vena cava
Right pulmonary artery
Right pulmonary veins
Pulmonary semilunar valve
Right atrium
Tricuspid valve
Right ventricle
Caudal vena cava

Aorta
Left pulmonary artery
Left pulmonary veins
Left atrium
Aortic semilunar valve
Bicuspid (mitral) valve
Left ventricle
Septum

FIGURE 8–2 Internal heart structures

of the heart is called the **atrium** (ā-trē-uhm). There are two superior chambers of the heart. The plural of atrium is **atria** (ā-trē-ah). All vessels coming into the heart enter here. **Atri/o** is the combining form for atria. The atria are separated into left and right parts by the **interatrial septum** (ihn-tər-ā-trē-ahl sehp-tuhm). A septum is a separating wall or partition.

The interior chambers of the heart are **ventricles** (vehn-trih-kuhlz). **Ventricul/o** is the combining form for ventricle. The ventricles are separated from each other by the **interventricular septum.** The interventricular septum in reptiles is not complete; hence, the ventricles are open to each other and only count as one heart chamber. The ventricles are the pumping chambers of the heart and all vessels leaving the heart leave via the ventricles.

The narrow tip of the heart is called the **apex** (ā-pehcks) or cardiac apex (Figure 8–1).

FOUR OF A KIND

Blood flow through the heart is controlled by valves (Figure 8–2 and Table 8–1). A **valve** (vahlv) is a mem-

branous fold. The combining form for valve is **valv/o** or **valvul/o.** There are four heart valves:

1. **right atrioventricular valve** (ā-trē-ō-vehn-trihck-yoo-lahr vahlv) or right AV valve—This valve controls the opening between the right atrium and right ventricle. It is also called the **tricuspid valve** (trī-kuhs-pihd vahlv) because it has three points or cusps (tri- = three; cusps = points).
2. **pulmonary semilunar valve** (puhl-mah-nār-ē sehm-ē-loo-nahr vahlv) or pulmonary valve—This valve is located between the right ventricle and the pulmonary artery and controls blood entering the lungs. Semilunar means half-moon, and this valve is shaped like a half moon.
3. **left atrioventricular valve** (ā-trē-ō-vehn-trihck-yoo-lahr vahlv) or left AV valve—This valve controls the opening between the left atrium and left ventricle. It is also called the **mitral valve** (mī-trahl vahlv) or **bicuspid** (bī-kuhs-pihd) because it has two points (bi- = two).
4. **aortic semilunar valve** (ā-ōr-tihck sehm-ē-loo-nahr vahlv) or aortic valve—The aortic valve is located

TABLE 8–1
Blood Flow Through the Heart

↓ The right atrium receives blood from all tissues, except the lungs, through the cranial and caudal venae cavae. Blood flows from here through the tricuspid valve into the right ventricle. (This is systemic circulation.)
↓ The right ventricle pumps the blood through the pulmonary semilunar valve and into the pulmonary artery, which carries it to the lungs. (This is pulmonary circulation.)
↓ The left atrium receives oxygenated blood from the lungs through the four pulmonary veins. The blood flows through the mitral valve into the left ventricle. (This is pulmonary circulation.)
↓ The left ventricle receives blood from the left atrium. From the left ventricle, blood goes out through the aortic semilunar valve and into the aorta and is pumped to all parts of the body, except the lungs. (This is systemic circulation.)
↓ Blood is returned by the venae cavae to the right atrium and the cycle continues.

between the left ventricle and the aorta and controls blood entering the arterial system. It is also half-moon shaped.

FASCINATIN' RHYTHM

A rhythm is the recurrence of an action or function at regular intervals. The heart's contractions are supposed to be rhythmic. The rate and regularity of the heart rhythm, termed the **heartbeat,** is determined by electrical impulses from nerves that stimulate the myocardium. The heartbeat or cardiac cycle is an alternating sequence of relaxation and contraction of the heart chambers. **Cardiac output** is the volume of blood pumped by the heart per unit time. To effectively pump blood throughout the body, contraction and relaxation of the heart must be synchronized accurately. These electrical impulses, also called the conduction system, are controlled by the sinoatrial node, atrioventricular node, bundle of His, and Purkinje fibers (Figure 8–3).

The **sinoatrial node** (sī-nō-ā-trē-ahl nōd) or SA node is located in the wall of the right atrium near the entrance of the superior vena cava. The SA node, along with atypical cardiac muscle cells called **Purkinje** (pər-kihn-jē) **fibers,** establishes the basic rhythm of the heart and is called the pacemaker of the heart. (Purkinje fibers are less developed in the atria and are usually associated with the ventricles.)

Electrical impulses from the SA node start waves of muscle contractions in the heart. The impulse in the right atrium spreads over the muscles of both atria, causing them to simultaneously contract. This contraction forces blood into the ventricles. Atrial contraction is termed **atrial systole** (sihs-stohl-ē). **Inotropy** (ihn-ō-trōp-ē) is the term meaning force of contraction.

These impulses continue to travel to the **atrioventricular node** (ā-trē-ō-vehn-trihck-yoo-lahr nōd) or AV node. The AV node is located in the interatrial septum. This slow contraction causes a pause after atrial contraction to allow the ventricles to fill with blood. The AV node transmits the electrical impulses to the bundle of His (also called the AV bundle).

The **bundle of His** (hihs) is located within the interventricular septum. The bundle of His continues on

through the ventricle as ventricular Purkinje fibers which carry the impulse through the ventricular muscle, causing the ventricles to contract. Ventricular contraction is termed **ventricular systole.** Ventricular contraction forces blood into the aorta and pulmonary arteries.

The normal heart rhythm is called the **sinus rhythm** because it starts in the sinoatrial node. If the SA node does not function properly and is unable to send the impulse to the rest of the heart, other areas of the conduction system can take over and initiate a heartbeat. The resulting abnormal rhythm is called an **arrhythmia** (ā-rihth-mē-ah) or **dysrhythmia** (dihs-rihth-mē-ah). Antiarrhythmic drugs are substances that control heartbeat irregularities (Table 8–2).

TABLE 8–2
Terms Relating to Rhythm

palpitation (pahl-pih-tā-shuhn)	pounding with or without irregularity in rhythm
fibrillation (fih-brih-lā-shuhn)	rapid, random, and ineffective heart contractions
flutter (fluht-tər)	cardiac arrhythmia in which atrial contractions are rapid but regular
bradycardia (brā-dē-kahr-dē-ah)	abnormally slow heartbeat
tachycardia (tahck-ē-kahr-dē-ah)	abnormally rapid heartbeat
paroxysm (pahr-ohck-sihzm)	sudden convulsion or spasm
normal sinus arrhythmia	irregular heart rhythm resulting from variation in vagal nerve tone as a result of respiration (a nonpathologic arrhythmia)
asystole (ā-sihs-tō-lē)	without contraction or lack of heart activity. "Flat line" on an EKG.
syncope (sihn-kō-pē)	temporary suspension of respiration and circulation
gallop (gahl-ohp)	low-frequency vibrations occurring during early diastole and late diastole

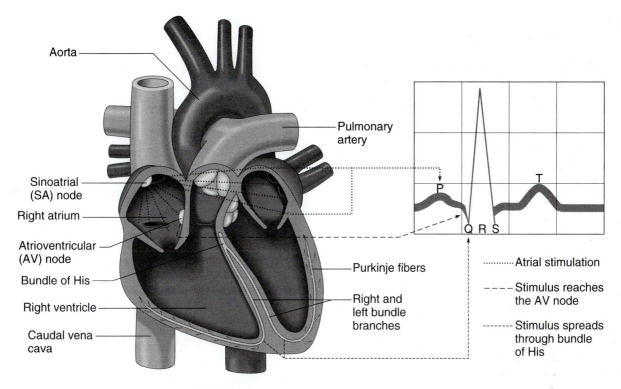

FIGURE 8–3 Conduction systems of the heart

What is Systole?
Systole (sihs-tō-lē) is a term that generally means contraction, derived from the Greek word to draw together. However, the term systole is used to denote ventricular contraction. Remember the atria and ventricles cannot contract together, although they follow in rapid succession. **Diastole** (dī-ah-stō-lē) means expansion, derived from the Greek word to draw apart. The term diastole is used to denote relaxation or the time when the chambers are expanded. During diastole, the atria fill with blood. Then the atria contract, forcing blood into the ventricles, and then the ventricles contract. If someone refers to systole and diastole without the modifiers atrial or ventricular, it is assumed that they mean ventricular contraction (systole) and ventricular relaxation (diastole).

Loading
When discussing workload of the heart it is divided into preload and afterload. **Preload** (prē-lōd) is the ventricular end-diastolic volume or the volume of blood entering the right side of the heart. **Afterload** (ahf-tər-lōd) is the impedance to ventricular emptying presented by aortic pressure. Preload problems are usually associated with right-sided heart disease, while afterload is associated with left-sided heart disease. Drugs used to treat heart disease may alter preload or afterload or both.

MINDING YOUR P'S AND Q'S (AND R'S AND S'S AND T'S)

The electrical events in the conduction system can be visualized by wave movement on an **electrocardiogram** (ē-lehck-trō-kahr-dē-ō-grahm). An electrocardiogram, abbreviated ECG or EKG, is the record of the electrical activity of the myocardium. The ECG or EKG is a tracing that shows the changes in voltage and polarity (positive and negative) over time (Figure 8–4). The process of recording the electrical activity of the myocardium is **electrocardiography** (ē-lehck-trō-kahr-dē-ohg-rah-fē).

Electrocardiography produces a tracing that represents the variations in electric potential caused by excitation of heart muscle and is detected at the body surface. These variations in electric potential are detected by conductors called **leads** (lēdz) (Figure 8–5).

DO YOU HEAR WHAT I HEAR?

Auscultation (ahws-kuhl-tā-shuhn) is listening to body sounds and usually involves the use of a stethoscope. A **stethoscope** (stehth-ō-skōp) is an instrument used to listen; however, if the term is broken down into its basic components **stetho-** means chest and **-scope** means instrument to visually examine or monitor. Traditionally, the stethoscope is considered to be a monitoring device of the chest area, but its use has been expanding to auscultation of other body parts.

When ausculting the heart, a lubb/dubb sound is heard. The lubb is the first sound heard (called the first heart sound) and is caused by closure of the AV valves. The dubb is the second sound heard (called the second heart sound) and is caused by closure of the semilunar valves. Systole or ventricular contraction occurs between the first and second heart sounds, while diastole or ventricular relaxation occurs between the second and first heart sounds.

A **heart murmur** (mər-mər) is an abnormal sound associated with the turbulent flow of blood. A murmur may be caused by a leak in a valve. A leak results in the inability (of the valve) to perform at the proper level, and this inability to perform at the proper level is termed **insufficiency** (ihn-sah-fihsh-ehn-sē). Narrowing of a valve may also result in turbulent blood flow, causing a murmur.

Murmurs are described as systolic (the swooshing noise occurring between the first and second heart sounds) or diastolic (the swooshing noise occurring between the second and first heart sounds). Murmurs may be further described as **holosystolic** (hō-lō-sihs-stohl-ihck) meaning it occurs during the entire ventricular contraction phase. Holosystolic is also called **pansystolic** (pahn-sihs-stohl-ihck). The prefixes **holo-** and **pan-** both mean all. These different phases in which a murmur occurs aids in the identification of the cause of the murmur. For example, murmurs ausculted during systole may be atrioventricular insufficiency or aortic or pulmonic valve **stenosis** (stehn-ō-sihs) or narrowing. Murmurs ausculted during diastole may be atrioventricular stenosis or aortic or pulmonic valve insufficiency.

In addition, murmurs can be described as crescendo and decrescendo. **Crescendo** (kreh-shehn-dō) **murmurs** are abnormal swooshing cardiac sounds that progressively increase in loudness, while **decrescendo** (deh-kreh-shehn-dō) **murmurs** progressively decrease in loudness.

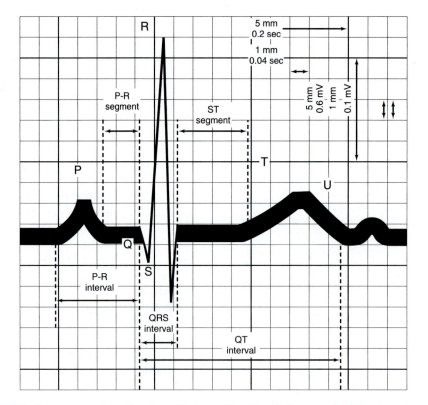

FIGURE 8–4 Anatomy of an electrocardiogram. The first deflection, the P wave, represents excitation (depolarization) of the atria. The P-R interval represents conduction through the atrioventricular valve. The QRS complex results from excitation of the ventricles. The Q-T interval represents ventricular depolarization and repolarization. The S-T segment represents the end of ventricular depolarization to the onset of ventricular repolarization. The T wave results from recovery (repolarization) of the ventricles. *Source:* From Glaze, K.: Basic electrocardiography—Part I. *Veterinary Technician* 17(10):721, 1996; with permission.

FIGURE 8–5 Electrocardiography in a dog. *Source:* Lodi Veterinary Hospital, S.C.

Anatomic location of where the murmur is ausculted is also helpful in the determination of its cause. Where the murmur is heard the loudest is the point of maximal intensity (which is abbreviated PMI). The PMI is usually located at the auscultation site of the "defective" valve. Occasionally, murmurs may result in vibrations felt on palpation of the chest. This vibration felt on palpation is termed **thrill** (thrihl).

Clicks may also be ausculted during an examination. Clicks may be a sign of mitral insufficiency or may be of unknown origin. Other things that may be ausculted are split heart sounds (heartbeat sounds that are divided), crackles (which may be associated with movement or respiratory sounds), and rumbles (most likely due to shivering).

What Are We Ausculting?

When ausculting the heart there are various things to focus on. One is heart rate or the speed at which the heart is beating. Tachycardia and bradycardia can affect how blood is transported through the body. Quality of the heartbeat should also be assessed. Is the heartbeat strong, weak, or thready? Abnormal sounds like murmurs, clicks, and splits should also be assessed. Rhythm also needs to be checked. Auscultation should also be done in conjunction with pulse detection. A pulse deficit is when the ventricles contract without enough force to propel blood to the periphery.

Blood Vessels

There are three major types of blood vessels: arteries, veins, and capillaries. The combining forms for vessel are **angi/o** and **vas/o**. These combining forms are generally used in reference to blood vessels, but may be used to describe other types of vessels as well.

The **lumen** (loo-mehn) is the opening within a vessel through which fluid flows. The diameter of the lumen is affected by **constriction** (kohn-strihckt-shuhn) or narrowing of the vessel diameter and **dilation** (dī-lā-shuhn) or widening of the vessel diameter. **Vasoconstrictors** (vahs-ō-kohn-strihck-tərz or vās-ō-kohn-strihck-tərz) are things that narrow a vessel's diameter; **vasodilators** (vahs-ō-dī-lāt-ərz or vās-ō-dī-lāt-ərz) are things that widen a vessel's diameter.

The **hilum** (hī-luhm) is the depression where vessels and nerves enter an organ.

The pumping action of the heart drives blood into the **arteries** (ahr-tər-ēz). An artery is a blood vessel that carries blood away from the heart. Blood in the arteries is usually oxygenated (the main exception is the pulmonary artery) and is bright red in color. The combining form for artery is **arteri/o.**

The **aorta** (ā-ōr-tah) is the main trunk of the arterial system and begins from the left ventricle of the heart. The combining form for aorta is **aort/o.** After leaving the left ventricle, the aorta arches dorsally and then progresses caudally. The aorta is located ventral to the vertebrae. The aorta branches into other arteries that supply many muscles and organs of the body. The branches from the aorta are usually named for the area in which they supply blood. For example, the **celiac** (sē-lē-ahck) **artery** supplies the liver, stomach, and spleen (**celi/o** is derived from the Greek term koilia meaning belly); the **renal arteries** supply the kidneys; and the **ovarian** (or **testicular**) **arteries** supply the ovaries (or testicles). Occasionally, arteries are named for their location as in the **subclavian** (suhb-klā-vē-ahn) **artery,** which is located under the collarbone (Figure 8–6).

The **arterioles** (ahr-tē-re-ōlz) are smaller branches of arteries. The combining form **arter/i** means vessel that carries blood away from the heart and the suffix **-ole** means small. Arterioles are smaller and thinner than arteries and carry blood to the capillaries.

The **capillaries** (cahp-ih-lār-ēz) are single cell thick vessels that connect the arterial and venous systems. Blood is able to flow rapidly through arteries and veins; however, blood flow is slower through the capillaries. This slower flow allows time for the exchange of oxygen, nutrients, and waste products. Blood in the alveolar capillaries picks up oxygen and gives off carbon dioxide. In the rest of the body, oxygen diffuses (passes through) from the capillaries into tissue, and carbon dioxide diffuses from tissue into the capillaries. Blood flow through tissues is called **perfusion** (pər-fū-shuhn). An indicator of perfusion is **capillary refill time** or CRT. A CRT can be obtained by applying pressure to mucous membranes and timing how long it takes for the pink color to return.

Capillaries connect with **venules** (vehn-yoolz), which are tiny blood vessels that carry blood to the veins. **Veins** (vānz) form a low-pressure collecting system that returns blood to the heart. Veins have thinner

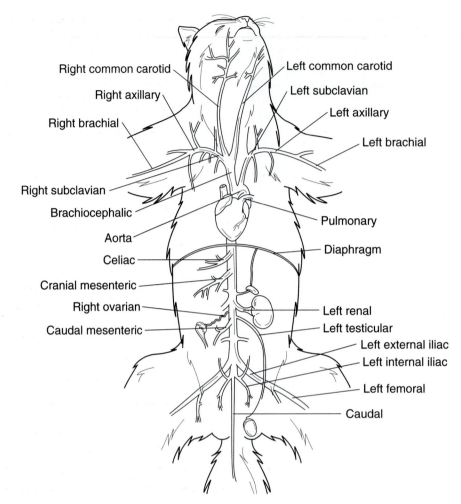

Right common carotid
Right axillary
Right brachial

Left common carotid
Left subclavian
Left axillary
Left brachial

Right subclavian
Brachiocephalic
Aorta
Celiac
Cranial mesenteric
Right ovarian
Caudal mesenteric

Pulmonary
Diaphragm

Left renal
Left testicular
Left external iliac
Left internal iliac
Left femoral
Caudal

FIGURE 8–6 Major arteries in the cat

walls and are less elastic than arteries, which have muscular walls to allow contraction and expansion to move blood throughout the body. Because the veins do not have muscular walls, contractions of the skeletal muscles cause the blood to flow through the veins toward the heart. Veins also have valves that permit blood flow toward the heart and prevent blood from flowing away from the heart. The combining forms for vein are **ven/o** and **phleb/o.**

Similar to arteries, veins are usually named for the area they take blood away from. The **jugular** (juhg-yoo-lahr) **vein** drains the head and neck area (jugulum is Latin for throat), **femoral veins** drain the legs, and **renal veins** drain the kidneys. An exception to this naming structure is the azygous vein. The **azygous** (ahz-ih-gihs) **vein** is a single vein that drains the chest wall and adjacent structures (the prefix **a-** means without, the combining form **zygon** means yoke or pair) and is named based on the fact that it is not paired in the body (Figure 8–7).

Vessels are used to distribute substances throughout the body. Many drugs are injected into vessels and transported to different areas of the body. Examples include:

▶ **intravenous** (ihn-trah-vē-nuhs) = within a vein. **Perivascular** (pehr-ih-vahs-kyoo-lahr) or pertaining to around the vessels is an undesired route of administration and is usually an error of intravenous injection.
▶ **intraarterial** (ihn-trah-ahr-tēr-ē-ahl) = within an artery

UNDER PRESSURE

Blood pressure is the tension exerted by blood on the arterial walls. It is dependent on the energy produced by the heart, the elasticity of the arterial walls, and the volume and **viscosity** (vihs-koh-siht-ē; viscosity means resistance to flow) of the blood. The **pulse** (puhlz) is the rhythmic expansion and contraction of an artery produced by pressure.

Blood pressure is measured by a **sphygmomanometer** (sfihg-mō-mah-nohm-eh-tər), which measures the amount of pressure exerted against the walls of the vessels. **Sphygm/o** is the combining form for pulse, **man/o** is the combining form for pressure, and **-meter** is the suffix meaning devise. **Systolic** (sihs-stohl-ihck) **pressure** occurs when the ventricles contract and is highest toward the end of the stroke output of the left ventricle. **Diastolic** (dī-ah-stohl-ihck) **pressure** occurs when the ventricles relax and is lowest late in ventricular dilation.

The combining form **tensi/o** means pressure or tension and is used when describing blood pressure. **Hypertension** is high blood pressure (**hyper** = excessive or above normal), while **hypotension** is low blood pressure (**hypo** = deficient or less than normal). Drugs used to lower blood pressure are called **antihypertensives** (ahn-tih-hī-pər-tehns-ihvs).

▶ Test Me: Cardiovascular System ◀

Diagnostic tests performed on the cardiovascular system include

- **electrocardiography** (ē-lehck-trō-kahr-dē-ohg-rah-fē) = process of recording the electrical activ-

ity of the heart. An **electrocardiogram** (ē-lehck-trō-kahr-dē-ō-grahm) is the record of the electrical activity of the heart and is abbreviated ECG or EKG. The machine that records the electrical activity of the heart is an **electrocardiograph** (ē-lehck-trō-kahr-dē-ō-grahf).

- **Holter monitor** (hōl-tər mohn-ih-tər) = 24-hour EKG that records the heart rates and rhythms onto a specialized tape recorder
- **radiography** (rā-dē-ohg-rah-fē) = procedure of imaging objects following exposure of sensitized film to X-rays. The resulting film is called a **radiograph** (rā-dē-ō-grahf) or X-ray. (This is an exception to the **-gram, -graph, -graphy** organization.)
- **angiography** (ahn-jē-ohg-rah-fē) = radiographic study of the blood vessels following injection of radiopaque material. An **angiogram** (ahn-jē-ō-grahm) is the film produced from this radiographic procedure.
- **cardiac catheterization** (kahr-dē-ahck kahth-eh-tər-ih-zā-shuhn) = radiographic study in which a

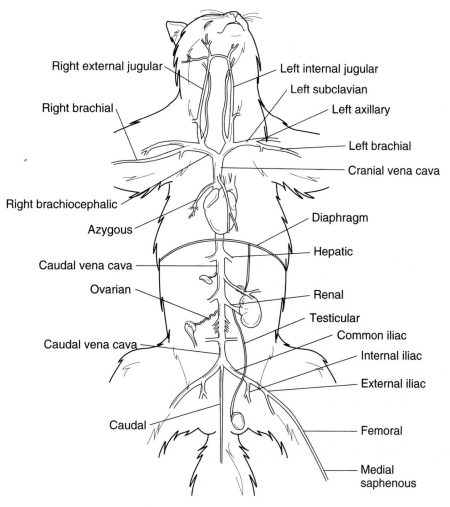

FIGURE 8–7 Major veins in the cat

catheter is passed into a blood vessel and is guided into the heart to detect pressures and patterns of blood flow.

- **echocardiography** (ehck-ō-kahr-dē-ohg-rah-fē) = process of evaluating the heart structures using sound waves. (**ech/o** is a combing form for sound). **Doppler** (dohp-lǝr) **echocardiography** uses the differences in frequency between sound waves and their echoes to measure the velocity of a moving object (Figure 8–8).
- **tourniquet** (toor-nih-keht) = constricting band applied to a limb to control bleeding or to assist in drawing blood
- **angiocardiography** (ahn-jē-ō-kahr-dē-ohg-rah-fē) = radiographic study of the blood vessels and heart using contrast material. The resulting film is an **angiocardiogram** (ahn-jē-ō-kahr-dē-ō-grahm).

▶ **Path of Destruction: Cardiovascular System** ◀

Pathologic conditions of the cardiovascular system include

- **atherosclerosis** (ahth-ǝr-ō-skleh-rō-sihs) = hardening and narrowing of the arteries. This may occur due to a **plaque** (plahck), which is a patch or raised area. **Ather/o** is the combining form for plaque or fatty substance.
- **infarct** (ihn-fahrkt) = localized area of necrosis caused by an interrupted blood supply
- **ischemia** (ihs-kē-mē-ah) = deficiency in blood supply (the combining form **isch/o** means hold back)

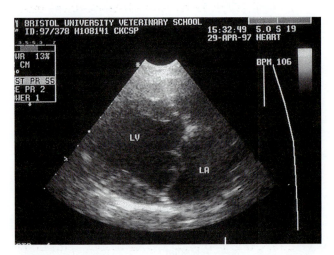

FIGURE 8–8 Echocardiogram of a canine heart *Source:* Courtesy of Mark Jackson, North Carolina State University

💡 Doppler echocardiography is based on the Doppler effect. An example of the Doppler effect (differences in frequency between sound waves and their echoes to measure the velocity of a moving object) is a police siren. A police siren seems to have a higher pitch when the police car is approaching than after it passes. Similarly, the pitch of a train whistle is higher when the train is approaching than after it passes. Medical use of the Doppler effect enables detection of blood flow based on various pitches after it passes by a transducer. Christian Doppler, an American mathematician, was the first person to describe this effect.

- **congestive** (kohn-gehs-tihv) **heart failure** = syndrome that reflects insufficient cardiac output to meet the body's needs; abbreviated CHF. **Congestion** (kohn-jehs-zhuhn), which is accumulation of fluid, and **edema** (eh-dē-mah), which is accumulation of fluid in the intercellular spaces, may be seen with CHF. **Ascites** (ah-sī-tēz) is fluid accumulation in the peritoneal cavity seen in dogs secondary to CHF and other diseases. **Pleural effusion** (ploor-ahl eh-fū-zhuhn) is abnormal fluid accumulation between the layers of the membrane encasing the lungs and is seen in cats secondary to CHF. Fluid accumulation can be relieved with the use of diuretics. **Diuretics** (dī-yoo-reht-ihcks) are substances that increase urine excretion.
- **carditis** (kahr-dī-tihs) = inflammation of the heart
- **pericarditis** (pehr-ih-kahr-dī-tihs) = inflammation of the pericardium
- **myocarditis** (mī-ō-kahr-dī-tihs) = inflammation of the myocardium
- **endocarditis** (ehn-dō-kahr-dī-tihs) = inflammation of the endocardium and sometimes the heart valves. Endocarditis may be further modified as in bacterial endocarditis.
- **mitral valve prolapse** (mī-trahl vahlv prō-lahps) = abnormal protrusion of the left atrioventricular valve that results in incomplete closure of the valve
- **mitral valve insufficiency** (mī-trahl vahlv ihn-sah-fihsh-ehn-cē) = inability of the left atrioventricular valve to perform at the proper level; may be caused by fibrosis, endocarditis, or other conditions that occur in the mitral valve area
- **vasculitis** (vahs-kyoo-lī-tihs) = inflammation of a blood or lymph vessel
- **aneurysm** (ahn-yoo-rihzm) = localized balloon-like enlargement of an artery
- **hematoma** (hē-mah-tō-mah) = collection of blood
- **hemangioma** (hē-mahn-jē-ō-mah) = benign tumor comprised of newly formed blood vessels
- **thrombus** (throhm-buhs) = blood clot attached to the interior wall of a vein or artery. A thrombosis (throhm-bō-sihs) is an abnormal condition in

which a blood clot develops within a blood vessel. Substances that slow and prevent blood clotting are called **anticoagulants** (ahn-tih-kō-āg-yoo-lahnts).

- **occlusion** (ō-kloo-shuhn) = blockage in a vessel or passageway in the body
- **embolus** (ehm-bō-luhs) = foreign object (such as a clot, air, tissue, etc) that is circulating in blood. An **embolism** (ehm-bō-lihzm) is blockage of a vessel due to a foreign object.
- **patent ductus arteriosus** (pā-tehnt duhck-tuhs ahr-tē-rē-ō-sihs) = persistence of the fetal communication between the left pulmonary artery and aorta that should close shortly after birth; abbreviated PDA. (Patent means remaining open.) A PDA may cause overloading of the left ventricle, which may lead to left ventricular failure. A continuous murmur is usually a sign of a PDA.
- **ventricular septal defect** (vehn-trihck-yoo-lahr sehp-tahl dē-fehckt) = opening in the wall dividing the right and left ventricles that may allow blood to shunt from the right ventricle to the left ventricle without becoming oxygenated. A **shunt** (shuhnt) means to bypass or divert. A shunt resulting from a ventricular septal defect would bypass the lungs. Ventricular septal defect is abbreviated VSD. A harsh holosystolic murmur is usually a sign of a VSD.
- **tetralogy of Fallot** (teht-rahl-ō-gē of fahl-ō) = multiple cardiac defect that includes pulmonary stenosis, ventricular septal defect, overriding aorta, and right ventricular hypertrophy
- **pulmonic stenosis** (puhl-mah-nihck stehn-ō-sihs) = narrowing of the opening and valvular area between the pulmonary artery and right ventricle
- **aortic insufficiency** (ā-ōr-tihck ihn-sah-fihsh-ehn-cē) = inability of the aortic valve to perform at the proper levels, which results in blood flowing back into the left ventricle from the aorta
- **cardiomegaly** (kahr-dē-ō-mehg-ah-lē) = heart enlargement
- **cardiomyopathy** (kahr-dē-ō-mī-ohp-ah-thē) = disease of heart muscle. May be further classified as **hypertrophic** (hī-pər-trō-fihck), which is excessive growth of the left ventricle or **dilated** (dī-lāt-ehd), which is characterized by a thin-walled left ventricle. Dilated cardiomyopathy is also known as **congestive** (kohn-gehs-tihv).
- **cor pulmonale** (kōr puhl-mah-nahl-ē) = alterations in the structure and/or function of the right ventricle caused by pulmonary hypertension; also called pulmonary heart disease. **Cor** means heart and **pulmon/o** is the combining form for lung.
- **hypocapnia** (hī-pō-kahp-nē-ah) = below-normal levels of carbon dioxide

- **hypercapnia** (hī-pər-kahp-nē-ah) = above-normal levels of carbon dioxide. Hypercapnia results in reduced levels of oxygen (**hypoxia** = hī-pohck-sē-ah) and may cause a bluish tinge to the skin and mucous membranes. This bluish tinge is termed **cyanosis** (the combining form **cyan/o** means blue)
- **hypoxia** (hī-pohck-sē-ah) = below-normal levels of oxygen
- **heart block** = interference with the electrical conduction of the heart. Heart block may be partial or complete and is graded out in degrees based on the characteristics of the block.
- **cardiac tamponade** (kahr-dē-ahck tehm-pō-nohd) = compression of the heart due to fluid or blood collection in the pericardial sac
- **regurgitation** (rē-gərg-ih-tā-shuhn) = backflow; used to describe backflow of blood due to imperfect closure of heart valves
- **angiopathy** (ahn-jē-ohp-ah-thē) = disease of vessels
- **dirofilariasis** (dī-rō-fihl-ahr-ē-ah-sihs) = heartworm infection. The scientific name of heartworm is *Dirofilaria immitis* (dī-rō-fihl-ahr-ē-ah ihm-ih-tihs), from which dirofilariasis is derived. Heartworm disease is currently found in dogs, cats, and ferrets. Heartworms mature and breed in the larger blood vessels. Mature heartworms produce tiny larvae called **microfilariae** (mī-krō-fihl-ahr-ē-ah). Mature heartworms may obstruct blood flow through the heart and blood vessels. A dead heartworm can cause pulmonary embolism. Obstruction of blood flow from the vena cava due to heavy heartworm infestation is termed **caval** (kā-vahl) **syndrome.** Heartworm disease can be prevented by the used of **prophylactic** (prō-fih-lahck-tihck) medication. **Prophylaxis** (prō-fih-lahck-sihs) means prevention. If an animal has heartworm disease treatment including use of an **adulticide** (ah-duhlt-ih-sīd) or substance that kills mature or adult (heart)worms and a **microfilaricide** (mī-krō-fihl-ahr-ih-sīd) or substance that kills larvae or juvenile (heart)-worms.
- **shock** (shohck) = inadequate tissue perfusion. There are different types of shock, but one type occurs after cardiac arrest or cessation of heartbeat. Treatment of shock includes **resuscitation** (reh-suhs-ih-tā-shuhn) or the restoration of life. Resuscitative measures include fluid administration, cardiac massage, and artificial respiration. **Cardiopulmonary resuscitation** (kahr-dē-ō-puhl-mohn-ār-ē) or CPR addresses only the cardiac and respiratory systems.

Terms Used to Describe Disease:

congenital
(kohn-jehn-ih-tahl)   present at birth
hereditary (hər-eh-dih-tahr-ē)   genetically transmitted from
                                        parent to offspring
anomaly (ah-nohm-ah-lē)   deviation from normal
idiopathic (ihd-ē-ō-pahth-ihck)   unknown cause
iatrogenic (ī-aht-rō-jehn-ihck)   produced by treatment

▶ Procede with Caution: Cardiovascular System ◀

Procedures performed on the cardiovascular system include

- **angioplasty** (ahn-jē-ō-plahs-tē) = surgical repair of blood or lymph vessels. Angioplasties may be **transluminal** (trahnz-loo-mehn-ahl), which means the procedure is done through the opening of vessel or **percutaneous** (pehr-kyoo-tā-nē-uhs), which means the procedure is done through the skin
- **valvotomy** (vahl-vah-tō-mē) = surgical incision into a valve or membranous flap
- **stent** (stehnt) = small expander implanted in a blood vessel to prevent it from collapsing (a stent is also a device to hold tissue in place or to provide support for a graft)
- **defibrillation** (dē-fihb-rih-lā-shuhn) = use of electrical shock to restore the normal heart rhythm.
- **transfusion** (trahnz-fū-shuhn) = introduction of whole blood or blood components into the bloodstream of the recipient (Figure 8–9).
- **hemostasis** (hē-mō-stā-sis) = control or stoppage of bleeding
- **central venous pressure** (sehn-trahl vē-nuhs prehs-sər) = tension exerted by blood in the cranial vena cava; abbreviated CVP. CVP is monitored by catheterization of the cranial vena cava via the jugular vein. The catheter is connected to a fluid-filled column and a syringe or bag that serves as a fluid source.

- **angiorrhaphy** (ahn-jē-ōr-ah-fē) = suture of vessel
- **arteriectomy** (ahr-teh-rē-ehck-tō-mē) = surgical removal of part of a blood vessel that carries blood away from the heart
- **arteriotomy** (ahr-tē-rē-oh-tō-mē) = incision of a blood vessel that carries blood away from the heart

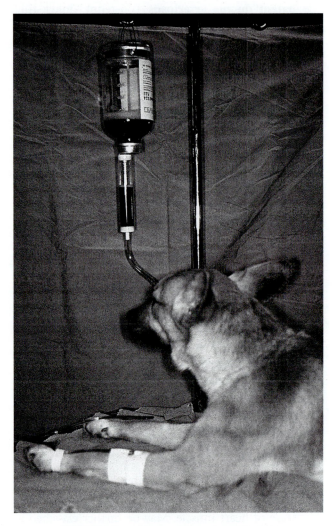

FIGURE 8–9 A dog receiving a transfusion

REVIEW EXERCISES

Multiple Choice—Choose the correct answer.

1. The right atrioventricular valve is also known as the
   a. mitral valve
   b. semilunar valve
   c. tricuspid valve
   d. bicuspid valve

2. The double-walled membranous sac enclosing the heart is the
   a. peritoneum
   b. pericardium
   c. perimyocardium
   d. pericardosis

3. A partition or wall separating something is called a
   a. septum
   b. valve
   c. lumen
   d. plaque

4. A localized area of necrosis caused by an interrupted blood supply is
   a. ischemia
   b. resuscitation
   c. pulse
   d. infarct

5. Introduction of whole blood or blood components into the bloodstream of the recipient is a/an
   a. embolism
   b. thrombus
   c. transfusion
   d. stent

6. To bypass or divert is called a/an
   a. preload
   b. shunt
   c. stent
   d. tourniquet

7. Cor means
   a. abnormality
   b. vessel
   c. heart
   d. valve

8. Disease of heart muscle is
   a. cardiopathy
   b. cor pulmonale
   c. cardiovalvopathy
   d. cardiomyopathy

9. Heart enlargement is
   a. cardiac swelling
   b. cardiac augmentation
   c. cardiac dilation
   d. cardiomegaly

10. Blood flow through tissue is
    a. ischemia
    b. infarct
    c. auscultation
    d. perfusion

Label the Diagrams in Figures 8–10 and 8–11.

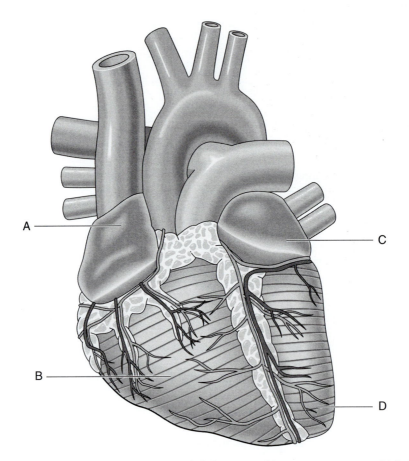

FIGURE 8–10 External heart structures. Label the external heart structures to which the lines are pointing.

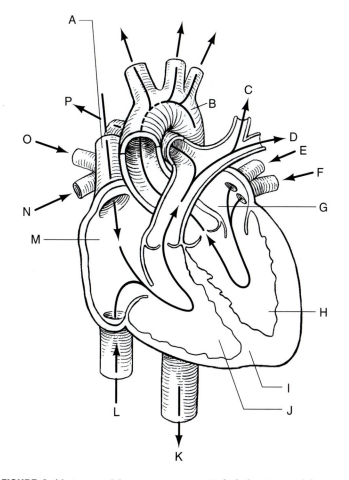

FIGURE 8–11 Internal heart structures. Label the internal heart structures to which the lines are pointing.

Case Studies—Define the underlined terms in each case study.

A 5 yr old M/N Doberman was presented with signs of <u>lethargy</u>, <u>syncope</u>, and <u>cyanotic</u> mucous membranes. After ausculting the heart, the veterinarian detected a <u>cardiac arrhythmia</u> and <u>tachycardia</u>. An <u>EKG</u> and <u>radiograph</u> of the heart were ordered. The radiograph revealed <u>cardiomegaly</u>, which helped to support the veterinarian's <u>diagnosis</u> of <u>cardiomyopathy</u>.

lethargy _____

syncope _____

cyanotic _____

cardiac arrhythmia _____

tachycardia _____

EKG _____

radiograph _____

cardiomegaly _____

diagnosis _____

cardiomyopathy _____

A farmer called the clinic because one of his cows was suddenly <u>off feed</u> and had not been producing as much milk as before. Upon arrival at the farm, the veterinarian noted that the cow was reluctant to move, had an arched back, and appeared tachypnic. <u>PE</u> revealed <u>tachycardia</u>, <u>tachypnea</u>, <u>dyspnea</u>, <u>pyrexia</u>, and abducted elbows. Auscultation of the <u>thorax</u> revealed muffled lung and heart sounds. The farmer was questioned as to his use of magnets to prevent metallic objects from staying in the <u>rumen</u> or reticulum. The owner did not use magnets as a <u>prophylactic</u> measure, so the veterinarian suspected <u>acute traumatic reticuloperitonitis</u> in this cow. <u>Acute traumatic reticuloperitonitis</u> is commonly called hardware disease and is seen when swallowed metallic objects fall into the <u>reticulum</u> of the <u>ruminant</u> stomach, pierce the reticulum wall, and cause contamination in the <u>peritoneal cavity</u>. Occasionally the object punctures the <u>diaphragm</u>, enters the thoracic cavity, and punctures the <u>pericardial sac</u> causing <u>pericarditis</u>. A magnet was placed in this cow via a <u>balling gun</u>, and antibiotics were initiated. If the cow does not improve, she may need to be shipped (sent to slaughter) because of the high cost of treating this disease.

off feed _____

PE _____

tachycardia _____

tachypnea _____

dyspnea _____

pyrexia _____

thorax _____

rumen _____

prophylactic _____

acute _____

traumatic _____

reticuloperitonitis _____

acute traumatic reticuloperitonitis _____

reticulum _____

ruminant_____

peritoneal cavity _____

diaphragm _____

pericardial sac_____

pericarditis _____

balling gun _____0

Fill in the blanks to trace the flow of blood through the heart.

Blood enters the heart through two large veins called the _____ and _____, which empty blood into the _____.

Blood flows from the right atrium to the right ventricle through the _____, which is also called the_____. After blood enters the right ventricle, it passes through the _____ and enters the lungs via the _____. Blood becomes oxygenated in the lungs and returns to the heart via the _____. Once blood enters the left atrium it passes through the _____ or _____ on its way to the _____. Blood passes through one last valve, the _____, before it enters the aorta and flows to various parts of the body.

A Breath of Fresh Air

FUNCTIONS OF THE RESPIRATORY SYSTEM

The **respiratory** (rehs-pih-tōr-ē or rehs-pih-rah-tōr-ē) **system** is the body system that brings oxygen from the air into the body for delivery via the blood to the cells. Once the blood has delivered the oxygen to the cells, it picks up carbon dioxide and carries it back to the lungs where this waste is expelled out into the air. Carbon dioxide has acid properties and therefore is also involved in maintaining the body's acid/base status.

The term **respiration** means the exchange of gases (oxygen and carbon dioxide) between the atmosphere and the cells of the body. The gas exchange between the blood and the cells is called internal or cellular respiration. External respiration is the absorption of atmospheric oxygen by the blood in the lungs and the diffusion of carbon dioxide from the blood in the lungs to atmospheric air.

Ventilation (vehn-tih-lā-shuhn) is a term that means to bring in fresh air. Ventilation is used to refer to breathing. Ventilation may be natural, as in normal breathing, or assisted, as in the use of a ventilator. Ventilators are devices to assist breathing and therefore should not be called respirators.

STRUCTURES OF THE RESPIRATORY SYSTEM

The respiratory tract is routinely divided into the upper and lower respiratory tracts. The **upper respiratory tract** (uhp-pər rehs-pih-tōr-ē trahck or uhp-pər rehs-pih-rah-tōr-ē trahck) is the part of the respiratory system that consists of the nose, mouth, pharynx, epiglottis, and larynx. The **lower respiratory tract** (lō-ər rehs-pih-tōr-ē trahck or lō-ər rehs-pih-rah-tōr-ē trahck) is the part of the respiratory system that consists of the trachea, bronchial tree, and lungs. The trachea is sometimes considered part of the upper respiratory tract (Figure 9–1).

Upper Respiratory Tract

SNIFFING IT OUT

Air enters and exits the body through the **nose. Nas/o** and **rhin/o** are combining forms for nose. In animals not all noses look alike. The rigidity of the nose in swine has lead to its being called the **snout** (snowt) because it is so different from other species' noses. The nose consists of **nostrils** or **nares** (nār-ēz), which are

119

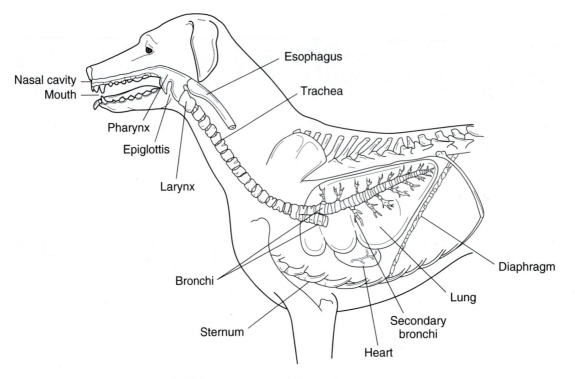

FIGURE 9–1 Structures of the respiratory system

the paired external openings of the respiratory tract. Nares vary from species to species and may have the ability to open widely as in equine or may remain relatively the same as in the canine.

In **endotherms** (ehn-dō-thərmz) or "warm-blooded" animals, the nasal passages contain **nasal turbinates** (tər-bih-nātz). The nasal turbinates are scroll-like bones covered with highly vascular mucous membranes. The nasal turbinates warm, humidify, and filter inspired air. There are two major nasal passages associated with the nasal turbinates: the dorsal nasal meatus and ventral nasal meatus (**meat/o** is the combining form for opening or passageway). A **nasogastric** (nā-zō-gahs-trihck) **tube** is a tube that passes through the nose and down to the stomach. A nasogastric tube is placed through the ventral nasal meatus.

Air passes from the nose through the nasal cavity. The front part of the nostrils and nasal cavity are referred to as the **vestibule** (vehs-tih-buhl). The nose is divided by a wall of cartilage called the nasal septum (nā-zahl sehp-tuhm). The combining form **sept/o** means partition.

The respiratory system is lined with **mucous membrane** (myoo-kuhs mehm-brān), which is a specialized form of epithelial tissue. The mucous membranes secrete **mucus** (myoo-kuhs). Mucus is a slime-like substance that is composed of glandular secretion, salts, cells, and leukocytes. Mucus helps to moisten, warm, and filter the air as it enters the nose. **Cilia** (sihl-ē-ah) are thin hairs located inside the nostrils. Cilia filter the air to remove **debris** (deh-brē).

The **olfactory** (ohl-fahck-tōr-ē) **receptors** are the receptors responsible for the sense of smell. The combining form **olfact/o** means smell. Olfactory receptors are nerve endings located in the mucous membranes in the nasal cavity.

The **tonsils** (tohn-sihlz) are lymphatic tissue that protects the nasal cavity and upper throat. The combining form for tonsil is tonsillo. The tonsils are covered in Chapter 15.

> **Rhin/o** is the Greek root for nose. Consider the term rhinoceros. Rhin/o means nose and **cer/o** means horn; therefore a rhinoceros is an animal with a horn on its nose. Where do you think a rhinovirus causes disease?

GIVE ME SPACE

A **sinus** (sīn-uhs) is an air-filled or fluid-filled space. A sinus in the respiratory system refers to an air-filled or fluid-filled space within bone. Sinuses have a mucous membrane lining. The functions of sinuses are to provide mucus, to make bone lighter, and to help to produce sound. The combining form for sinus is **sinus/o** (Table 9–1).

BACK OF THE THROAT

Air passes through the nasal cavity to the **pharynx** (fār-ihnks). The pharynx is commonly called the throat. The pharynx is the common passageway for the upper respiratory and gastrointestinal tracts. The pharynx ex-

TABLE 9–1
Sinuses

Sinus	Species Found	Location
frontal (frohn-tahl)	all domestic species	dorsal part of skull between nasal cavity and orbit
maxillary (mahx-ihl-ār-ē)	all domestic species (maxillary recess in carnivores)	maxilla with nasal cavity on each side
sphenoid (sfehn-oyd)	feline, bovine, equine, swine	sphenoid bone; opens to nasal cavity
palatine (pahl-eh-tehn)	ruminants, equine	palatine bone; communicates with maxillary sinus
lacrimal (lahck-rih-mahl)	swine, ruminants	lacrimal bone
conchal (kohn-kahl)	swine, ruminants, equine	formed by enclosure of conchae

tends from the back of the nasal passages and mouth to the larynx and connects the nasal passages to the larynx and the mouth to the esophagus. The combining form for throat is **pharyng/o.**

The pharynx has three divisions:

▶ **nasopharynx** (nā-zō-fār-ihnkz) = portion of the throat posterior to the nasal cavity and above (dorsal to) the soft palate
▶ **oropharynx** (ō-rō-fār-ihnkz) = portion of the throat between the soft palate and epiglottis
▶ **laryngopharynx** (lah-rihng-gō-fār-ihnkz) = portion of the throat below the epiglottis which opens into the voice box and esophagus.

The nasopharynx is the passageway for air entering through the nose, while the oropharynx and laryngopharynx are passageways for air entering through the nose and food entering through the mouth. During swallowing, the **soft palate** (pahl-aht) moves dorsally and caudally to close off the nasopharynx to prevent food from going up into the nose. Palate means roof of the mouth and the combining form for palate is **palat/o.** The **epiglottis** (ehp-ih-gloht-ihs) acts like a lid and covers the larynx during swallowing. The covering of the larynx by the epiglottis does not allow food to enter the trachea and go into the lungs. The combining form for epiglottis is **epiglott/o** (Figure 9–2).

MAKE A SOUND

The **larynx** (lār-ihnkz) is the part of the respiratory tract located between the pharynx and trachea. The larynx is commonly referred to as the **voice box.** The larynx contains the **vocal cords** (vō-kahl kōrdz), which are paired membranous bands in the larynx that help produce sound. **Laryng/o** is the combining form for the voice box.

The vocal apparatus is the **glottis** (gloh-tihs). The glottis consists of vocal cords and the space between them. Air passing through the glottis causes a vibration that produces sound. The combining form for

glottis is **glott/o.** The vocal apparatus of avian species is the **syrinx** (sehr-ihnks) and is located between the trachea and bronchi.

Lower Respiratory Tract

WINDPIPE

Air passes from the larynx to the **trachea** (trā-kē-ah). The trachea is commonly called the **windpipe** and extends from the neck to the chest. The trachea attaches to the larynx in the neck and passes into the **thorax** (thōr-ahcks) or chest cavity through the **thoracic** (thoh-rahs-ihck) **inlet.** The trachea is located ventral to the esophagus and is held open by a series of C-shaped cartilaginous rings. The open part of the C's are typically along the dorsal aspect, which is adjacent to the esophagus. This allows for easier expansion of the esophagus when the animal swallows. The trachea is also lined with cilia, which help filter debris. The combining form for the windpipe is **trache/o.**

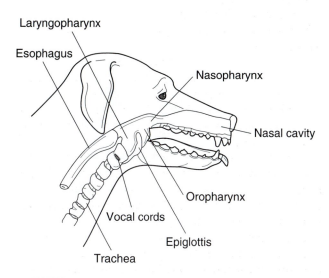

FIGURE 9–2 Structures of the nasal cavity and larynx

BRANCHING OFF

The bottom of the trachea (distal end of the trachea) divides into two branches which are called the **tracheal bifurcation** (trā-kē-ahl bī-fər-kā-shuhn). The branches from the trachea are called **bronchi** (brohng-kī). The combining form for bronchi is **bronch/o.** Each **bronchus** (brohng-kuhs) leads to a separate lung (right or left) and continues to divide. This continual division appears similar to a tree and its branches; therefore, the bronchi and its branches are sometimes referred to as the **bronchial tree.** Each bronchus that leads to a separate lung is called a **principal bronchus** (right principal or left principal bronchus). The principal bronchi divide into smaller branches called **secondary** (sehc-ohnd-ār-ē) or lobar (lō-bahr) **bronchi.** The secondary bronchi divide into **tertiary** (tər-shē-ār-ē) or **segmental** (sehg-mehn-tahl) **bronchi.** The tertiary bronchi are smaller units and are also called **bronchioles** (brohng-kē-ōlz) or **bronchiolus** (brohng-kē-ō-luhs). The suffix **-ole** means small, so it can be derived that bronchioles are smaller than bronchi. The combining form for bronchiole is **bronchiol/o.** Bronchioles contain no cartilage or glands. The bronchioles continue to divide. The terminal bronchioles are the last portion of a bronchiole that does not contain alveoli. The respiratory bronchioles are the final branches of the bronchioles. The respiratory bronchioles have alveolar outcroppings and branch into alveolar ducts (Figure 9–3).

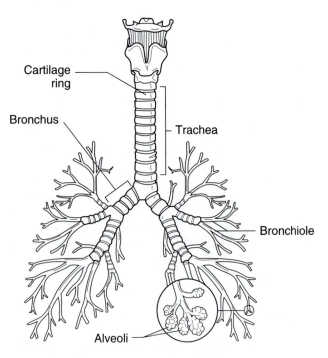

Cartilage ring

Bronchus

Trachea

Bronchiole

Alveoli

FIGURE 9–3 Lower respiratory tract structures

GIVE ME AIR

Alveoli (ahl-vē-ō-lī) are air sacs in which most of the gas exchange occurs. An **alveolus** (ahl-vē-ō-luhs), which is Latin for small hollow thing, is a small grape-like cluster at the end of each bronchiole. The alveolus is connected to the bronchiole via an alveolar duct. The combining form **alveol/o** means small sac.

Alveoli have thin, flexible membrane walls that are surrounded by a network of microscopic capillaries. Gas exchange occurs across the alveolar membrane. Oxygen diffuses into the blood in alveolar capillaries and binds to the hemoglobin in erythrocytes. Carbon dioxide diffuses from the plasma across the alveolar membrane into the alveolus. When an animal exhales, much of this air is pushed back out of the alveolus, back up through the respiratory tract, and out the nose or mouth.

Alveoli contain liquid that reduces alveolar surface tension. This liquid is called **surfactant** (sihr-fahck-tehnt). Surfactant prevents collapse of the alveoli during expiration.

WHAT'S HOLDING US TOGETHER?

The thoracic cavity is contained within the ribs. The combining form **cost/o** means ribs. Intercostal means pertaining to between the ribs.

The lungs are located within the thoracic cavity. The thoracic cavity also protects the lungs. The combining form **thorac/o** and the suffix **-thorax** both mean chest cavity or chest.

The **lung** (luhng) is the main organ of respiration. There are two lungs (right and left) that are composed of divisions called **lobes** (lōbz). A lobe is a well-defined portion of an organ and is used in describing areas in the lung, liver, and other organs. The combining form **lob/o** means well-defined portion. There is species variation as to the number and names of lung lobes. The combining forms **pneum/o, pneumon/o,** and **pneu** mean lungs or air; the combining forms **pulm/o** and **pulmon/o** mean lung.

The term **parenchyma** (pahr-ehnk-ih-mah) refers to the functional elements of an organ, as opposed to its framework or **stroma** (strō-mah). The functional elements of the lung are referred to collectively as the lung parenchyma.

The region between the lungs is called the **mediastinum** (mē-dē-ahs-tī-nuhm). The mediastinum is the space between the lungs that houses the heart, aorta, lymph nodes, esophagus, trachea, bronchial tubes, nerves, thoracic duct, and thymus (Figure 9–4).

AROUND AND AROUND

Each lung is encased in a membranous sac called the **pleura** (ploor-ah). The combining form **pleur/o** means

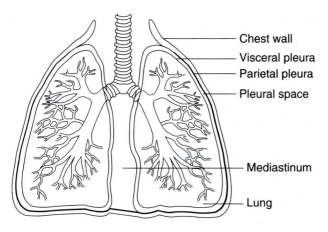

FIGURE 9–4 Structures of the thoracic cavity

Chest wall
Visceral pleura
Parietal pleura
Pleural space
Mediastinum
Lung

membrane surrounding the lung; the plural form of pleura is **pleurae.**

▶ **parietal pleura** (pah-rī-eh-tahl ploor-ah) is the outer layer of the membrane lining the inner wall of the thoracic cavity

▶ **visceral pleura** (vihs-ər-ahl ploor-ah) is the inner layer of the membrane lining the outside of the lung

▶ **pleural** (ploor-ahl) **space** is the airtight space found between the parietal pleura and visceral pleura. The pleural space or pleural cavity contains a small amount of lubricating fluid to prevent friction when the membranes rub together during respiration.

A WALL ACROSS

The thoracic and peritoneal cavities are separated from each other by the **diaphragm** (dī-ah-frahm). The prefix **dia-** means across and the combining form **phragm/o** means wall. The diaphragm is a muscle and contraction of the diaphragm causes air pressure within the lungs to drop below atmospheric pressure. This produces a vacuum within the thoracic cavity to draw in air. When the diaphragm relaxes, this negative pressure is no longer generated and air is forced out of the lung. The combining forms **diaphragmat/o** and **phren/o** mean diaphragm. A **diaphragmatic hernia** (dī-ah-frahg-mah-tihck hər-nē-ah) is an abnormal displacement of organs through the muscle separating the chest and abdomen. The phrenic nerve innervates the diaphragm.

> The combining form **phren/o** also is a term used to refer to the mind. The ancient Greeks once thought that the spleen and kidneys were the organs responsible for emotion, and since the diaphragm sits across the top of these organs phren/o was used for both the diaphragm and mind.

A BREATH OF FRESH AIR

Breathing is the inhalation and exhalation of air. **Inhalation** (ihn-hah-lā-shuhn) is the drawing in of a breath. Inspiration is another term for the drawing in of a breath. **Exhalation** (ehcks-hah-lā-shuhn) is the release of a breath. **Expiration** is another term for the release of a breath. Spirare in Latin means to breathe. The combining form **spir/o** also means breath or breathing. Most medical terms use the Greek root **-pnea** to refer to breathing.

▶ **apnea** (ahp-nē-ah) = absence of breathing
▶ **dyspnea** (dihsp-nē-ah) = difficult or labored breathing
▶ **bradypnea** (brād-ihp-nē-ah) = abnormally slow respiratory rates
▶ **tachypnea** (tahck-ihp-nē-ah) = abnormally rapid respiratory rates
▶ **hyperpnea** (hī-pərp-nē-ah) = abnormal increase in the rate and depth of respirations
▶ **hypopnea** (hī-pōp-nē-ah) = abnormally slow or shallow respirations
▶ **hyperventilation** (hī-pər-vehn-tih-lā-shuhn) = abnormally rapid deep breathing which results in decreased levels of cellular carbon dioxide.
▶ **agonal breathing** (āg-uh-nuhl) = respirations near death or extreme suffering

Respiration involves the exchange of oxygen (O_2) and carbon dioxide (CO_2). The combining forms **ox/i, ox/o,** and **ox/y** refer to O_2, and **capn/o** refers to CO_2. **Hypoxia** (hī-pohck-sē-ah) refers to an inadequate supply of oxygen to tissue despite an adequate blood supply. **Hypercapnia** (hī-pər-kahp-nē-ah) refers to excessive amounts of carbon dioxide in the blood. **Hyperventilation** (hī-pər-vehn-tih-lā-shuhn), an abnormal increase in the rate or depth of breathing, may lead to **hypocapnia** (hī-pō-kahp-nē-ah), which is a decrease in the carbon dioxide levels in the blood.

When carbon dioxide dissolves in water, some of it reacts with the water to form carbonic acid (H_2CO_3). Since carbon dioxide breaks down into a weak acid, it affects the blood pH. An excessive amount of carbon dioxide in the blood can lower the pH of blood; this is termed **respiratory acidosis** (ah-sih-dō-sihs). If carbon dioxide levels are abnormally low, **respiratory alkalosis** (ahl-kah-lō-sihs) may result. Changes in blood pH can also result from metabolic factors and are then termed **metabolic acidosis** or **metabolic alkalosis** (Figure 9–5 and Table 9–2).

▶ **Test Me: Respiratory System** ◀

Diagnostic procedures performed on the system include

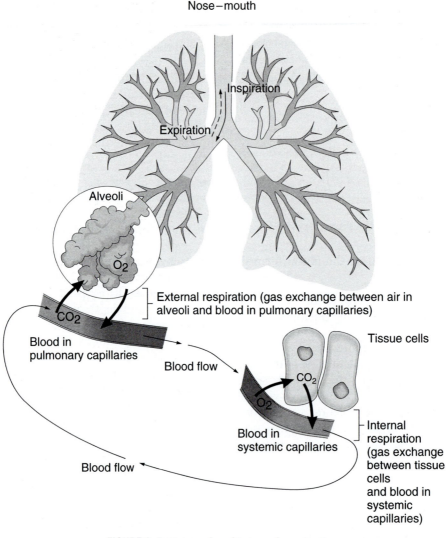

Nose—mouth

Inspiration

Expiration

Alveoli

O_2

CO_2

External respiration (gas exchange between air in alveoli and blood in pulmonary capillaries)

Blood in pulmonary capillaries

Blood flow

Tissue cells

CO_2

O_2

Blood in systemic capillaries

Internal respiration (gas exchange between tissue cells and blood in systemic capillaries)

Blood flow

FIGURE 9–5 External and internal respiration

- **auscultation** (ahws-kuhl-tā-shuhn) = act of listening. The respiratory tract is auscultated with a **stethoscope** (stehth-ō-skōp). Respiratory rhythm, rate, and sound are evaluated upon auscultation. Pathologic respiratory sounds are called **adventitious** (ahd-vehn-tish-uhs) **sounds.**

Things to listen for include

  - **respiratory rate** (RR) = number of respirations per minute. One inspiration and one expiration form a single respiration. RR varies with species.
  - **bubbling** = sound of popping bubbles that suggests fluid accumulation
  - **crepitation** (krehp-ih-tā-shuhn) = fine or coarse interrupted crackling noises coming from collapsed or fluid-filled alveoli during inspiration; also called **rales** (rahlz) or crackles

- **decreased lung sounds** = less or no sound of air movement suggests consolidation
- **rhonchi** (rohn-kī) = abnormal, continuous, musical, high-pitched whistling sound heard during inspiration; also called **wheezes** (wē-zehz)
- **stridor** (strī-dōr) = snoring, squeaking, or whistling that suggests airway narrowing
- **vesicular sounds** (vehs-ihck-yoo-lahr) = sound resulting from air passing through small bronchi and alveoli.
- **percussion** (pər-kush-uhn) = diagnostic procedure used to determine density in which sound is produced by tapping various body surfaces with the finger or an instrument. The sound produced over the chest where air is present varies from an area where fluid is present.

- **transtracheal wash** (trahnz-trā-kē-ahl) = sterile collection of fluid or mucus from the trachea via a catheter inserted through the skin into the trachea to assess respiratory disease; abbreviated TTW. Fluid may be used for cytologic and microbiologic examination (Figure 9–6).
- **tracheal wash** (trā-kē-ahl) = collection of fluid or mucus from the trachea via an endotracheal tube to assess respiratory disease. Fluid may be used for cytologic and microbiologic examination.
- **bronchoalveolar lavage** (brohng-kō-ahl-vē-ō-lahr lah-vahj) = collection of fluid or mucus from the lower respiratory tract (bronchi or alveoli) via an endoscope or through an endotracheal tube inserted as far down the trachea as possible before infusing fluid and aspirating a sample. Fluid may be used for cytologic examination.
- **phlegm** (flehm) = thick mucus secreted by the respiratory lining. Phlegm ejected through the

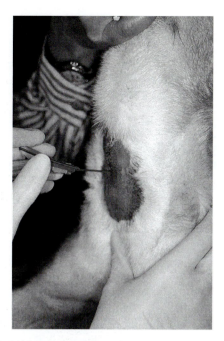

FIGURE 9–6 Transtracheal wash in a dog. A TTW is initiated with insertion of a catheter in the trachea of a dog. A syringe filled with saline is attached to the opposite end of the catheter so that a small volume of saline may be introduced into the trachea and then quickly aspirated along with mucus. This sample is then ready for cytologic and microbiologic assessment.

TABLE 9–2
Lung Volume Terminology

tidal volume	amount of air exchanged during normal respiration (air inhaled and exhaled in one breath)
inspiratory reserve volume or **complemental air**	amount of air inspired over the tidal volume (extra amount that could be inhaled after normal inspiration)
expiratory reserve volume or **supplemental air**	amount of air expired over the tidal volume (extra amount that could be exhaled after normal expiration)
residual volume	air remaining in the lungs after a forced expiration (amount of air trapped in alveoli)
dead space	air in the pathway of the respiratory system (is termed dead because this air is not currently participating in gas exchange)
minimal volume	amount of air left in alveoli after the lung collapses
vital capacity	largest amount of air that can be moved in the lung (tidal volume + inspiratory and expiratory reserve volumes)

mouth is called **sputum** (spyou-tum). Sputum may be used for cytologic examination.

- **bronchoscopy** (brohng-kohs-kō-pē) = visual examination of the tube located between the trachea and bronchioles (bronchus). Bronchoscopy may be used to examine the bronchi for disease or foreign objects. A **bronchoscope** (brohng-kō-skōp) is an instrument used to visually examine the bronchus.
- **laryngoscopy** (lahr-ihng-gohs-kō-pē) = visual examination of the voice box. Laryngoscopy is used to examine the larynx for disease, tissue repair, or foreign objects. A **laryngoscope** (lahr-ihng-ō-skōp or lahr-ihn-gō-skōp) is an instrument used to visually examine the voice box.
- **spirometer** (spər-oh-mē-tər) = instrument used to measure air taken in and out of the lungs.
- **thoracocentesis** (thō-rah-kō-sehn-tē-sihs) = puncture of the chest wall with a needle to obtain fluid from the pleural cavity. This fluid may be used for cytologic and microbiologic examination. Thoracocentesis may also be performed to drain pleural effusions or to reexpand a collapsed lung. Thoracocentesis is also called **thoracentesis** (thō-rah-sehn-tē-sihs) (Figure 9–7).

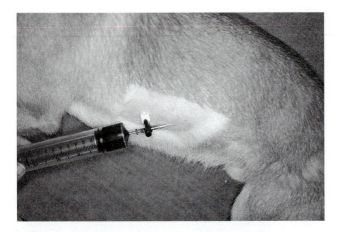

FIGURE 9–7 Thoracocentesis in a dog. Thoracocentesis is used to collect fluid or gas from the chest cavity. Fluid is collected ventrally, and gas is aspirated dorsally.

- **radiography** (rā-dē-ohg-rah-fē) = image of internal structures created by exposure of sensitized film to X-rays (Figure 9–8). Ultrasound does not work well for the respiratory system because the ultrasound beam cannot pass through a gas-containing structure to provide information about the internal structures.
- **trephination** (trē-fīn-ā-shuhn) = insertion of a hole boring instrument (trephine) into a sinus to establish fluid drainage or to allow access to the roots of teeth

▶ Path of Destruction: Respiratory System ◀

Pathologic conditions of the respiratory tract include
- **chronic obstructive pulmonary disease** (krohn-ihck ohb-struhck-tihv puhl-mah-nār-ē dih-zēz) = general term for abnormal conditions in equine species in which expiratory flow is slowed; commonly called heaves and is abbreviated COPD. Horses with heaves may have a heave line, which is increased abdominal musculature associated with increased expiratory effort in a horse with COPD.
- **stenotic nares** (stehn-ah-tihck nār-ēz) = narrowed nostrils that result in reduced airway flow
- **asphyxiation** (ahs-fihck-sē-ā-shuhn) = interruption of breathing resulting in lack of oxygen; also called suffocation
- **anoxia** (ā-nohck-sē-ah) = absence of oxygen (almost complete lack of oxygen)
- **cyanosis** (sī-ah-nō-sihs) = abnormal condition of blue discoloration. Cyanosis is cause by inadequate oxygen levels.
- **sinusitis** (sī-nuh-sī-tihs) = inflammation of a sinus
- **upper respiratory infection** = invasion of the nose, mouth, pharynx, epiglottis, larynx, and/or

trachea by pathogenic organisms; abbreviated URI. Signs of URI include cough, nasal and ocular discharge, dyspnea, and respiratory noise. A **cough** (kowf) is a sudden noisy expulsion of air from the lungs. Coughs may be **paroxysmal** (pahr-ohck-sihz-mahl), which means spasmlike and sudden. **Tuss/i** is the combining form for cough.

- **rhinitis** (rī-nī-tihs) = inflammation of the nasal mucous membranes
- **aspiration** (ahs-pih-rā-shuhn) = inhaling a foreign substance into the upper respiratory tract
- **epistaxis** (ehp-ih-stahck-sihs) = nosebleed
- **pharyngitis** (fār-ihn-jī-tihs) = inflammation of the throat
- **inflammation** (ihn-flah-mā-shuhn) = localized protective response to destroy, dilute, or wall off injury; classic signs are heat, redness, swelling, pain, and loss of function
- **laryngitis** (lahr-ihn-jī-tihs) = inflammation of the voice box
- **laryngoplegia** (lahr-ihng-gō-plē-jē-ah) = paralysis of the voice box
- **laryngospasm** (lah-rihng-ō-spahzm) = sudden fluttering and/or closure of the voice box
- **phonation** (fō-nā-shuhn) = act of producing sound. Phonation may be affected by pathology as is described with the appropriate prefix. **Aphonation** (ā-foh-nā-shuhn) is the inability to produce sound.
- **tracheitis** (trā-kē-ī-tihs) = inflammation of the windpipe
- **asthma** (ahz-mah) = chronic allergic disorder
- **bronchiectasis** (bronhg-kē-ehck-tah-sihs) = dilation of the bronchi. Bronchiectasis may be a **sequela** (sē-qwehl-ah) of inflammation or obstruction. Sequela is a condition following as a consequence of a disease.

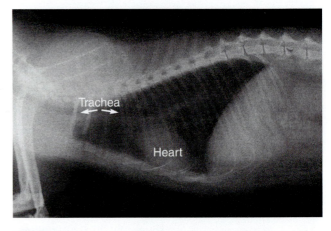

FIGURE 9–8 Thoracic radiograph in a cat. This lateral thoracic radiograph reveals normal thoracic structures. *Source:* Lodi Veterinary Hospital, S.C.

- **bronchitis** (brohng-kī-tihs) = inflammation of the bronchi. Bronchitis may be **acute** (ah-kūt), which means occurring over a short course with a sudden onset, or **chronic** (krohn-ihck), which means occurring over a long course with a progressive onset.
- **emphysema** (ehm-fih-zē-mah) = chronic lung disease caused by enlargement of the alveoli or changes in the alveolar wall
- **pulmonary edema** (puhl-mohn-ār-ē eh-dē-mah) = accumulation of fluid in the lung tissue. **Edema** (eh-dē-mah) is abnormally large amounts of fluid in the intercellular tissue spaces.
- **atelectasis** (aht-eh-lehk-tah-sihs) = incomplete expansion of the alveoli; may also mean collapse of a lung
- **pleurisy** (ploor-ih-sē) = inflammation of the pleura; also called **pleuritis** (ploor-ī-tihs)
- **pneumothorax** (nū-mō-thōr-ahckz) = abnormal accumulation of air or gas in the chest cavity (Figure 9–9).
- **pleural effusion** (ploor-ahl eh-fū-zhuhn) = abnormal accumulation of fluid in the pleural space. **Effusion** (eh-fū-zhuhn) is fluid escaping from blood or lymphatic vessels into tissues or spaces. A small amount of lubricating fluid in this space is normal.
- **pyothorax** (pī-ō-thō-rahcks) = accumulation of pus in the chest cavity. **Pus** is a fluid product of inflammation composed of leukocytes, **exudate** (ehcks-yoo-dāt) (high-protein fluid), and cell debris.
- **hemoptysis** (hē-mohp-tih-sihs) = spitting of blood from the lower respiratory tract
- **pneumonia** (nū-mō-nē-ah) = abnormal condition of the lung that usually involves inflammation and congestion of the lung. **Congestion** (kohn-gehs-zhuhn) is the abnormal accumulation of fluid. **Interstitial** (ihn-tər-stih-shahl) pertains to the area between the cells; **interstitial pneumonia** is an abnormal lung condition with increased fluid between the alveoli and a decrease in lung function
- **bronchopneumonia** (brohng-kō-nū-mō-nē-ah) = abnormal condition of the bronchi and lung
- **pleuropneumonia** (ploor-ō-nū-mō-nē-ah) = abnormal condition of the pleura and the lung (usually involves inflammation and congestion)
- **snuffles** (snuhf-uhlz) = common term for upper respiratory disease of rabbits caused by *Pasteurella multocida*
- **tracheitis** (trā-kē-ī-tihs) = inflammation of the windpipe
- **rhinorrhea** (rī-nō-rē-ah) = nasal discharge
- **pulmonary fibrosis** (puhl-mō-nār-ē fī-brō-sihs) = abnormal formation of fibers in the alveolar walls

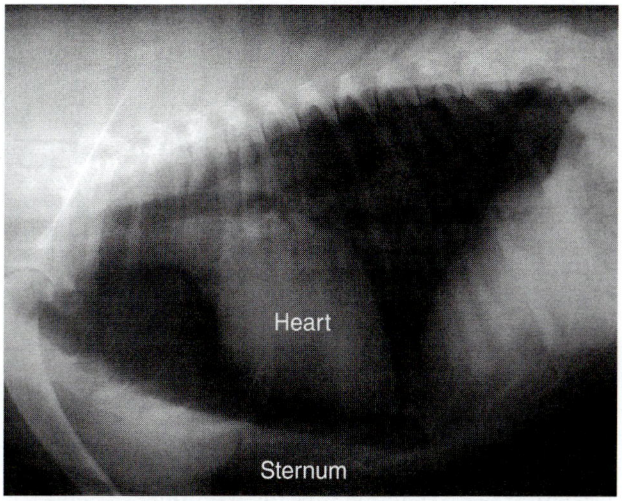

FIGURE 9–9 Pneumothorax in a dog. Lateral projection of a pneumothorax demonstrates dorsal displacement of the heart from the sternum. *Source:* University of Wisconsin Veterinary Hospital—Radiology.

- **equine laryngeal hemiplegia** (ē-qwīn lahr-ihn-jē-ahl hehm-ih-plē-jē-ah) = disorder of horses that is characterized by abnormal inspiratory noise during exercise associated with degeneration of the left recurrent laryngeal nerve and atrophy of the laryngeal muscles; also called **left laryngeal hemiplegia** or **roaring**
- **polyp** (pohl-uhp) = growth or mass protruding from a mucous membrane (usually benign)
- **rhinopneumonitis** (rī-nō-nū-moh-nī-tihs) = inflammation of the nasal mucous membranes and lungs

▶ Procede With Caution: Respiratory System ◀

Procedures performed on the respiratory system include
- drugs used on the respiratory system include **bronchoconstrictors** (brohng-kō-kohn-strihck-tərz), which are substances that narrow the openings into the lung, and **bronchodilators** (brohng-kō-dī-lā-tərz), which are substances that expand the openings into the lung. **Mucolytics** (mū-kō-lih-tihckz) are substances used to break down (**-lysis** means break down or separate) mucus or the slime-like discharge from glandular secretion. **Antitussives** (ahn-tih-tuhs-ihvz) are substances used to control or prevent coughing. The prefix **anti-** means against and the combining form **-tussi** means cough.
- **sinusotomy** (sī-nuhs-oht-ō-mē) = surgical incision into a sinus
- **pharyngoplasty** (fār-rihng-ō-plahs-tē) = surgical repair of the throat

- **pharyngostomy** (fār-ihng-ohs-tō-mē) = surgical creation of an opening into the throat. A **stoma** (stō-mah) is an opening on a body surface and may occur naturally or may be created surgically.
- **pharyngotomy** (fār-ihng-oht-ō-mē) = surgical incision into the throat
- **laryngectomy** (lār-ihn-jehck-tō-mē) = surgical removal of the voice box
- **laryngoplasty** (lah-rihng-ō-plahs-tē) = surgical repair of the voice box
- **tracheoplasty** (trā-kē-ō-plahs-tē) = surgical repair of the windpipe
- **tracheotomy** (trā-kē-oht-ō-mē) = surgical incision into the windpipe
- **tracheostomy** (trā-kē-ohs-tō-mē) = surgical creation of an opening into the windpipe (usually involves insertion and placement of a tube)
- **endotracheal intubation** (ehn-dō-trā-kē-ahl ihn-too-bā-shuhn) = passage of a tube through the oral cavity, pharynx, and larynx into the windpipe. An endotracheal tube provides a **patent** (pā-tehnt) airway for administration of anesthetics or for critical care patients. Patent means open, unobstructed, or not closed.
- **pneumonectomy** (nū-mō-nehck-tō-mē) = surgical removal of lung tissue
- **lobectomy** (lō-behck-tō-mē) = surgical removal of a lobe
- **pleurectomy** (ploor-ehck-tō-mē) = surgical removal of all or part of the pleura
- **thoracotomy** (thō-rah-koht-o-mē) = surgical incision into the chest wall
- **chest tube** = hollow device inserted into the thoracic cavity to remove fluid or gas. Chest tubes are passed when animals are severely dyspnic due to pressure on the lungs.

REVIEW EXERCISES

Multiple Choice—Choose the correct answer.

1. The wall that divides the nasal cavity is called the
   a. nasodivision
   b. nares
   c. nasal septum
   d. nasal meatus

2. Parts of the respiratory tract contain thin hairs called
   a. flagella
   b. naris
   c. surfactant
   d. cilia

3. An abnormal condition of blue discoloration is termed
   a. bluing
   b. cyanosis
   c. xanthochromia
   d. erythemia

4. Inhaling a foreign substance into the upper respiratory tract is called
   a. asphxiation
   b. effusion
   c. atelectasis
   d. aspiration

5. Hypoxia is
   a. below normal levels of oxygen
   b. above normal levels of oxygen
   c. below normal levels of carbon dioxide
   d. below normal levels of carbon dioxide and oxygen

6. Liquid that reduces alveolar surface tension is called
   a. surfactant
   b. mucus
   c. rhinorrhea
   d. mucorrhea

7. A condition following as a consequence of disease is a/an
   a. chronic condition
   b. acute condition
   c. sequela
   d. consequensosis

8. A substance that works against, controls, or stops a cough is a/an
   a. bronchoconstrictor
   b. bronchodilator
   c. mucolytic
   d. antitussive

9. Abnormal accumulation of blood in the pleural cavity is
   a. hemothorax
   b. hemoptysis
   c. hemopleuritis
   d. hemopneumonia

10. A growth or mass protruding from a mucous membrane is a
    a. nasogastric
    b. polyp
    c. bifurcation
    d. stridor

Label the diagram in Figure 9–10.

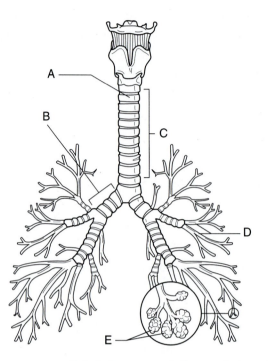

FIGURE 9–10 Respiratory tract. Label the parts of the respiratory tract as indicated by arrows. Provide the combining form for each of these anatomic parts.

Matching—Match the common term with the proper anatomic term.

_____ windpipe	pharynx
_____ throat	thorax
_____ voice box	larynx
_____ chest	trachea
_____ nostril	naris

Case Studies—Define the underlined terms in each case study.

An 8 mo old F DSH was presented to the clinic for routine surgery. The cat was given an IV drug to sedate it so that an endotracheal tube could be placed. The endotracheal tube (ET tube) would serve as a delivery route for inhalant anesthesia. Upon attempting to intubate the cat, the larynx started to swell and spasm due to trauma. The cat had inspiratory dyspnea, a severe cough, cyanotic mucous membranes, and an elevated pulse rate. The diagnosis of laryngeal spasm was made. A drug was placed in the laryngeal area to control the spasms (lidocaine), and endotracheal intubation was again attempted. After a few attempts, the ET tube was inserted and the cat's breathing and mucous membrane color improved. If the ET tube would not have been able to have been placed, a tracheotomy would have been performed.

endotracheal tube _____

intubate _____

inspiratory dyspnea _____

cough _____

cyanotic _____

diagnosis _____

laryngeal spasm _____

tracheotomy _____

A group of beef cattle were moved into a feedlot about 2 wk ago. Some of the cattle are now experiencing anorexia, are pyrexic, and have mucopurulent nasal discharge. Some of the cattle have rapid shallow breathing and a cough. Auscultation of the lungs revealed moist rales. Due to the history and respiratory signs the veterinarian suspected bovine pneumonic pasteurellosis or shipping fever, a severe respiratory disease seen in younger animals after shipping or stress. The affected cattle were isolated and treated with antibiotics. Management practices, such as immunization and stress reduction, were discussed with the owner.

anorexia _____

pyrexic _____

mucopurulent nasal discharge _____

auscultation _____

rales _____

bovine _____

pneumonic _____

Skin Deep

► Identify the structures of the integumentary system
► Describe the functions of the integumentary system
► Recognize, define, spell, and pronounce the terms used to describe the diagnosis, pathology, and treatment of the integumentary system

FUNCTIONS OF THE INTEGUMENTARY SYSTEM

The **integumentary** (ihn-tehg-yoo-mehn-tah-rē) **system** consists of skin and its appendages (appendages include glands, hair, fur, wool, feathers, scales, claws, beaks, horns, hooves, and nails). Considered one of the largest organ systems in the body, the integumentary system is involved in many processes.

Skin plays a role in the immunologic system, waterproofs the body, prevents fluid loss, provides species-specific coloration, and provides a site for vitamin D synthesis. Exocrine glands, both sebaceous and sweat, are located in the integumentary system. Sebaceous glands lubricate the skin and discourage bacterial growth on the skin. Sweat glands regulate body temperature and excrete wastes through sweat. Hair and nails are other components of the integumentary system. Hair helps control body heat loss and is a sense receptor. Nails protect the dorsal surface of the distal phalanx.

STRUCTURES OF THE INTEGUMENTARY SYSTEM

Skin

Skin covers the external surfaces of the body. The skin is composed of **epithelial** (ehp-ih-thē-lē-ahl) **tissue** and is sometimes referred to as the **epithelium** (ehp-ih-thē-lē-uhm). The combining forms for skin are **cutane/o, derm/o,** and **dermat/o;** the suffix **-derma** means skin. **Dermatology** (dər-mah-tohl-ō-jē) is the study of skin.

SKIN STRATIFICATION

The skin is made up of three layers: the epidermis, dermis, and subcutaneous layers (Figure 10–1).

The outermost or most superficial layer of skin is the **epidermis** (ehp-ih-dər-mihs). The prefix **epi-** means above and **dermis** means skin. The epidermis is several layers thick and does not contain blood vessels. The epidermis is sometimes called the avascular layer due to the absence of blood vessels. The epidermis is dependent on the deeper layers for nourishment.

The thickness of the epidermis varies greatly from region to region in any animal. The thickest layers of the epidermis are found in the areas of greatest exposure, such as the foot pads and teats.

The epidermis is made up of squamous epithelium and the basal layer. **Squamous epithelium** (skwā-muhs ehp-ih-thē-lē-uhm) is composed of flat, plate-like cells. Because these flat, plate-like cells are arranged in many layers, it is called stratified squamous epithelium.

The basal layer is the deepest layer of the epidermis. Cells layer, multiply, and push upward in the basal layer. As the cells move superficially, they die and become filled with **keratin** (kehr-ah-tihn). Keratin is a protein that provides skin with its waterproofing properties. The

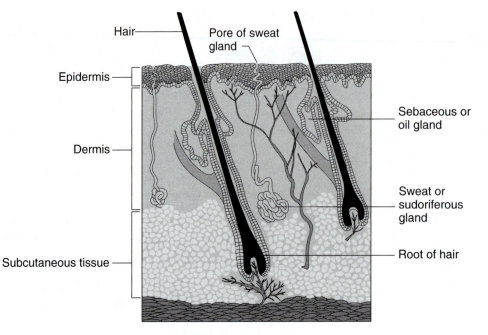

FIGURE 10–1 Skin layers

combining form for keratin is **kerat/o.** Kerat/o also means horny or cornea. The basal layer also contain **melanocytes** (mehl-ah-nō-sītz), which produce and contain a black pigment. This black pigment is called **melanin** (mehl-ah-nihn). The combining form **melan/o** means black or dark. Melanin protects the skin from some of the harmful rays of the sun and is responsible for skin pigmentation. The absence of normal pigmentation is called **albinism** (ahl-bih-nihz-uhm); true albinism means hair, skin, and eyes have no pigmentation (Table 10–1).

The **dermis** (dər-mihs) is the layer directly deep to the epidermis. The dermis is also called the **corium** (kō-rē-uhm). The dermis is composed of blood and lymph vessels, nerve fibers, and the accessory organs of the skin. The dermis also contains connective tissue which is composed of the following cells:

▶ **fibroblasts** (fī-brō-blahsts) = fiber-producing cells. **Collagen** (kohl-ah-jehn) is the major fiber in the dermis.

▶ **collagen** (kohl-ah-jehn) = tough, flexible, fibrous protein found in skin, bone, cartilage, tendons, and ligaments. Kolla in Greek means glue and **-gen** means to produce.

▶ **histiocytes** (hihs-tē-ō-sīts) = phagocytic cells that engulf foreign substances; also called tissue **macrophages** (mahck-rō-fahjs) or (mahck-rō-fājs).

TABLE 10–1
Layers of the Epidermis

There are five layers of the epidermis (from dermis to most superficial): stratum means layer or sheetlike mass
stratum basale (strah-tuhm bā-sahl) or **stratum germinativum** (gər-mihn-ā-tihv-uhm) = deepest or basal layer that continually multiply to replenish cells lost from the epidermal surface. **Cuboidal** (qū-boy-dahl) or cube-like cells arranged in row(s).
stratum spinosum (strah-tuhm spī-nō-suhm) = layer immediately superficial to the stratum basale, which is thickest in hairless regions and areas of high wear and tear. **Keratinization** (kehr-ah-tihn-ah-zā-shuhn) and **desquamation** (dehs-kwah-mā-shuhn) begin in this layer. Keratinization is the development of the hard, protein constituent of hair, nails, epidermis, horny structures, and tooth enamel. Desquamation is the process in which cell organelles gradually dissolve. The stratum spinosum is also called the **prickle** or **spinous layer.**
stratum granulosum (strah-tuhm grahn-yoo-lō-suhm) = layer immediately superficial to the stratum spinosum. Cells contain keratin granules in their cytoplasm.
stratum lucidum (strah-tuhm loo-sih-duhm) = layer immediately superficial to the stratum granulosum, which is clear due to the accumulation of keratin fibers in cell cytoplasm. This layer is not present in all species, but when present it is found in areas of high wear and tear like the foot pads.
stratum corneum (strah-tuhm kohr-nē-uhm) = most superficial layer of the epidermis, which consists of layers of dead, highly keratinized, and flattened cells; also called the **horny layer**

▶ **mast cells** = cells that respond to insult by producing and releasing histamine and heparin

> **histamine** (hihs-tah-mēn) = chemical released in response to allergens that causes itching
> **heparin** (hehp-ah-rihn) = anticoagulant chemical released in response to injury

▶ **perception** (pər-cehp-shuhn) = ability to recognize sensory stimuli. Perception is received by nerve impulses that recognize temperature, touch, pain, and pressure. **Tactile** (tahck-tīl) perception is the ability to recognize touch sensation.

The **subcutaneous** (suhb-kyoo-tahn-ē-uhs) **layer** or **hypodermis** (hī-pō-dər-mihs) is located deep to or under the dermis and is composed of connective tissue. The subcutaneous layer contains a large amount of **fat** or **lipid** (lihp-ihd). **Adipocytes** (ahd-ih-pō-sīts) are fat cells that produce lipid. **Adip/o** is the combining form for fat (Figure 10–2).

Skin Association

Appendages or structures associated with the skin include glands, hair, fur, wool, feathers, scales, claws, beaks, horns, hooves, and nails.

GLANDS

There are two main categories of skin glands: sebaceous and sweat glands. **Sebaceous** (seh-bā-shuhs) **glands** or **oil glands** secrete an oily substance called **sebum** (sē-buhm). **Seb/o** is the combining form that means sebum or oily substance. Sebaceous glands are located in the dermis and are closely associated with hair follicles. Sebum is released from its gland through **ducts** (duhcks) that open into the hair follicles. Ducts are tube-like passages; tiny ducts are called **ductules.** Sebum moves from the hair follicle to the skin surface where it lubricates the skin. Sebum is slightly acidic and retards bacterial growth on the skin. Sebaceous glands are considered **holocrine** (hō-lō-krihn) glands because the secreting cells and its secretion make up the discharge produced. Sebaceous glands are found in the anal sacs, glands that produce musk, and circumoral and supracaudal glands which are used by cats to mark territory when they groom and rub their tail, respectively.

Sweat (sweht) or **sudoriferous** (soo-dohr-ihf-ohr-uhs) **glands** are aggregations of cells that are located in the dermis. Sweat glands are divided into **eccrine** (ē-krihn) **glands** and **apocrine** (ahp-ō-krihn) **glands.** Eccrine sweat glands produce and secrete water, salt,

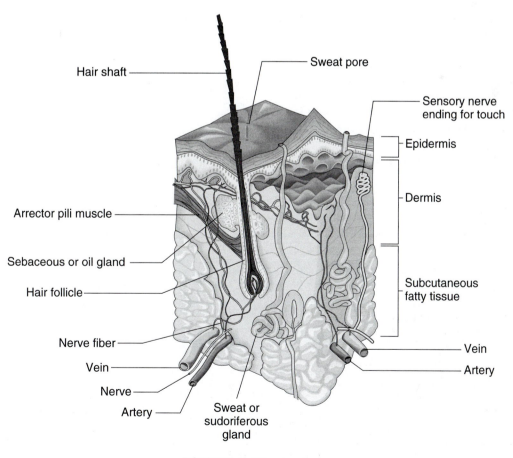

Hair shaft
Sweat pore
Sensory nerve ending for touch
Epidermis
Dermis
Arrector pili muscle
Sebaceous or oil gland
Hair follicle
Subcutaneous fatty tissue
Nerve fiber
Vein
Vein
Artery
Nerve
Artery
Sweat or sudoriferous gland

FIGURE 10–2 Skin structures

and waste (sweat) and are located in various regions of the body depending on the species of animal. Eccrine sweat glands are tiny, coiled glands that have ducts that open directly onto the skin surface through pores. Apocrine glands produce and secrete a strong-smelling substance into the hair follicles. Apocrine glands are found throughout the body. Apocrine glands get their name from the fact that the free end or apical end of the cell is cast off along with the secretory products to make the substance released. Sweat glands help regulate body temperature against **hyperthermia** (hī-pər-thər-mē-ah) or increased body temperature and **hypothermia** (hī-pō-thər-mē-ah) or decreased body temperature.

Hidrosis (hī-drō-sihs) is the production and excretion of sweat. The combining form **hidr/o** means sweat. **Anhidrosis** (ahn-hī-drō-sihs) is the abnormal reduction of sweating; **hyperhidrosis** (hī-pər-hī-drō-sihs) is excessive sweating.

Ceruminous (seh-roo-mihn-uhs) glands are modified sweat glands that are located in the ear canal. The ceruminous glands secrete **cerumen** (seh-roo-mehn), a waxy substance of varying colors depending on the species, that is commonly called **earwax.**

HAIR

Hair is rod-like fibers made of dead protein cells filled with keratin. The combining forms for hair are **pil/i, pil/o,** and **trich/o.** The hair shaft is the portion of hair extending beyond the skin surface and is composed of the **cuticle** (kū-tih-kuhl), **cortex** (kōr-tehckz), and **medulla** (meh-doo-lah). The cuticle is one cell layer thick and appears scaly. The cortex is the main component of the hair shaft, is several layers thick, and is responsible for coat color. The medulla is the innermost component of the hair shaft.

Hair **follicles** are sacs that hold the hair fibers. The **arrector pili** (ah-rehck-tər pī-lī) is a tiny muscle attached to the hair follicle that causes the hair to stand erect in response to cold temperatures or stressful situations. When a dog contracts the arrector pili along the dorsal side of the neck and down the spine it is called "raising the hackles." **Piloerection** (pī-lō-ē-rehck-shuhn) is the condition of the hair standing straight up.

What is Hair?

Animals have many different types of hair; therefore, there are many different terms to describe the types of hair animals may have.

► **fur** = short, fine, soft hair
► **pelt** = skin plus fur or hair
► **guard hairs** = long, straight, stiff hairs that form the outer coat; also called **primary hairs** or **topcoat.** Guard hairs include tail and mane hair, bristly hair of swine, and most of the fur hair.
► **secondary hairs** = finer, softer, and wavy hair; also called the **undercoat.** Secondary hairs include wool and wavy hair located near the skin of rabbits.
► **tactile** (tahck-tīl) **hair** = long, brittle, extremely sensitive hairs usually located on the face; also called **vibrissae** (vī-brihs-ā), which are technically longer tactile hairs. An example of vibrissae are cat whiskers.
► **cilia** (sihl-ē-ah) = thin, short hairs. An example of cilia are the eyelashes.
► **simple pattern hair growth** = guard hairs grow from separate follicular openings as in cattle
► **compound pattern hair growth** = multiple guard hairs grow from single follicles as in dogs
► **shedding** = normal hair loss due to temperature, hormones, photoperiod (light), nutrition, and other nondisease causes

Feathers and scales will be covered in Chapter 23, which includes exotic animals and zoologic terminology.

Things tend to be named for a reason. Sometimes bacteria, viruses, and fungi are named for their size, shape, or organism they affect. One example is the fungi, *Trichophyton.* **Trich/o** means hair and **phyt/o** means plant. **Fungi** are vegetative organisms (plants) that grow as branched, hairlike filaments (trich/o).

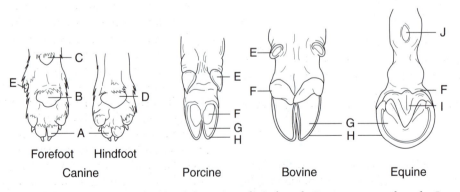

FIGURE 10–3 Comparison of animal feet. A = digital pad, B = metacarpal pad, C = carpal pad, D = metatarsal pad, E = dewclaw, F = bulb or heel, G = sole, H = wall, I = frog, J = ergot

NAILS, CLAWS, AND HOOVES

The distal phalanx of animals is covered by nails, claws, or hooves. Nails, claws, and hooves all have a wall, sole, and pad although they may be called different things. **Walls** are usually located dorsal and lateral to the distal phalanx. The **sole** is located ventral to the distal phalanx and is usually flakey. **Footpads** or **tori** (tohr-ē) provide cushioning and protection for the bones of the foot. Pads are usually thick and composed of keratinized epithelium. The pad has a subcutaneous layer that contains a large number of adipose cells and elastic connective tissue. Sweat glands are also found in most mammalian foot pads (Figure 10–3).

Dogs and cats have **digital pads** on the palmar and plantar surfaces of the phalanges. **Metacarpal** and **metatarsal pads** are singular pads located on the palmar and plantar surface of the metacarpal/metatarsal area. **Carpal pads** are located on the palmar surface of each carpus. Carpal pads do not bear weight when the animal is standing. Dogs and cats are called **digitigrade** (dihg-iht-ih-grād) animals because they walk on their toes, with only the digital and metacarpal/metatarsal pads making contact with the ground. **Plantigrade** (plahnt-ih-grād) animals have well-developed footpads as those in primates. Plantigrade animals walk with phalanges, metacarpals/metatarsals, and carpal/tarsal bones making contact with the ground.

The equine hoof is divided into various regions: the coronary band, periople, wall, bars, sole, bulb, and frog (Figure 10–4).

► **coronary** (kohr-ō-nār-ē) **band** = region where hoof meets the skin; analogous to the cuticle of the human nail. The coronary band is the site of hoof wall growth; also called the **coronet**.

► **periople** (pehr-ē-ō-puhl) = flaky tissue band located at junction of the coronary band and the hoof wall and extends distally. The periople widens at the heel to cover the bulbs of the heels.

► **wall** = epidermal tissue that includes the toe (front), quarters (sides), and heels (back)

► **bars** (bahrz) = raised V-shaped structure on ventral surface of hoof. Bars are located on either side of the frog.

► **sole** (sōl) = softer hoof tissue located on the ventral surface of the hoof (bottom of the hoof)

► **frog** (frohg) = V-shaped pad of soft horn located in the central region of the ventral hoof surface of equine (located between the bars). When weight is put on the frog, blood is forced out of the foot to aid in circulation of blood throughout the foot.

► **bulbs** (buhlbz) **of heel** = upward thickening of the frog above the heels of the wall

---

In dogs and cats **nails** and **claws** are keratin plates covering the dorsal surface of the distal phalanx. The dorsal and lateral surface of the claw is covered by the wall and the ventral surface is the sole. Beneath the wall and sole is the connective tissue dermis which contains

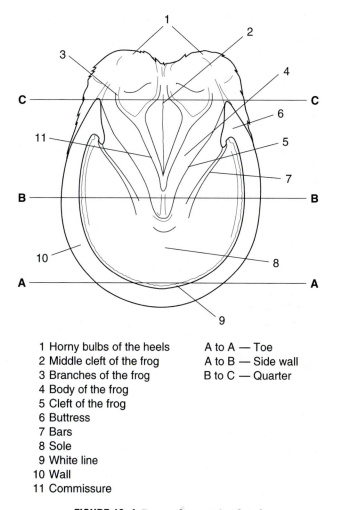

1 Horny bulbs of the heels
2 Middle cleft of the frog
3 Branches of the frog
4 Body of the frog
5 Cleft of the frog
6 Buttress
7 Bars
8 Sole
9 White line
10 Wall
11 Commissure

A to A — Toe
A to B — Side wall
B to C — Quarter

FIGURE 10–4 Parts of an equine hoof

numerous blood vessels and nerve endings. This sensitive tissue is known as the **quick. Quicking** is the term used to describe trimming the nail or claw to the level of the dermis. Quicking results in bleeding and pain. The combining form for claw in **onych/o.**

Hooves are the horny covering of the distal phalanx in **ungulate** (uhng-yoo-lātz) or **hooved animals** like equine, ruminants, and swine. Some ungulates have a solid hoof as in equine and some have cloven or split hooves as in ruminants and swine. The ventral surface of the hoof is the sole and is relatively large in equine, while it is smaller in ruminants and swine. The combining form for hoof is **ungul/o.**

The pads of ungulates vary with the species. In ruminants and swine, the footpad is called the bulb or heel. The pad of equine is called the frog and along with the bulb provides shock absorption. The **corium** (kōr-ē-uhm)is the dermis of the hoof and is located under the epidermal surface of the hoof wall, sole, and frog. The corium corresponds to the quick.

VESTIGIAL STRUCTURES

Vestigial or rudimentary structures of the integumentary system include dewclaws, chestnuts, and ergots. **Dewclaws** (doo-klaw) are rudimentary bones. The dewclaw in dogs is the first digit, while in cloven-hoofed animals the dewclaws are digits II and V. Dewclaws in dogs are usually found in the forepaw (although they occasionally are seen in the hindpaw) and may be removed within a few days of birth to avoid trauma.

Chestnuts and ergots are vestigial pads in equine. **Chestnuts** are located on the medial surface of the leg; in the front leg they are located above the knee and in the hind leg they are located below the hock. Chestnuts correspond to carpal pads in the dog. No two chestnuts are alike, and they do not change in size or shape throughout an equine's life. **Ergots** are located within a tuft of hair on the fetlock joint. Ergots correspond to metacarpal/metatarsal pads in the dog. Digital pads of dogs are replaced by the bulbs of the heel (and the frog in equine) in ungulates.

What Is the Difference between Horns and Antlers?

Horns and antlers are protective structures located in the head region of animals. **Horns** are permanent structures that grow continuously after birth. Horns grow from the frontal skull bones and originate from keratinized epithelium. **Cornification** (kohr-nih-fih-kā-shuhn) is the conversion of epithelium into keratin or horn. Horns may be located in different positions as can be seen in the different ruminant species. Breeds that are naturally hornless are called **polled.**

Antlers are not permanent structures and are shed and re-grown annually. Antlers grow from the skull as do horns. Antlers are initially covered with skin called **velvet,** which is rubbed off after the skin dies. When the velvet is rubbed off, the bone is exposed, the antlers lose their blood supply and are eventually shed.

▶ Test Me: Integumentary System ◀

Diagnostic procedures performed on the integumentary system include

- **biopsy** (bī-ohp-sē) = removal of living tissue for examination of life. The combining form **bi/o** means life; the suffix **-opsy** means view of. An **incisional biopsy** (ihn-sih-shuhn-ahl bī-ohp-sē) is the removal of a piece of a tumor or lesion for examination (Figure 10–5). **Incision** (ihn-sih-shuhn) means to cut into. An **excisional biopsy** (ehcks-sih-shuhn-ahl bī-ohp-sē) is the removal of an entire tumor or lesion plus a margin surrounding tissue for examination. **Excision** (ehcks-sih-shuhn) means to cut out of. A needle biopsy is insertion of a sharp instrument (needle) into a tissue for examination. Examination of biopsies involves the use of a microscope.

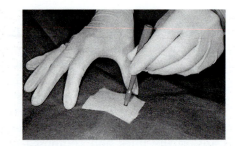

FIGURE 10–5 Punch biopsy of the skin. A punch biopsy is a type of incisional biopsy commonly used to obtain full thickness skin samples. The punch has a circular opening that comes in different sizes.

- **exfoliative cytology** (ehcks-fōl-ē-ah-tihv sī-tohl-ō-jē) = scraping of cells from tissue and examining them under a microscope. Exfoliative is the term pertaining to falling off.
- **skin scrape** = microscopic examination of skin for the presence of mites; skin is sampled by taking a scalpel blade across an area of skin that is squeezed or raised so that the sample contains a deep skin sample (Figure 10–6).
- **tissue culture** = diagnostic or research procedure that takes epithelial cells and grows them in a medium
- **intradermal** (ihn-trah-dər-mahl) **skin testing** = injection of test substances into the skin layer to observe a reaction (Figure 10–7); used for diagnosis of **atopy** (ah-tō-pē) with the injection of

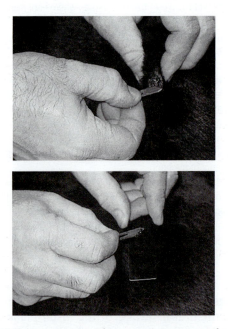

FIGURE 10–6 Skin scrape. Skin scrapings are used to detect mites. *Source:* Courtesy of Ron Fabrizius, DVM, Diplomat ACT

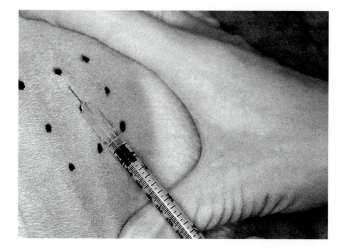

FIGURE 10–7 Intradermal skin testing. Intradermal skin testing is used to assess atopy.

multiple allergens or for tuberculosis testing by injecting tuberculin into the skin layer and observing the injection site for a 24-, 36-, and 72-hour postinjection reaction (tuberculosis testing is referred to as **purified protein derivative** or PPD testing)
- fungal and bacterial cultures will be covered in Chapter 16.

▶ Path of Destruction: Integumentary System ◀

Pathologic conditions of the integumentary system include
- **seborrhea** (sehb-ō-rē-ah) = skin condition characterized by overproduction of sebum (oil)
- **comedo** (kōm-eh-dō) = blackhead or buildup of sebum and keratin in a pore. Plural is **comedones** (kōm-eh-dō-nehz)
- **sebaceous cyst** (seh-bā-shuhs sihst) = closed sac of yellow, fatty material. A **cyst** (sihst) is a closed sac containing fluid or semisolid material
- **alopecia** (ahl-ō-pē-shah) = hair loss resulting in hairless patches or complete lack of hair. **Shedding** is normal hair loss due to various causes (Figure 10–8).
- **lesion** (lē-zhuhn) = pathologic change of tissue; used to describe abnormalities in many locations
- **contusion** (kohn-too-zhuhn) = injury that does not break the skin; characterized by pain, swelling, and discoloration.
- **papule** (pahp-yool) = small, raised skin lesion less than 0.5 cm in diameter
- **plaque** = solid raised lesion greater than 0.5 cm in diameter

- **macule** (mahck-yool) = flat, discolored lesion less than 1 cm in diameter; also called **macula** (mahck-yoo-lah)
- **nodule** (nohd-yoo-uhl) = small knot protruding above the skin
- **patch** = localized skin color change greater than 1 cm in diameter
- **scale** = flake
- **crust** = collection of dried sebum and cell debris
- **wheal** (whēl) = smooth, slightly raised swollen area that itches
- **purpura** (pər-pə-rah) = condition characterized by hemorrhage into the skin that causes bruising. The two types of purpura are ecchymosis and petechia.
- **ecchymosis** (ehck-ih-mō-sihs) = purplish nonelevated patch of bleeding into the skin; also called a bruise; pleural is **ecchymoses** (ehck-ih-mō-sēz)
- **petechiae** (peh-tē-kē-ah) = small, pinpoint hemorrhages. Singular is **petechia** (peh-tē-kē-ā).
- **pallor** (pahl-ohr) = skin paleness.
- **verrucae** (veh-roo-sē) = warts
- **hemangioma** (hē-mahn-jē-ō-mah) = benign tumor composed of newly formed blood vessels
- **pustule** (puhs-tyuhl) = small, circumscribed, pus-filled skin elevation. **Circumscribed** (sehr-kuhm-skrībd) means contained within a limited area.
- **vesicle** (vehs-ih-kuhl) = contained skin elevation filled with fluid that is greater than 0.5 cm in diameter; also called a **blister, bulla** (buhll-ah), or **bleb**
- **bullae** (buhl-ā) = multiple contained skin elevations filled with fluid that are greater than 0.5 cm in diameter. The singular form is **bulla** (buhl-ah).
- **abscess** (ahb-sehsz) = localized collection of pus

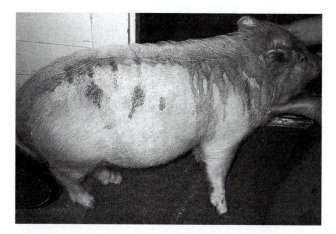

FIGURE 10–8 Alopecia in a pig with sarcoptic mange. *Source:* Courtesy of Ron Fabrizius, DVM, Diplomat ACT

(a) Surface lesions

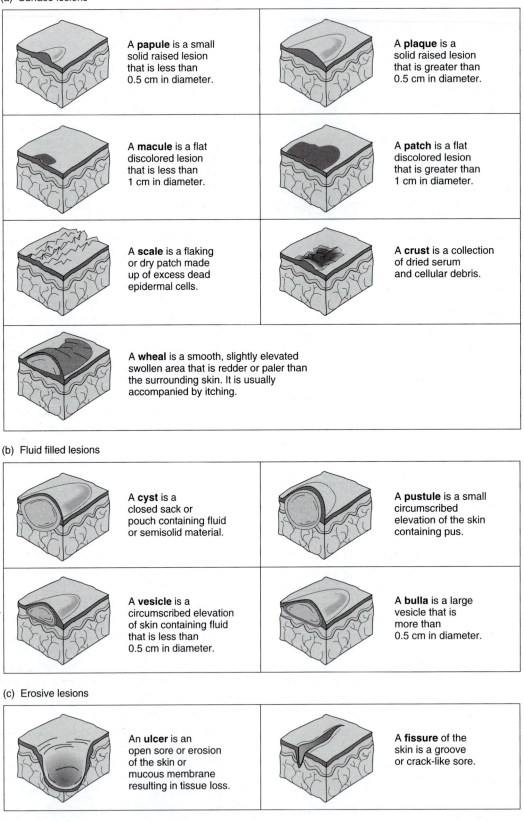

A **papule** is a small solid raised lesion that is less than 0.5 cm in diameter.

A **plaque** is a solid raised lesion that is greater than 0.5 cm in diameter.

A **macule** is a flat discolored lesion that is less than 1 cm in diameter.

A **patch** is a flat discolored lesion that is greater than 1 cm in diameter.

A **scale** is a flaking or dry patch made up of excess dead epidermal cells.

A **crust** is a collection of dried serum and cellular debris.

A **wheal** is a smooth, slightly elevated swollen area that is redder or paler than the surrounding skin. It is usually accompanied by itching.

(b) Fluid filled lesions

A **cyst** is a closed sack or pouch containing fluid or semisolid material.

A **pustule** is a small circumscribed elevation of the skin containing pus.

A **vesicle** is a circumscribed elevation of skin containing fluid that is less than 0.5 cm in diameter.

A **bulla** is a large vesicle that is more than 0.5 cm in diameter.

(c) Erosive lesions

An **ulcer** is an open sore or erosion of the skin or mucous membrane resulting in tissue loss.

A **fissure** of the skin is a groove or crack-like sore.

FIGURE 10–9 Skin lesions. Skin lesions may be (a) raised, discolored, (b) fluid filled, and/or (c) erosive in nature.

- **laceration** (lahs-ər-ā-shuhn) = accidental cut into the skin
- **ulcer** (uhl-sihr) = erosion of skin or mucous membrane. **Decubital ulcers** (dē-kyoo-bih-tahl uhl-sihrz) are erosions of skin or mucous membranes as a result of prolonged pressure; also called **bedsores.**
- **fissure** (fihs-sər) = cracklike sore (Figure 10–9)
- **fistula** (fihs-tyoo-lah) = abnormal passage from an internal organ to the body surface or between two internal organs. Plural is **fistulae** (fihs-tyoo-lā).
- **dermatosis** (dər-mah-tō-sihs) = abnormal skin condition. Plural is **dermatoses** (der-mah-tō-sēz).
- **keratosis** (kehr-ah-tō-sihs) = abnormal condition of epidermal overgrowth and thickening. Plural is **keratoses** (kehr-ah-tō-sēs).
- **dermatitis** (dər-mah-tī-tihs) = inflammation of the skin. **Contact dermatitis** is inflammation of the skin caused by touching an irritant.
- **pruritus** (proo-rī-tuhs) = itching
- **urticaria** (ər-tih-kā-rē-ah) = localized areas of swelling that itch; also called **hives**
- **erythroderma** (eh-rihth-rō-dər-mah) = abnormal redness of skin occurring over a widespread area. Combining forms for red are **erythr/o, erythem/o,** and **erythemat/o.**
- **erythema** (ehr-ih-thē-mah) = skin redness. **Erythematous** (ehr-ih-thehm-ah-tuhs) means pertaining to redness.
- **infestations** (ihn-fehs-tā-shuhns) are occupation and dwelling of a parasite on the external surface of tissue. **Ectoparasites** (ehck-tō-pahr-ah-sīts) live on the external surface; **ecto-** means outside. A **louse** (lows) is a wingless parasitic insect; plural is **lice** (lyse). **Pediculosis** (pehd-ih-koo-lō-sihs) is lice infestation. A **mite** is an insect with a hard exoskeleton and paired, jointed legs. **Mange** (mānj) is a common term for skin disease caused by mites. There are different types of mange such as **sarcoptic** (sahr-kohp-tihck) or **demodectic** (deh-mō-dehck-tihck) depending upon the type of mite involved. **Chiggers** (chihg-gərs) is infestation by mite larvae that results in severe pruritus. **Acariasis** (ahs-kahr-ī-ah-sihs) is infestation with ticks or mites. **Maggots** (mā-gohts) are insect larvae especially found in dead or decaying tissue. **Myiasis** (mī-ī-ah-sihs) is infestation by fly larvae.
- **abrasion** (ah-brā-shuhn) = injury in which superficial layers of skin are scraped
- **cellulitis** (sehl-yoo-lī-tihs) = inflammation of connective tissue. Inflammation may be **diffuse** (dih-fuhs) meaning widespread or **localized** (lō-kahl-īzd) meaning within a well-defined area.

- **dermatocellulitis** (dər-mah-tō-sehl-yoo-lī-tihs) = inflammation of the skin and connective tissue
- **pyoderma** (pī-ō-dər-mah) = skin disease containing pus. **Pus** (puhs) is an inflammatory product made up of leukocytes, cell debris, and fluid. **Purulent** (pər-ū-lehnt) means containing or producing pus. **Puppy pyoderma** is a skin disease in puppies characterized by pus containing lesions. **Juvenile pyoderma** is a skin disease in puppies that progresses to a systemic disease characterized by fever, anorexia, and enlarged and abscessing lymph nodes; juvenile pyoderma is referred to as **puppy strangles.**
- **furuncle** (fyoo-ruhng-kuhl) = localized skin infection in a gland or hair follicle; also called a **boil. Furunculosis** (fyoo-ruhng-kuh-lō-sihs) is the abnormal condition of persistent boils over a period of time.
- **carbuncle** (kahr-buhng-kuhl) = cluster of furuncles
- **dermatomycosis** (dər-mah-tō-mī-kō-sihs) = abnormal skin condition due to superficial fungus; also called **dermatophytosis** (dər-mah-tō-fī-tō-sihs). **Dermatophytes** (dər-mah-tō-fītz) are superficial fungi that are found on the skin. An example of a dermatophyte is the fungus that causes ringworm.
- **gangrene** (gahng-grēn) = necrosis associated with loss of circulation. **Necrosis** (neh-krō-sihs) is condition of dead tissue; **necrotic** (neh-krō-tihck) means pertaining to dead tissue. Decay that produces a foul-smell is called **putrefaction** (pyoo-treh-fahck-shuhn).
- **granuloma** (grahn-yoo-lō-mah) = small area of healing tissue
- **skin tag** = small growths that hang from the body by stalks
- **polyp** = growth from mucous membranes
- **papilloma** (pahp-ih-lō-mah) = benign epithelial growth that is lobed
- **lipoma** (lī-pō-mah) = benign growth of fat cells; also called **fatty tumor**; commonly seen in older dogs
- **burn** = tissue injury caused by heat, flame, electricity, chemicals, or radiation
- **sarcoma** (sahr-kō-mah) = malignant neoplasm of soft tissue arising from connective tissue
- **carcinoma** (kahr-sih-nō-mah) = malignant neoplasm of epithelial tissue
- **flea allergy dermatitis** (dər-mah-tī-tihs) = inflammation of the skin caused by an allergic reaction to flea saliva; abbreviated FAD. An **allergen** (ahl-ər-jehn) is a substance that produces an allergic response.
- **atopy** (ah-tō-pē) = hypersensitivity reaction in animals involving pruritus with secondary dermatitis;

commonly called **allergies** or **allergic dermatitis.**
Hypersensitization is an increased response to an
allergen. **Hyposensitization** is a decreased response
to an allergen. Animals with atopy may undergo a
series of hyposensitization injections to decrease
their response to a specific allergen(s).

- **ecthyma** (ehck-thih-mah) = skin infection with
shallow eruptions caused by a virus (Figure
10–10).
- **footrot** = bacterial (*Fusobacterium* sp) hoof disease that spreads from the interdigital skin to the
deeper foot structures
- **onychomycosis** (ohn-kē-ō-mī-kō-sihs) = superficial fungal infection of the claw
- **paronychia** (pahr-ohn-kē-ah) = bacterial or viral
infection of the claw
- **hyperkeratosis** (hī-pər-kehr-ah-tō-sihs) = increased growth of the horny layer of the skin; also
called **acanthokeratodermia** (ā-kahn-thō-kehr-
ah-tō-dər-mah)
- **dyskeratosis** (dihs-kehr-ah-tō-sihs) = abnormal
alteration in keratinization
- **scar** (skahr) = mark left by a healing lesion where
excess collagen was produced to replace injured
tissue; also called **cicatrix** (sihck-ah-trihcks) or **ci-
catrices** (sihck-ah-trih-sēz), which are multiple
scars
- **melanoma** (mehl-ah-nō-mah) = tumor or growth
of pigmented skin cells. **Malignant melanoma** is the
term used to describe cancer of the pigmented skin
cells. One form of melanoma is **amelanotic
melanoma,** which is an unpigmented malignant
melanoma.

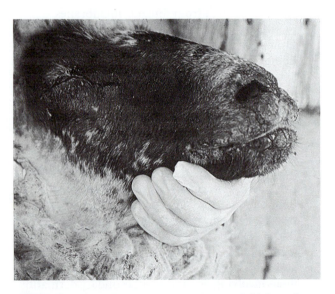

FIGURE 10–10 Contagious ecthyma (soremouth) in a sheep.
Source: Courtesy of Ron Fabrizius, DVM, Diplomat ACT

- **exanthema** (ehcks-ahn-thē-mah) = cutaneous
rash due to fever or disease. Singular is **exanthem**
(ehcks-ahn-thuhm).
- **acne** (ahck-nē) = skin inflammation caused by
plugged sebaceous glands and comedone development from papules and pustules. **Chin acne** is a
common condition seen in cats in which acne develops on the chin and lip area.
- **eosinophilic granuloma** (ē-ō-sihn-ō-fihl-ihck
grahn-yoo-lō-mah) **complex** = collective name
for autoimmune lesion of eosinophilic ulcer,
eosinophilic plaque, and linear granuloma found
in cats and rarely in dogs. This complex of diseases
affects the skin, mucocutaneous junctions, and
oral mucosa of cats, involving raised, ulcerated
plaques. These lesions are named for their location: **eosinophilic ulcer** or **rodent ulcer** is located
on the lower lip and oral mucosa of cats;
eosinophilic plaques are raised pruritic lesions on
the ventral abdomen of cats; **linear granulomas**
are located in a line usually on the caudal aspect of
the hindlimb of cats.
- **feline miliary dermatitis** (mihl-ē-ahr-ē dər-mah-
tī-tihs) = skin disease of cats in which multiple
crusts and bumps are present predominantly on
the dorsum that can be associated with many
causes
- **discoid lupus erythematosus** (dihs-koyd loo-pihs
eh-rih-thehm-ah-tō-sihs) = canine autoimmune
disease in which the bridge of the nose (and sometimes the face and lips) exhibit depigmentation,
erythema, scaling, and erosions; abbreviated DLE;
may have been referred to as collie nose or solar
dermatitis in the past
- **eczema** (ehcks-zeh-mah) = general term for inflammatory skin disease characterized by erythema, papules, vesicles, crusts, and scabs either
alone or in combination
- **frostbite** = tissue damage caused by extreme cold
or contact with chemicals with extreme temperature (i.e., liquid nitrogen)
- **parakeratosis** (pahr-ah-kehr-ah-tō-sihs) = lesion
characterized by thick scales, cracking, and red
raw surface caused by the persistence of ker-
atinocyte nuclei in the horny layer of skin
- **pemphigus** (pehm-fih-guhs) = group of immune
mediated skin diseases characterized by vesicles,
bullae, and ulcers. The most common form is
pemphigus vulgaris (pehm-fih-guhs vuhl-gahr-
ihs), which consists of shallow ulcerations frequently involving the oral mucosa and mucocutaneous junctions. Pemphix is Greek for blister.
- **acute moist dermatitis** (ah-kūt moyst dər-mah-tī-
tihs) = bacterial skin disease that is worsened by
licking and scratching; also called **hot spot**

▶ Procede With Caution: Integumentary System ◀

Procedures performed on the integumentary system include
- **debridement** (dē-brīd-mehnt) = removal of tissue and foreign material to aide healing
- **cauterization** (kaw-tər-ī-zā-shuhn) = destruction of tissue using electric current, heat, or chemicals
- **cryosurgery** (krī-ō-sihr-gər-ē) = destruction of tissue using extreme cold
- **laser** (lā-zər) = device that transfers light into an intense beam for various purposes; acronym for light amplification by stimulated emission of radiation. An acronym is a word formed by initial letters of compound terms.

REVIEW EXERCISES

Multiple Choice—Choose the correct answer.

1. Pruritus is commonly called
   a. hair loss
   b. dry skin
   c. itching
   d. pus

2. Skin redness is termed
   a. cellulitis
   b. erythema
   c. scleroderma
   d. scarring

3. Hypersensitivity reaction in animals involving pruritus with secondary dermatitis is called
   a. atrophy
   b. allergen
   c. antigen
   d. atopy

4. Hair loss resulting in hairless patches or complete lack of hair is termed
   a. shedding
   b. lesion
   c. alopecia
   d. plaque

5. Occupation and dwelling of parasites on the external skin surfaces is called
   a. parasitism
   b. ectoparasites
   c. infestation
   d. myiasis

6. A skin disease containing pus is
   a. pyometra
   b. pyoderma
   c. pyoerythema
   d. pyosis

7. Producing or containing pus is referred to as
   a. abscess
   b. purulent
   c. mucocutaneous
   d. polyp

8. Removal of tissue and foreign material to aid healing is
   a. skin scraping
   b. tissue culture
   c. biopsy
   d. debridement

9. Putrefaction is
   a. foul-smelling decay
   b. homogenation
   c. granulomatous
   d. cellulitis

10. Large tactile hair is
    a. sensogenic
    b. plantigrade
    c. vibrissa
    d. cerumen

Matching—Match the common dermatologic term with its medical term.

_____hive	polled
_____hot spot	cicatrix
_____fatty tumor	furuncle
_____pale	urticaria
_____blister	abrasion
_____scar	acute moist dermatitis
_____oil	pruritus
_____scrape	atopy
_____boil	lipoma
_____redness	scale
_____itching	erythema
_____flake	pallor
_____warts	sebum
_____allergic dermatitis	verrucae
_____hornless	fissure
_____crack	vesicle

Label the diagram in Figure 10–11.

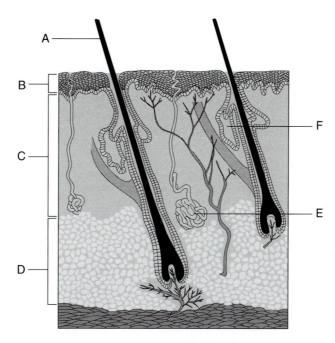

A

B

C

D

F

E

FIGURE 10–11 Skin structures. Label the parts of the skin indicated by the arrows.

Case Study—Define the underlined terms in the following case study.

A 3 yr old F/S Golden retriever was presented with clinical signs of <u>pruritus</u>, abdominal <u>dermatitis</u>, and <u>otitis</u>. <u>Skin scrapes</u> were negative for external parasites. Ear cytology revealed a large number of yeast. The dog was referred to a <u>dermatologist</u>, who diagnosed <u>atopy</u> via <u>intradermal skin testing</u>. The dog was put on a <u>hypoallergenic</u> diet and was given <u>hyposensitization</u> injections. Medications were prescribed to control the pruritus and secondary <u>pyoderma</u>.

pruritus _____

dermatitis _____

otitis _____

skin scrapes _____

dermatologist _____

atopy _____

intradermal skin testing _____

hypoallergenic _____

hyposensitization _____

pyoderma _____

11 The Great Communicator

Objectives *In this chapter, you should learn to:*

► Identify and describe the major structures and functions of the endocrine system
► Describe the role of the hypothalamus and pituitary gland in hormone secretion
► Recognize, define, spell, and pronounce terms related to the diagnosis, pathology, and treatment of the endocrine system

FUNCTIONS OF THE ENDOCRINE SYSTEM

The endocrine (ehn-dō-krihn) system is composed of ductless glands that secrete chemical messengers called hormones into the bloodstream. The prefix **endo-** means within and the suffix **-crine** means to secrete or separate. Hormones enter the bloodstream and are carried throughout the body to affect a variety of tissues and organs. Tissues and organs the hormones act upon are called **target organs. Hormone** (hōr-mōn) is a Greek term meaning impulse or to set in motion; hormones may either excite or inhibit a motion or action.

STRUCTURES OF THE ENDOCRINE SYSTEM

The glands of the normal endocrine system include (Figure 11–1):

► one pituitary gland (with two lobes)
► one thyroid gland (right and left lobes fused ventrally)
► four parathyroid glands
► two adrenal glands
► one pancreas
► one thymus
► one pineal gland
► two gonads (ovaries in females, testes in males)

The Master Gland

The **pituitary** (pih-too-ih-tār-ē) **gland** is located at the base of the brain just below the hypothalamus. The pituitary gland is also called the **hypophysis** (hī-poh-fī-sihs) because it is a growth located beneath or ventral to the hypothalamus or part of the brain. The pituitary gland is known as the master gland because it secretes many hormones that control or master other endocrine glands. The combining form for pituitary gland is **pituit/o.**

The pituitary gland acts in response to stimuli from the **hypothalamus** (hī-pō-thahl-ah-muhs). The hypothalamus is located below the thalamus in the brain and secretes releasing and inhibiting factors that affect the release of substances from the pituitary gland (Figure 11–2). The hypothalamus is connected to the pituitary gland via a stalk called the **infundibulum** (ihn-fuhn-dihb-yoo-luhm). Infundibulum means funnel-shaped passage or opening.

The pituitary gland is made up of two lobes: the anterior and posterior. The anterior lobe is also known as the **adenohypophysis** (ahd-ehn-ō-hī-pohf-ih-sihs) because it produces hormones (that is, has glandular function; **aden/o** = gland). Hormones released from the anterior pituitary gland are sometimes referred to as **indirect-acting hormones** because they cause their target organ to produce a second hormone. The posterior

147

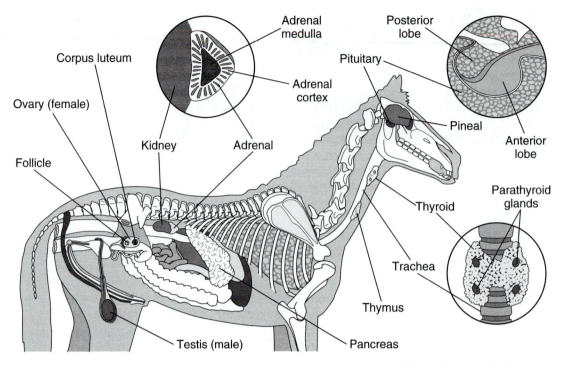

FIGURE 11–1 Endocrine gland locations. Relative locations of the endocrine glands

lobe is also known as the **neurohypophysis** (nū-rō-hī-pohf-ih-sihs) because it responds to a neurologic stimulus and does not produce hormones, but rather stores and secretes them. Hormones released from the posterior pituitary gland are sometimes referred to as **direct-acting hormones** because they produce the desired affect directly in the target organ (Figure 11–3).

SECRETIONS OF THE ANTERIOR PITUITARY GLAND

▶ **thyroid** (thī-royd) **-stimulating hormone** = augments growth and secretions of the thyroid gland; abbreviated TSH

▶ **adrenocorticotropic** (ahd-rēn-ō-kōr-tih-kō-trō-pihck) **hormone** = augments the growth and secretions of the adrenal cortex; abbreviated ACTH

▶ **follicle** (fohl-lihck-kuhl) **-stimulating hormone** = augments the secretion of estrogen and growth of eggs in the ovaries (female) and the production of sperm in the testes (male); abbreviated FSH. FSH is a type of **gonadotropic** (gō-nahd-ō-trō-pihck) **hormone.** Gonadotropic can be divided into **gonad/o,** which means gamete-producing gland (ovary or testes) and **-tropic,** which means having an affinity for.

▶ **luteinizing** (loo-tehn-īz-ihng) **hormone** = augments ovulation and aids in the maintenance of pregnancy in females; abbreviated LH. **Lute/o** is the combining form for yellow. Luteinizing hormone transforms an ovarian follicle into a corpus lu-

teum or yellow body. LH is a type of gonadotropic hormone.

▶ **interstitial** (ihn-tər-stihsh-ahl) **cell-stimulating hormone** = stimulates testosterone secretion in males (abbreviated ICSH). ICSH is now considered to be LH.

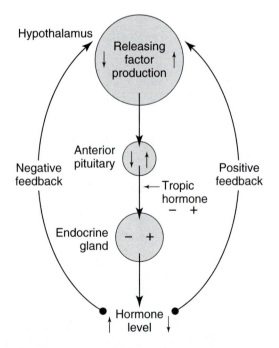

FIGURE 11–2 Secretions of the pituitary gland

- **prolactin** (prō-lahck-tihn) = augments milk secretion and influences maternal behavior; also known as **lactogenic hormone** or **luteotropin** (the combining form **lact/o** means milk)
- **growth hormone** = accelerates body growth; abbreviated GH; also known as **somatotropin** (the combining form **somat/o** means body)
- **melanocyte** (mehl-ah-nō-sit) **-stimulating hormone** = augments skin pigmentation; abbreviated MSH

SECRETIONS OF THE POSTERIOR PITUITARY GLAND

- **antidiuretic hormone** (ahn-tih-dī-yoo-reht-ihck) = maintains water balance in the body by augmenting water reabsorption in the kidneys (**anti-** means against, **diuretic** means pertaining to increased urine secretion); abbreviated ADH; also known as **vasopressin** (**vas/o** is the combining form for vessel and **press/i** is the combining form for tension; vasopressin affects blood pressure)

- **oxytocin** (ohcks-ē-tō-sihn) = stimulates uterine contractions during parturition and milk letdown from the mammary ducts

The Thyroid Gland

The **thyroid** (thī-royd) gland is a butterfly-shaped gland due to the right and left lobes being fused ventrally by an isthmus (the isthmus may be rudimentary in horses and dogs). The thyroid gland is located on either side of the larynx. The thyroid gland regulates metabolism and iodine uptake. The combining forms for thyroid gland are **thyr/o** and **thyroid/o.**

SECRETIONS OF THE THYROID GLAND

- **triiodothyronine** (trī-ī-ō-dō-thī-rō-nēn) = one of the thyroid hormones that regulates metabolism; abbreviated T_3
- **thyroxine** (thī-rohcks-ihn) = one of the thyroid hormones that regulates metabolism; abbreviated T_4

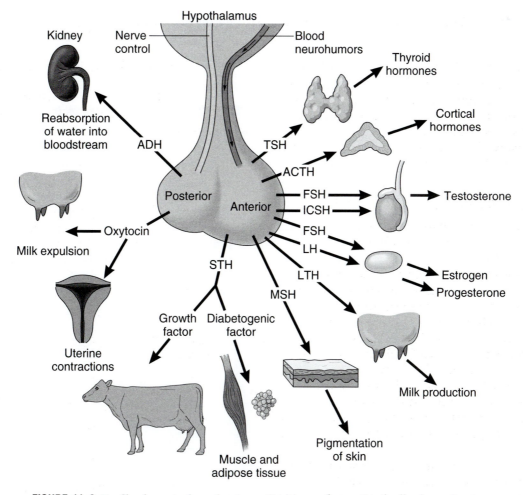

FIGURE 11–3 Feedback control mechanisms. Positive and negative feedback mechanisms control the levels of a particular hormone in the blood by either secreting releasing or inhibiting factors that affect hormone release.

▶ **calcitonin** (kahl-sih-tō-nihn) = thyroid hormone that promotes calcium absorption into bones from blood

Near the Thyroid

The **parathyroid** (pahr-ah-thī-royd) **glands** are glands located on the surface of the thyroid gland that secrete parathyroid hormone or **parathormone** (pahr-ah-thōr-mōn) or PTH. The prefix **para-** means near or before. PTH helps regulate blood calcium and phosphorus levels. PTH increases blood calcium levels by reducing bone calcium levels. Calcium is regulated in the body by the antagonistic actions of PTH and calcitonin. PTH also regulates phosphorus content of blood and bones. The combining form for the parathyroid glands is **parathyroid/o**.

Toward or Above the Kidneys

Two small glands, known as the **adrenal** (ahd-rē-nahl) or **suprarenal** (soop-rah-rē-nahl) glands, are located cranial to each kidney. The prefix **ad-** means toward, the combining form **ren/o** means kidney, and the prefix **supra-** means above. The adrenal gland regulates electrolytes, metabolism, sexual functions, and the body's response to injury. The combining forms **adren/o** and **adrenal/o** refer to the adrenal glands.

Each adrenal gland consists of two parts: the **adrenal cortex** (ahd-rē-nahl kōr-tehcks) or outer portion and the **adrenal medulla** (ahd-rē-nahl meh-doo-lah) or inner portion. The combining form **cortic/o** means outer and the combining form **medull/o** means inner or middle portion (Figure 11–4).

SECRETIONS OF THE ADRENAL CORTEX

The adrenal cortex hormones are classified as steroids. A **steroid** (stehr-oyd) is a substance that has a specific chemical structure. **Corticosteroids** (kōr-tih-kō-stehr-oydz) are produced by the adrenal cortex.

▶ **mineralocorticoids** (mihn-ər-ahl-ō-kōr-tih-koydz) = group of corticosteroids that regulates electrolyte and water balance by effecting ion transport in the kidney. The principal mineralocorticoid is **aldosterone** (ahl-dohs-tər-own), which is a hormone that regulates sodium and potassium.

▶ **glucocorticoids** (gloo-kō-kōr-tih-koydz) = group of corticosteroids that regulate carbohydrate, fat, and protein metabolism, resistance to stress, and immunologic functioning. An example of a glucocorticoid is **hydrocortisone** (hī-drō-kōr-tih-zōn) or **cortisol** (kōr-ih-zohl), which regulates carbohydrate, fat, and protein metabolism.

▶ **androgens** (ahn-drō-jehnz) = group of corticosteroids that aid in the development and maintenance of male sex characteristics. The combining form **andr/o** means male and the suffix **-gen** means producing. **Anabolic** (ahn-ah-bohl-ihck) **steroids** are synthetic medications similar in structure to testosterone that are used to increase strength and muscle mass. The prefix **ana-** means up or excessive and the combining form **bol/o** means throwing. Anabolic means pertaining to throwing (or building) up. The opposite of anabolism is catabolism. The prefix **cata-** means down; therefore, **catabolism** (kaht-ah-bōl-ihz-uhm) means throwing (or breaking) down.

SECRETIONS OF THE ADRENAL MEDULLA

▶ **epinephrine** (ehp-ih-nehf-rihn) = **catecholamine** (kaht-ih-kōl-ih-mēn) that stimulates the sympathetic nervous system (the fight or flight system) and increases blood pressure, heart rate, and blood glucose. Epinephrine is also known as **adrenaline** (ahd-rehn-ah-lihn). Because epinephrine increases blood pressure, it is known as a vasopressor. A **vasopressor** (vahs-ō-prehs-ōr) is a substance that stimulates blood vessel contraction and increases blood pressure.

▶ **norepinephrine** (nōr-ehp-ih-nehf-rihn) = catecholamine that stimulates the sympathetic nervous system

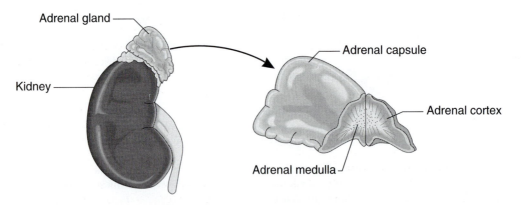

Adrenal gland

Kidney

Adrenal capsule

Adrenal cortex

Adrenal medulla

FIGURE 11–4 Portions of the adrenal gland

(the fight or flight system) and increases blood pressure, heart rate, and blood glucose. Norepinephrine is also known as **noradrenaline.**

What is in a name?

Sometimes it seems as though medical terminology is a bunch of long words strung together with no relationship to each other. However, consider the names of the hormones released from the adrenal medulla:

epinephrine = the prefix **epi-** means above. The combining form **nephr/o** means kidney. From its name it can be derived that epinephrine is made in the gland above the kidney, which is known as the adrenal gland.

adrenaline = the prefix **ad-** means toward. The combining form **ren/o** means kidney. From its name it can be derived that adrenaline is made in the gland toward the kidney, which is known as the adrenal gland

norepinephrine = the prefix **nor-** means normal and denotes the parent compound in a pair of related substances. From its name it can be derived that norepinephrine is a substance very similar in chemical structure and function to epinephrine.

The Dual Function Gland

The **pancreas** (pahn-krē-ahs) is an aggregation of cells located near the proximal duodenum that has both exocrine and endocrine functions. The exocrine function of the pancreas involves the secretion of digestive enzymes and is covered in Chapter 6 on the digestive system. The endocrine function of the pancreas involves the secretion of blood glucose-regulating hormones. (Blood glucose is also regulated in conjunction with other hormones such as thyroid hormone.) The combining form for pancreas is **pancreat/o.**

Specialized cells in the pancreas called the **islets of Langerhans** (ī-lehts ohf lahng-ər-hahnz) secrete the hormones that help regulate blood glucose.

ENDOCRINE SECRETIONS OF THE PANCREAS

▶ **insulin** (ihn-suh-lihn) = hormone that transports blood glucose into body cells
▶ **glucagon** (gloo-kah-gohn) = hormone that increases blood glucose

The Thymus

The **thymus** (thī-muhs) is a gland predominant in young animals located near midline in the cranioventral portion of the thoracic cavity. The thymus gland has immunologic function. The combining form for the thymus is **thym/o.** Thym/o also refers to the soul;

some people believed that the thymus gland was the seat of the soul since it lies in such close proximity to the heart.

SECRETION OF THE THYMUS

▶ **thymosin** (thī-mō-sihn) = augments maturation of lymphocytes; therefore, it plays a role in the immune system, which is covered in Chapter 15 on the hematologic, immune, and lymphatic systems.

The Pineal Gland

The **pineal gland** or **pineal body** (pī-nē-ahl) is an aggregation of cells located in the central portion of the brain. The pineal gland's functions are not fully understood, but one function is has is that it secretes hormones that affect circadian rhythm. The combining form **pineal/o** refers to the pineal gland.

SECRETION OF THE PINEAL GLAND

▶ **melatonin** (mehl-ah-tōn-ihn) = controls circadian rhythm and reproductive timing. **Circadian** can be divided into **circa** (around) and **diem** (day). Circadian refers to the events occurring within a 24-hour period.

Gonads

The **gonads** (gō-nahdz) are gamete producing glands. A **gamete** (gahm-ēt) is a sex cell. Gamete-producing glands are the **ovaries** in females and **testes** in males. **Gonad/o** is the combining form for gonad or gamete-producing gland. **Gonadotropic hormones** stimulate the gonads.

SECRETIONS OF THE OVARY

Secretions of the ovary are stimulated by **human chorionic** (kōr-ē-ohn-ihck) **gonadotropin** or **hCG.** hCG is secreted by the placenta after pregnancy has occurred. Other gonadotropins also influence ovarian secretion.

▶ **estrogen** (ehs-trō-jehn) = hormone that aids in the development of secondary sex characteristics (an example is mammary gland development) and regulates ovulation in females
▶ **progesterone** (prō-gehs-tər-own) = hormone that aids in the maintenance of pregnancy. Progesterone is also secreted from the corpus luteum and placenta.

SECRETION OF THE TESTES

▶ **testosterone** (tehs-tohs-tər-own) = augments the development of secondary sex characteristics. Examples of secondary sex characteristics in animals include horns in rams (not horns in all species), boar tusks, and shoulder girth in cattle and horses. Testosterone is also thought to be secreted from the ovaries and adrenal cortex, although these secretions are in very small amounts.

▶ Test Me: Endocrine System ◀

Diagnostic procedures performed on the endocrine system include

- **assays** (ahs-āz) = laboratory technique used to determine the amount of a particular substance in a sample. Assays will be covered in Chapter 17 on laboratory, diagnostic, and surgical terminology.
- **ACTH stimulation test** = blood analysis for cortisol levels after administration of synthetic adrenocorticotropic hormone; used to differentiate pituitary-dependent hyperadrenocorticism versus adrenal-dependent hyperadrenocorticism
- **dexamethasone** (dehx-ah-mehth-ah-zōn) **suppression test** = blood analysis for cortisol levels after administration of synthetic glucocorticoid (dexamethasone); used to differentiate pituitary-dependent hyperadrenocorticism versus adrenal-dependent hyperadrenocorticism. Also called **dex suppression test.** Dex suppression tests may either be high-dose or low-dose.
- **thyroid stimulation test** = blood analysis for thyroid hormone levels after administration of synthetic thyroid-stimulating hormone; used to differentiate pituitary dependent versus thyroid dependent dysfunction. **Synthetic** (sihn-theh-tihck) means pertaining to artificial production.
- **radioactive** (rā-dē-ō-ahck-tihv) **iodine uptake test** = analysis of thyroid function after induction of radioactive iodine that has been given orally or intravenously. Absorption of the radioactive iodine is measured with a counter for a specific time period.

▶ Path of Destruction: Endocrine System ◀

Pathologic conditions of the endocrine system include
- **endocrinopathy** (ehn-dō-krih-nohp-ah-thē) = disease of the hormone-producing system
- **hypercrinism** (hī-pər-krī-nihzm) = condition of excessive gland secretion
- **hypocrinism** (hī-pō-krī-nihzm) = condition of deficient gland secretion
- **pituitarism** (pih-too-ih-tār-ihzm) = any disorder of the pituitary gland
- **hypophysitis** (hī-poh-fī-sī-tihs) = inflammation of the pituitary gland
- **hyperpituitarism** (hī-pər-pih-too-ih-tahr-ihzm) = condition of excessive secretion of the pituitary gland

- **acromegaly** (ahck-rō-mehg-ah-lē) = enlargement of the extremities caused by excessive secretion of growth hormone after puberty. **Acr/o** means extremitites
- **diabetes insipidus** (dī-ah-bē-tēz ihn-sihp-ih-duhs) = insufficient antidiuretic hormone production or the inability of the kidneys to response to ADH stimuli. A **stimulus** (stihm-yoo-luhs) is an agent, act, or influence that produces a reaction. The plural of stimulus is **stimuli.**
- **thyroiditis** (thī-roy-dī-tihs) = inflammation of the thyroid gland
- **hypothyroidism** (hī-pō-thī-royd-ihzm) = condition of thyroid hormone deficiency. Signs of hypothyroidism include decreased metabolic rate, poor hair coat, lethargy, and increased sensitivity to cold. **Euthyroidism** (yoo-thī-royd-ihzm) is the condition of normal thyroid function. The prefix **eu-** means good, well, or easily.
- **hyperthyroidism** (hī-pər-thī-royd-ihzm) = condition of thyroid hormone excess. Signs of hyperthyroidism include increased metabolic rate, weight loss, and polyuria and polydipsia (Figure 11–5).
- **thyrotoxicosis** (thī-rō-tohck-sih-kō-sihs) = abnormal life-threatening condition of excessive poisonous quantities of thyroid hormone. The combining form **toxico** means poison.
- **thyromegaly** (thī-rō-mehg-ah-lē) = enlargement of the thyroid gland
- **hypoparathyroidism** (hī-pō-pahr-ah-thī-royd-ihzm) = abnormal condition of deficient parathyroid hormone secretion resulting in hypocalcemia.

FIGURE 11–5 Signs of increased metabolic rate, weight loss, polyuria and polydipsia may indicate hyperthyroidism or diabetes mellitus in cats. *Source:* Courtesy of Mark Jackson, North Carolina State University

FIGURE 11–6 Dog with hyperadrenocorticism. Hyperadrenocorticism or Cushing's disease results in polyuria, polydipsia, and redistribution of body fat. *Source:* Courtesy of Mark Jackson, North Carolina State University

Hypocalcemia (hī-pō-kahl-sē-mē-ah) is abnormally low blood calcium levels. **Hypo-** is the prefix for deficient, **calc/i** is the combining form for calcium, **-emia** is the suffix for blood condition. Hypercalcemia is abnormally high blood calcium levels because **hyper-** is the prefix meaning excessive.

- **hyperparathyroidism** (hī-pər-pahr-ah-thī-royd-ihzm) = abnormal condition of excessive parathyroid hormone secretion resulting in hypercalcemia
- **adrenopathy** (ahd-rēn-ohp-ah-thē) = disease of the adrenal glands
- **Addison's** (ahd-ih-sohnz) **disease** = disorder caused by deficient adrenal cortex function; also called **hypoadrenocorticism** (hī-pō-ahd-rēn-ō-kōr-tih-kihz-uhm)
- **aldosteronism** (ahl-doh-stər-ōn-ihzm) = disorder caused by excessive secretion of aldosterone by the adrenal cortex resulting in electrolyte imbalance. An electrolyte (ē-lehck-trō-līt) is a charged substance found in blood.
- **Cushing's** (kuhsh-ihngz) **disease** = disorder caused by excessive adrenal cortex production of glucocorticoid resulting in increased urination, drinking, and redistribution of body fat; also called **hyperadrenocorticism** (hī-pər-ahd-rēn-ō-kōr-tih-kihz-uhm) (Figure 11–6)
- **pheochromocytoma** (fē-ō-krō-mō-sī-tō-mah) = tumor of the adrenal medulla resulting in increased secretion of epinephrine and norepinephrine. **Phe/o** is the combining form for dusky; **chrom/o** is the combining form for color; **cyt/o** is the combining form for cell; **-oma** is the suffix meaning tumor. A pheochromocytoma is a tumor that takes on a dark (dusky) color be-

cause it is composed of chromaffin (or colored) cells.

- **pancreatitis** (pahn-krē-ah-tī-tihs) = inflammation of the pancreas (Figure 11–7)
- **hyperinsulinism** (hī-pər-ihn-suh-lin-ihzm) = disorder of excessive hormone that transports blood glucose into body cells
- **hyperglycemia** (hī-pər-glī-sē-mē-ah) = abnormally elevated blood glucose
- **hypoglycemia** (hī-pər-glī-sē-mē-ah) = abnormally low blood glucose
- **insulinoma** (ihn-suh-lihn-ō-mah) = tumor of the islet of Langerhans of the pancreas
- **diabetes mellitus** (dī-ah-bē-tēz mehl-ih-tuhs) = metabolic disorder of inadequate secretion of insulin or recognition of insulin by the body resulting in increased urination, drinking, and weight loss. Severe insulin deficiency may result in **ketoacidosis** (kē-tō-ah-sih-dō-sihs), which is an abnormal condition of low pH accompanied by ketones (by-products of fat metabolism). **Acidosis** (ah-sih-dō-sihs) is an abnormal condition of low pH
- **thymoma** (thī-mō-mah) = tumor of the thymus
- **pinealopathy** (pīn-ē-ah-lohp-ah-thē) = disorder of the pineal gland
- **hypergonadism** (hī-pər-gō-nahd-ihzm) = abnormal condition of excessive hormone secretion by the sex glands (ovaries in females; testes in males)
- **hypogonadism** (hī-pō-gō-nahd-ihzm) = abnormal condition of deficient hormone secretion by the sex glands (ovaries in females; testes in males)
- **gynecomastia** (gī-neh-kō-mahs-tē-ah) = condition of excessive mammary development in males

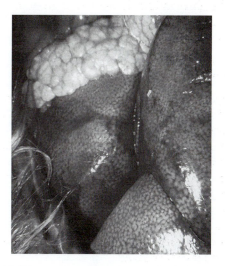

FIGURE 11–7 Pancreas of a dog with diabetes mellitus. *Source:* Courtesy of Mark Jackson, North Carolina State University

▶ Procede With Caution ◀

Procedures performed on the endocrine system include

- **hypophysectomy** (hī-pohf-ih-sehck-tō-mē) = surgical removal of the pituitary gland
- **chemical thyroidectomy** (thī-royd-ehck-tō-mē) = administration of radioactive iodine to suppress thyroid function; also called radioactive iodine therapy
- **thyroidectomy** (thī-royd-ehck-tō-mē) = surgical removal of all or part of the thyroid gland
- **lobectomy** (lō-behck-tō-mē) = surgical removal of a lobe or well-defined portion of an organ
- **parathyroidectomy** (pahr-ah-thī-royd-ehck-tō-mē) = surgical removal of one or more parathyroid glands
- **adrenalectomy** (ahd-rē-nahl-ehck-tō-mē) = surgical removal of one or both adrenal glands
- **pancreatectomy** (pahn-krē-ah-tehck-tō-mē) = surgical removal of the pancreas
- **pancreatotomy** (pahn-krē-ah-toht-ō-mē) = surgical incision into the pancreas
- **thymectomy** (thī-mehck-tō-mē) = surgical removal of the thymus
- **pinealectomy** (pīn-ē-ahl-ehck-tō-mē) = surgical removal of the pineal gland

REVIEW EXERCISES

Multiple Choice—Choose the correct answer.

1. The gland known as the master gland that helps maintain the appropriate levels of hormone in the body is the
   a. hypothalamus
   b. pituitary gland
   c. thyroid gland
   d. pancreas

2. The chemical substance that is secreted by the posterior pituitary gland that causes water reabsorption in the kidneys is known as
   a. prolactin
   b. luteinizing hormone
   c. ACTH
   d. antidiuretic hormone

3. The regulator of the endocrine system is the
   a. thyroid gland
   b. calcitonin
   c. hypothalamus
   d. parathyroid gland

4. Thyromegaly is
   a. enlargement of the thyroid gland
   b. augmentation of the thymus
   c. dissolution of the parathyroid glands
   d. radioactive iodine treatment of the thyroid gland

5. Surgical removal of a well-defined portion of an organ is a/an
   a. sacculectomy
   b. lumpectomy
   c. lobectomy
   d. cystectomy

6. An aggregation of cells specialized to secrete or excrete materials not related to their own function is a/an
   a. hormone
   b. gland
   c. hypoadenum
   d. hyperadenum

7. Hypoadrenocorticism, a disorder caused by deficient adrenal cortex production of glucocorticoid, is also known as
   a. Cushing's disease
   b. Grave's disease
   c. Addison's disease
   d. Langerhans' disease

8. A tumor of the islets of Langerhans of the pancreas is called a/an
   a. diabetes mellitus
   b. diabetes insipidus
   c. ketoacidosis
   d. insulinoma

9. The chemical substance that helps control circadian rhythm is
   a. circadianin
   b. pinealin
   c. melatonin
   d. thymin

10. Excessive mammary development in males is called
    a. feminum
    b. gynecomastia
    c. gyneconium
    d. feminomastia

Matching—Match the gland in Column I to its correct function or structure in Column II.

Column I

_____adrenals

_____thyroid

_____thymus

_____pancreas

_____pituitary

_____parathyroid

_____pineal

_____gonads

Column II

a. butterfly-shaped gland on either side of the larynx

b. gland located dorsal to the sternum

c. contains specialized cells that secrete hormones that affect sugar and starch metabolism

d. secretes melatonin

e. gamete-producing glands

f. small gland at the base of the brain

g. secretes hormone that reduces bone calcium levels and regulates phosphorus

h. two small glands located on top of each kidney

Label the diagram in Figure 11–8.

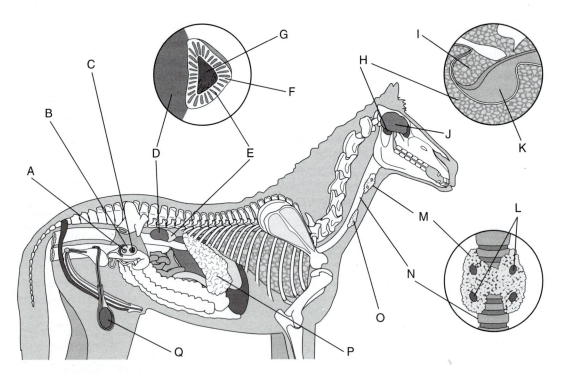

FIGURE 11–8 Endocrine organs. Label the endocrine organs and related structures. Provide the combining form for each endocrine gland.

Word Scramble—Use the definitions to unscramble the endocrine system terms.

condition of building up	mlsioaanb
condition of breaking down	mlsioaactb
abnormally elevated blood glucose	aieyeyhprglcm
inflammation of the pancreas	iittseaacrnp
gamete-producing gland	naodg
surgical removal of the thyroid gland	dyorihtetcmoy
study of the hormonal system	glnooyndeocri
production of new glucose	ssgiluncoeogeen
artificial production	nttheicys
abnormal condition of deficient thyroid hormone secretion	smidioryhtophy

Case Studies—Define the terms underlined in each case study.

A 7 <u>yr</u> old <u>M/N</u> Golden retriever, Goldie, was presented to the clinic for bite wounds from another dog in the household. After the dog's vaccination status was checked, he was examined. The dog's vital signs included <u>T</u> = 103 °F, <u>HR</u> = 120 <u>BPM</u>, RR = 42 breaths/min. Upon examination it was noted that the dog weighed 85<u>#</u>, the haircoat had areas of <u>alopecia</u>, and the dog was <u>lethargic</u>. The wounds from the dog fight consisted of 2 lacerations; 1 laceration was 4 <u>cm</u> long located on the <u>lateral</u> side of the L forelimb <u>proximal</u> to the <u>elbow</u> and the other laceration was 2 cm long located on the <u>medial</u> surface of the L <u>hock</u>. The owner was questioned as to whether these two dogs usually got along in the house and the owner stated that they had prior to Goldie's recent history of lethargy and unwillingness to go out for walks during the past month. The owner felt that Goldie did not want to go out on walks due to the change of seasons and the cold, damp weather. Goldie was admitted to the clinic and <u>anesthetized</u>, and the lacerations were cleaned and sutured. Antibiotics were dispensed 2 T <u>BID</u> for 7 <u>d</u>. Goldie came back for suture removal in 10 days. The lacerations had healed, but hair had not begun to regrow in the areas of the lacerations, and more areas of alopecia were noted. A thyroid panel was recommended to assess Goldie's thyroid function. Blood was drawn and a thyroid panel was run. The thyroid panel revealed that Goldie was <u>hypothyroid</u>, and thyroid supplementation was begun.

yr _____

M/N _____

T _____

HR _____

BPM _____

_____

alopecia _____

lethargic _____

cm _____

lateral _____

proximal _____

elbow _____

medial _____

hock _____

anesthetized _____

BID _____

d _____

hypothyroid _____

A 15 yr old F/S DSH was brought into the clinic for weight loss. Examination revealed the cat was tachycardic, emaciated, and lethargic. Vital signs were T = 102 °F, HR = 260 BPM, RR = 60 breaths/min. Abdominal palpation revealed a moderately full urinary bladder, slightly small kidneys, and gas-filled loops of bowel. Palpation of the neck revealed enlargement of thyroid glands. Blood was collected from the jugular vein for a CBC, chem screen, and T_4. Blood work revealed a slightly elevated BUN and creatinine, hypercholesterolemia, and elevated T_4 levels. It was recommended that the cat receive radioactive iodine treatments; however, the options of thyroidectomy and pharmaceutical management were also discussed. The cat was scheduled for radiation treatments at a referral center.

yr _____

F/S _____

DSH _____

tachycardic _____

emaciated _____

lethargic _____

T _____

HR _____

BPM _____

RR _____

palpation _____

vein _____

CBC _____

chem screen _____

T_4 _____

BUN _____

creatinine _____

hypercholesterolemia _____

radioactive iodine treatments _____

thyroidectomy _____

12

1 + 1 = 3 (or More)

Objectives _In this chapter, you should learn to:_

► Identify and describe the functions and structures of the male and female reproductive systems
► Recognize, define, spell, and pronounce the terms related to the diagnosis, pathology, and treatment of the reproductive systems
► Describe the estrous cycle in females using medical terminology

THE REPRODUCTIVE SYSTEM

The **reproductive** (rē-prō-duhck-tihv) **system** is responsible for the act of producing offspring. The reproductive system needs both male- and female-specific organs to complete offspring production. The term **theriogenology** (thēr-ē-ō-jehn-ohl-ō-jē) is used to describe animal reproduction or the study of producing beasts. **Theri/o** is the combining form for beast; **gen/o** is the combining form for producing, and **-logy** is the suffix meaning to study.

The reproductive organs, whether male or female, are called the **genitals** (jehn-ih-tahlz) or **genitalia** (jehn-ih-tā-lē-ah). The combining form **genit/o** refers to the organs of reproduction. The genitalia include external and internal organs.

The male reproductive system will be covered first followed by the female reproductive system.

FUNCTIONS OF THE MALE REPRODUCTIVE SYSTEM

The functions of the male reproductive system are to produce and deliver sperm to the egg to create life.

STRUCTURES OF THE MALE REPRODUCTIVE SYSTEM

Scrotum

The **scrotum** (skrō-tuhm) or **scrotal sac** is the external pouch that encloses and supports the testes. The scrotum encloses the testes outside of the body so that the testes are at a temperature lower than body temperature. This lower temperature is needed for sperm development. Scrotum is the Latin term for bag. The combining form **scrot/o** means scrotum.

The area between the scrotum and the anus in males is referred to as the **perineum** (pehr-ih-nē-uhm). **Perine/o** is the combining form for the area between the scrotum (or vulva in females) and anus.

Testes

The **testes** (tehs-tēz) or **testicles** (tehs-tih-kuhlz) are the male sex glands that produce spermatozoa. Testes refers to glands and **testis** (tehs-tihs) refers to a single gland. The orientation of the testes within the scrotum varies among species. Some species have open inguinal rings that allow the testes to be withdrawn from the scrotum and into the abdomen. Sex glands are called **gonads** (gō-nahdz). The combining forms for testes are **orch/o, orchi/o, orchid/o, test/o,** and **testicul/o** (Figure 12–1).

The testes develop in the fetal abdomen and descend into the scrotum before birth. The testes are suspended in the scrotum by the spermatic cord. The testicle is divided into compartments that contain coiled tubes called the **seminiferous tubules** (seh-mih-nihf-ər-uhs too-buhlz) and cells between the spaces are called **interstitial** (ihn-tər-stih-shahl) cells. The interstitial cells of the testes are called the **Leydig's** (lih-dihgz) **cells.** Leydig's cells have endocrine function. **Sertoli** (sihr-tō-lē) **cells** are specialized cells in the testes that support and nourish sperm growth. The seminiferous tubules

are channels in the testes in which sperm are produced and through which the sperm leave the testes. **Sperm** (spərm) or **spermatozoa** (spər-mah-tō-zō-ah) are the male gametes or sex cells. **Spermatozoon** (spər-mah-tō-zō-uhn) is one gamete. The combining forms for spermatozoa are **sperm/o** and **spermat/o.**

Ejaculated semen is occasionally evaluated microscopically to determine if the spermatozoa have normal morphology and motility and if they are present in adequate numbers. A spermatozoon has a head, midpiece, and tail (Figure 12–2). The head contains the nucleus. At the top of the head is a structure called the **acrosome** (ahk-rō-zōm), which contains enzymes that allow the spermatozoa to penetrate the ovum. The midpiece contains mitochondria to provide energy to

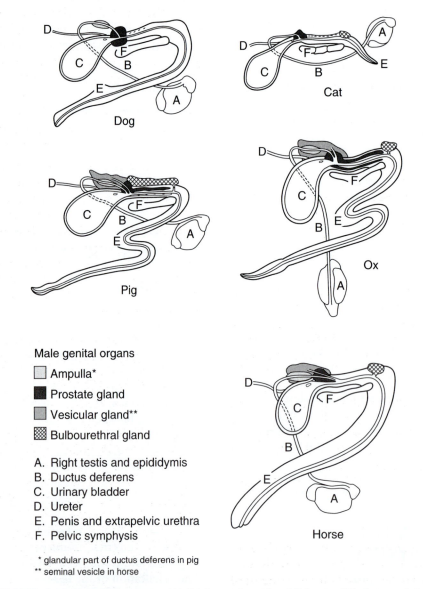

Male genital organs

- ⬜ Ampulla*
- ⬛ Prostate gland
- ▨ Vesicular gland**
- ▧ Bulbourethral gland

A. Right testis and epididymis
B. Ductus deferens
C. Urinary bladder
D. Ureter
E. Penis and extrapelvic urethra
F. Pelvic symphysis

* glandular part of ductus deferens in pig
** seminal vesicle in horse

FIGURE 12–1 Male genital organs. (A) Right testis and epididymis, (B) ductus deferens, (C) urinary bladder, (D) ureter, (E) penis and extrapelvic urethra, (F) pelvic symphysis

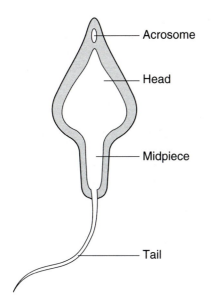

FIGURE 12–2 Parts of sperm

the sperm. The tail is actually a flagellum, providing movement for the spermatozoon to reach the ovum. **Spermatogenesis** (spər-mah-tō-jehn-eh-sihs) is production of male gametes.

> Gamete comes from the Greek terms gametes, which means husband, and gamete, which means wife. Gamein is Greek for to marry. Johann Gregor Mendel was the first to apply the term gamete in biology to mean sex cells. Gametes refer to the sperm in males and ovum (egg) in females.

Epididymis

The seminiferous tubules join together to form a cluster. Ducts emerge from this cluster and enter the epididymis. The **epididymis** (ehp-ih-dihd-ih-mihs) is the tube at the upper part of each testis that secretes part of the semen, stores semen before ejaculation, and provides a passageway for sperm. The epididymis is divided into head (or caput), body, and tail portions. The epididymis runs down the length of the testicle, turns upward, and becomes a narrower tube called the vas deferens (Figure 12–3). Sperm are collected in the epididymis where they become **motile** (mō-tihl). Motile means capable of spontaneous motion.

The combining form for epididymis is **epididym/o.**

Vas Deferens

The **vas deferens** is a tube connected to the epididymis that carries sperm into the pelvic region toward the urethra. Each vas deferens is encased by the spermatic cord (the spermatic cord also encases nerves and blood and lymph vessels along with the vas deferens). The end of the vas deferens is the **ductus deferens** (duhck-tuhs dehf-ər-ehnz). The ductus deferens is the excretory duct of the testes. The ductus deferens in swine has a glandular portion called the **ampulla** (ahmp-yoo-lah). An ampulla is an enlarged part of a tube or canal.

Accessory Sex Glands

The accessory sex glands include the seminal vesicles, prostate gland, and bulbourethral glands. Not all glands are present in all species (Table 12–1). The accessory sex glands add secretions to the sperm and flush urine from the urethra before sperm enter it.

The **seminal vesicles** (sehm-ih-nahl vehs-ih-kuhl) or **vesicular** (vehs-ih-koo-lahr) **glands** are two glands that open into the ductus deferens as it join the urethra. The seminal vesicles secrete a thick, yellow substance that nourishes sperm and adds volume to the ejaculated semen. **Semen** (sē-mehn) is the ejaculatory fluid that contains sperm and the secretions of the accessory sex glands. **Semin/i** is the combining form for semen.

> Seminal vesicles is the term used in horses; vesicular gland is used in the other species if the gland is present.

The **ejaculatory** (ē-jahck-yoo-lā-tōr-ē) **duct** is formed by the union of the ductus deferens and the duct from the seminal vesicles. The ejaculatory duct passes through the prostate and enters the urethra.

The **prostate** (proh-stāt) **gland** is a single gland that surrounds or is near the urethra and may be well defined or diffuse, depending upon the species. The prostate gland secretes a thick fluid that aids in the motility of sperm. The combining form **prostat/o** means prostate gland.

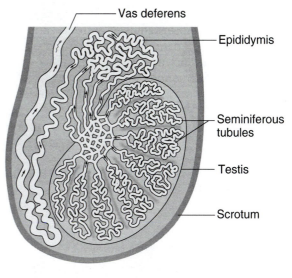

FIGURE 12–3 Cross-section of scrotal structures

TABLE 12–1
Male Accessory Glands of Different Species

	Male Accessory Sex Glands		
	Prostate	Vesicular gland/ seminal vesicle	Bulbourethral
dog	+	–	–
cat	+	–	+
pig	+	+	+
ruminants	+	+	+
horse	+	+	+

The **bulbourethral** (buhl-bō-yoo-rē-thrahl) **glands** are two glands located on either side of the urethra. The bulbourethral glands secrete a thick mucus that acts as a lubricant for sperm. These glands are called **Cowper's** (cow-pərz) **glands** in primates.

Urethra

The **urethra** (yoo-rē-thrah) is a tube passing through the penis to the outside of the body that serves both reproductive and urinary systems. The combining form for urethra is **urethr/o.**

Penis

The **penis** (pē-nihs) is the male sex organ that carries reproductive and urinary products out of the body. The glans penis is the distal part of the penis on which the urethra opens. The **prepuce** (prē-pyoos) is the retractable fold of skin covering the glans penis. The prepuce is sometimes called the **foreskin.** Dogs have an **os penis** (ohs pē-nihs), which is a bone encased in the penile tissue. All species except the cat have a cranioventrally directed penis. The combining forms **pen/i** and **priap/o** mean penis.

The penis is composed of erectile tissue that upon sexual stimulation fills with blood (under high pressure) and causes an erection. Some species, like the ruminants and swine, achieve an erection by straightening of the **sigmoid flexure** (sihg-moyd flehck-shər), an S-shaped bend in the penis. Other species, like equine and canine, have a penis with almost all erectile tissue. Erection in these animals is caused by blood engorgement of the erectile tissue (Figure 12–4).

FUNCTIONS OF THE FEMALE REPRODUCTIVE SYSTEM

The functions of the female reproductive system are to create and support new life.

STRUCTURES OF THE FEMALE REPRODUCTIVE SYSTEM

Ovaries

The **ovaries** (ō-vah-rēz) are a small pair of organs located in the lower abdomen (Figure 12–5). An ovary is the female gonad that produces estrogen, progesterone,

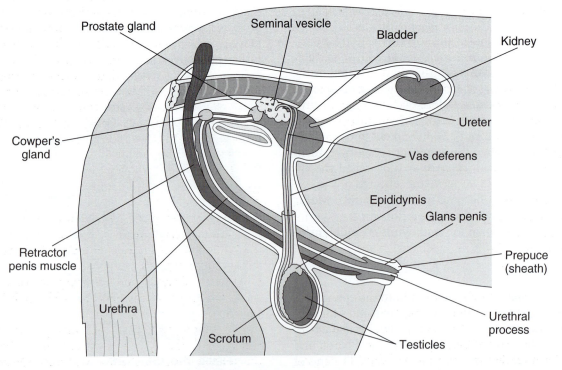

FIGURE 12–4 Reproductive tract of a stallion

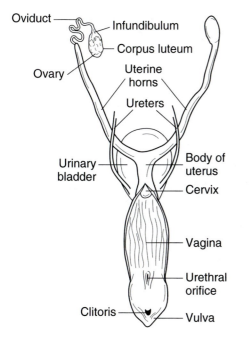

FIGURE 12–5 Reproductive tract of a bitch

Uterus

The **uterus** (yoo-tər-uhs) is a thick-walled, hollow organ with muscular walls and a mucous membrane lining. The uterus is situated dorsal to the urinary bladder and ventral to the rectum. The combining forms **hyster/o, metr/o, metri/o,** and **uter/o** mean uterus. The uterus consists of three parts:

▶ **cornus** (kōr-nuhs) = horn. The distal end of the uterus has two horns that travel toward the oviducts. **Cornu** means horn. Some animals are **bicornuate** (bī-kōrn-yoo-āt), which means having two uterine horns

▶ **corpus** (kōr-puhs) = body. The middle portion of the uterus. **Corpu** means body.

▶ **cervix** (sihr-vihckz) = neck. The lower portion of the uterus that extends into the vagina. **Cervic/o** means neck.

The uterus is composed of three major tissue types:

▶ **perimetrium** (pehr-ih-mē-trē-uhm) = membranous outer layer of the uterus. **Peri-** is the prefix for surrounding.

▶ **myometrium** (mī-ō-mē-trē-uhm) = muscular middle layer of the uterus. **My/o** is the combining form for muscle.

▶ **endometrium** (ehn-dō-mē-trē-uhm) = inner layer of the uterus. **Endo-** is the prefix meaning within.

Cervix

The **cervix** (sihr-vihckz) is the caudal continuation of the uterus and the cranial continuation of the vagina. The cervix contains ring like smooth muscle called **sphincters.** The main function of the cervix is to prevent foreign substances from entering the uterus. The cervix is usually closed tightly except during estrus, when it relaxes to allow entry of sperm. During pregnancy the cervix is closed with a mucus plug. The mucus plug is released near parturition to allow fetal passage. The combining form **cervic/o** means neck or neck like structure.

Vagina

The **vagina** (vah-jī-nah) is the muscular tube lined with mucosa that extends from the cervix to outside the body. The vagina accepts the penis during copulation and serves as a passage for semen into the body and excretions and offspring out of the body. The combining forms **colop/o** and **vagin/o** mean vagina.

A membranous fold of tissue may partially or completely cover the external vaginal orifice. This fold is called the **hymen** (hī-mehn). An **orifice** (ōr-ih-fihs) is an entrance or outlet from a body cavity.

and **ova** (ō-vah) or **eggs.** The ovaries contain many small sacs called **graafian follicles** (grahf-ē-ahn fohl-lihck-kuhlz). Each graafian follicle contains an ovum. The ova develop within the ovaries and are expelled (ovulated) when the egg matures. The combining forms **ovari/o** and **oophor/o** mean female gonad. The combining forms **oo/o, ov/i,** and **ov/o** mean egg. An egg cell is an **oocyte** (ō-ō-sīt).

Oviducts

The **oviducts** (ō-vih-duhckts) are paired tubes that extend from the cranial portion of the uterus to the ovary (although they are not attached to the ovary). The oviducts are also referred to as the **fallopian tubes** (fah-lō-pē-ahn) and **uterine tubes.** The combining form for oviducts is **salping/o,** which means tube.

The distal end of each oviduct is a funnel-shaped opening called the **infundibulum** (ihn-fuhn-dihb-yoo-luhm). The infundibulum contains fringed extensions called **fimbriae** (fihm-brē-ah) that catch ova when they leave the ovary. The fimbriae are not attached to the ovaries.

The proximal end of each oviduct is connected to the uterine horns. The oviducts serve as ducts to carry ova from the ovary to the uterus. The oviducts also transport sperm traveling up from the vagina and uterus. **Fertilization** (fər-tihl-ih-zā-shuhn) or egg and sperm union usually occurs in the oviduct.

Vulva

The **vulva** (vuhl-vah), also known as the female external genitalia or **pudendum** (pyoo-dehn-duhm), consists of the vaginal orifice, vestibular glands, clitoris, hymen, and urethral orifice. The combining forms for the vulva are **vulv/o** and **episi/o.** The **perineum** (pehr-ih-nē-uhm) is the region between the vaginal orifice and anus in females.

The **labia** (lā-bē-ah) are the fleshy borders or edges of the vulva and are occasionally referred to as the **lips.** In animals the vulva contains simple lips as opposed to humans who have major and minor labia. The **vaginal orifice** is the entrance from the vagina to outside the body. The **vestibular glands** (also known as **Bartholin's glands** in primates) are found in bovine, feline, and occasionally ovine species. The vestibular glands secrete mucus to lubricate the vagina. The **clitoris** (kliht-ō-rihs) is the sensitive, erectile tissue lo-cated in the ventral portion of the vulva. The clitoris is the analog of the glans penis of the male. The **urethral orifice** is found where the vagina and vulva join and is sometimes associated with a vestigial hymen (Figure 12–6).

Mammary Glands

The **mammary** (mahm-mah-rē) **glands** are milk-producing glands in females. The number of mammary glands varies with the species: the mare, ewe, and doe (goat) have two; cows have four; sows have six or more pairs; bitches and queens have four or more pairs. In litter-bearing species the glandular structures are usually paired, located on the ventral surface and are called mammary glands or **mammae** (mahm-ā). Each singular gland (mamma) is associated with one nipple. In large animals the mammary gland is called an **udder** (uh-dər), is located in the inguinal area, and

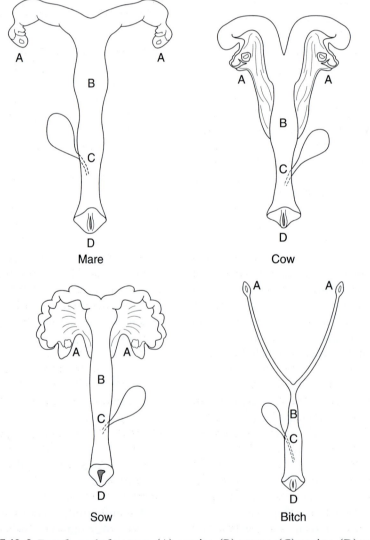

FIGURE 12–6 Female genital organs. (A) ovaries, (B) uterus, (C) vagina, (D) vulva

has either two or four functional teats. The nipple area is called a **teat** (tēt) in large animals. In cows the four mammae are called **quarters** (Figure 12–7).

Mammary glands are composed of connective and adipose tissue organized into lobes and lobules that contain milk-secreting sacs called **alveoli** (ahl-vē-ō-lī). Each lobe drains towards the teat or papilla via a **lactiferous** (lahck-tihf-ər-uhs) **duct**. **Lact/i** means milk. The lactiferous ducts come together to form the lactiferous or teat sinus. The **lactiferous sinus** is composed of the gland cistern (within the gland) and the teat cistern (within the teat). Milk travels from the gland cistern into the teat cistern. From the teat cistern, milk empties into the **papillary duct,** which is commonly called the **streak canal.**

The combining forms for mammary glands are **mamm/o** and **mast/o.**

The Estrous Cycle

The **estrous** (ehs-truhs) **cycle,** which is sometimes referred to as the heat cycle, occurs at the onset of puberty and continues throughout an animal's life. The ability to sexually reproduce begins at puberty and varies among species. The estrous cycle prepares the uterus to accept a fertilized ovum.

The female reproductive systems functions on cyclic intervals. Hormones secreted from the anterior pituitary gland and ovary control the estrous cycle. Although there is species variation during estrous cycle phases among animals, the basic patterns are the same.

The estrous cycle starts when ova develop within ovarian follicles. One or more follicles continue to develop until they reach the ripened follicle size (termed the graafian follicle). The graafian follicle(s) ruptures, which is termed ovulation. The ovum is expelled from the ovary into the oviduct. The ruptured follicle continues to grow and becomes filled with a yellow substance. The yellow ruptured follicle is now called the **corpus luteum** (kōr-puhs loo-tē-uhm) or yellow body. The corpus luteum is abbreviated CL. The CL secretes progesterone. If the ovum is fertilized, the CL will continue to secrete progesterone to prevent future estrous cycling. If the ovum is not fertilized, the CL shrinks, reduces its progesterone secretion, and a new estrous cycle begins. The stage of the estrous cycle in which the graafian follicle is present is called the **follicular** (fohl-ihck-yoo-lahr) **phase.** Estrogen is the predominant hormone during the follicular phase. The stage of the estrous cycle in which the corpus luteum is present is called the **luteal** (loo-tē-ahl) **phase.** Progesterone is the predominant hormone during the luteal phase.

The estrous cycle is also divided into phases. The phases of the estrous cycle are:

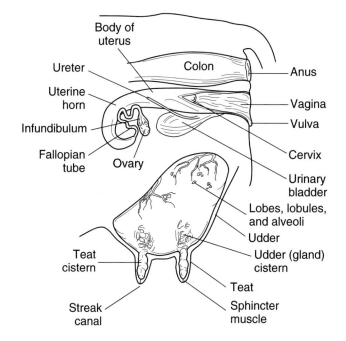

FIGURE 12–7 Cross-section of cow reproductive system

proestrus (prō-ehs-truhs) = period of the estrous cycle before sexual receptivity. The prefix **pro-** means before. Proestrus involves the secretion of follicle-stimulating hormone (FSH) by the anterior pituitary gland, which causes the follicles to develop in the ovary. FSH stimulates ovarian release of estrogen, which helps prepare the reproductive tract for pregnancy.

estrus (ehs-truhs) = period of the estrous cycle in which the female is receptive to the male. During estrus FSH levels decrease and luteinizing hormone (LH) levels increase, causing the graafian follicle to rupture and release its egg (ovulation). **Ovulation** (ohv-yoo-lā-shuhn) occurs and the animal is said to be in **heat.** This is also called **standing heat.**

metestrus (meht-ehs-truhs) = period of the estrous cycle after sexual receptivity. The CL forms and produces progesterone during this phase. Progesterone ensures proper implantation and maintenance of pregnancy. If an animal is not pregnant, the CL decreases in size and becomes a **corpus albicans** (kōr-puhs ahl-bih-kahnz) or white body. Metestrus may be followed by diestrus, estrus, pregnancy, or false pregnancy.

diestrus (dī-ehs-truhs) = period of the estrous cycle after metestrus. This short phase of inactivity and quietness is seen in polyestrous animals before the onset of proestrus.

anestrus (ahn-ehs-truhs) = period of the estrous cycle when the animal is sexually quiet. This long phase of quietness is seen in seasonally polyestrous and seasonally monestrous animals.

What do the following "estrous terms" mean?

monestrous (mohn-ehs-truhs) = pertaining to having one estrous or heat cycle per year.

polyestrous (pohl-ē-ehs-truhs) = pertaining to having more than one estrous or heat cycle per year.

spontaneous ovulators (spohn-tā-nē-uhs oh-vū-lā-tōrz) = ovum release occurs cyclically

induced ovulators (ihn-doosd oh-vū-lā-tōrz) = ovum is released only after copulation; also called **reflex ovulators**. Examples include cats, rabbits, ferrets, llamas, and mink.

seasonally = pertaining to a specific time of year. For example, queens are seasonally polyestrus. The queen will have multiple cycles from the seasons January through October (varies depending on photoperiod and geographic location).

PUTTING THE PARTS TOGETHER

Mating

For reproduction to occur the male and female of the species must **copulate** (kohp-yoo-lāt) to allow the sperm from the male to be transferred into the female. Copulation and **coitus** (kō-ih-tuhs) are terms that mean sexual intercourse. **Intromission** (ihn-trō-mihs-shuhn) is insertion of the penis into the vagina. During coitus, the male **ejaculates** (ē-jahck-yoo-lātz) into the female's vagina. **Ejaculat/o** means to throw or hurl out. Sperm travel through the vagina, into the uterus, and into the oviduct. When a sperm penetrates the ovum that is descending down the oviduct, fertilization occurs. **Fertilization** (fər-tihl-ih-zā-shuhn) is the union of ovum and sperm. If more than one ovum is passing down the oviduct when sperm are present, multiple fertilizations can take place.

Afterbirth

The **placenta** (plah-sehn-tah) is the female organ of mammals that develops during pregnancy and joins mother and offspring for exchange of nutrients, oxygen, and waste products. The placenta is also called the **afterbirth.** The **umbilical** (uhm-bihl-ih-kuhl) **cord** is the structure that forms where the fetus communicates with the placenta. The **umbilicus** (uhm-bihl-ih-kuhs) is the structure that forms on the abdominal wall where the umbilical cord was connected to the fetus.

Duo

Twins are two offspring born from the same gestational period. Fraternal twins are two offspring born during the same labor resulting from fertilization of separate ova by separate sperm. Maternal twins are two offspring born during the same labor resulting from fertilization of a single ovum by a single sperm (the fertilized egg separates into two parts).

Copulation and coitus are two words used to describe sexual intercourse. Other terms used in reference to mating include:

mount (mownt) 5 preparatory step to animal mating that involves one animal climbing on top of another animal or object; used as an indicator of heat (Figure 12–8)

tie (tī) 5 period of copulation between a male and female canine during which the two animals are locked together due to penile erectile tissue

conception (kohn-sehp-shuhn) is the beginning of a new individual resulting from fertilization. The fertilized egg is called a **zygote** (zī-gōt). Cell division of sex cells in which the cell receives half the chromosomes from each parent is called **meiosis** (mī-ō-sihs). **Implantation** (ihm-plahn-tā-shuhn) is the attachment and embedding of the zygote within the uterus. The developing zygote after implantation is called an **embryo** (ehm-brē-ō). An unborn animal is called a **fetus** (fē-tuhs), and this term is used more towards the end of pregnancy. Pregnancy is the time period between conception and parturition.

The umbilicus is also called the **navel** (nā-vuhl). **Umbilic/o** is the combining form for navel.

The placenta and its associated structures are referred to as the fetal membranes. The innermost membrane enveloping the embryo in the uterus is called the **amnion** (ahm-nē-ohn). The amnion forms the

The embryo has distinct layers that give rise to various tissue types. Layers of the embryo include the **ectoderm** (ehck-tō-dərm) or outer layer of the embryo, the **mesoderm** (mē-sō-dərm) is the middle layer of the embryo, and the **endoderm** (ehn-dō-dərm) is the inner layer of the embryo.

FIGURE 12–8 Cattle mounting. *Source:* Courtesy of Ron Fabrizius, DVM, Diplomat ACT

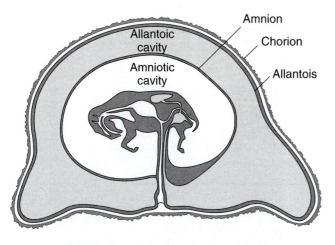

FIGURE 12–9 Equine placenta and fetus

amniotic cavity and protects the fetus by engulfing it in amniotic fluid. The amnion may be referred to as the **amniotic sac** or **bag of waters.** The **allantois** (ahl-ahn-tō-ihs) is the innermost layer of the placenta and forms a sac between itself and the amnion where fetal waste products accumulate. The **chorion** (kōr-ē-ohn) is the outermost layer of the placenta (Figure 12–9).

The ruminant placenta has elevations on it that are located on the maternal or fetal surface. The **cotyledon** (koht-eh-lē-dohn) is the elevation of the ruminant placenta that is on the fetal surface and adheres to the maternal caruncle. Cotyledons are also called **buttons.** The **caruncle** (kahr-uhnk-uhl) is the fleshy mass on the maternal ruminant placenta that attaches to the fetal cotyledon. The caruncle and cotyledon together form the **placentome** (plahs-ehn-tohm) (Figure 12–10).

Pregnancy

Pregnancy (prehg-nahn-sē) is the condition of having a developing fetus in the uterus and is the time period between conception and parturition. The combining form for pregnancy is **pregn/o. Cyesis** (sī-eh-sihs) also means pregancy. **Gestation** (jehs-tā-shuhn) is the period of development of the fetus in the uterus from conception to parturition and is the term more commonly used in reference to animals. The combining forms for gestation are **gest/o** and **gestat/o.** Gestation periods vary in length among animal species. A fetus is said to be **viable** (vī-ah-buhl) when it is capable of living outside of the mother. Viability depends on the species and age/weight of the fetus.

TERMS RELATED TO GESTATIONAL/PARTURITIONAL STATUS

gravid/o (grahv-ih-dō) = combining form for pregnant
-para (pahr-ah) = suffix meaning bearing a live fetus

nulligravida (nuhl-ih-grahv-ih-dah) = never been pregnant; **nulli-** means none; **nullipara** (nuhl-ih-pahr-ah) = female who has never borne a viable fetus
primigravida (preh-mih-grahv-ih-dah) = first pregnancy; **primi-** means first; **primipara** (prī-mihp-ah-rah) = female who has borne one offspring
multigravida (muhl-tih-grahv-ih-dah) = multiple pregnancies; **multi-** means many; **multiparous** (muhl-tihp-ah-ruhs) = female who has borne multiple offspring during different gestations. A **litter** (liht-tər) is multiple offspring born during the same labor.
viviparous (vī-vihp-ahr-uhs) = bearing live young. **Vivi-** is the prefix for live.
oviparous (ō-vihp-ahr-uhs) = bearing eggs. **Ovi-** means egg.

Birth

The act of giving birth is called **parturition** (pahr-tyoo-rihsh-uhn). Parturition is also referred to as **labor. Part/o** is the combining form for giving birth. The period before the onset of labor is the **antepartum** (ahn-tē-pahr-tuhm) period; the period right after labor is the **postpartum** (pōst-pahr-tuhm) period.

Parturition is divided into stages. The first stage of labor involves **dilation** (dī-lā-shuhn) of the cervix. Dilation is the act of stretching. **Dilatation** (dihl-ah-tā-shuhn) is stretching beyond normal. The second stage of labor involves uterine contractions of increasing frequency and strength and expulsion of the fetus. Expulsion of the fetus is termed **delivery** (deh-lihv-ər-ē). The third stage of labor involves separation of the placenta from the uterus.

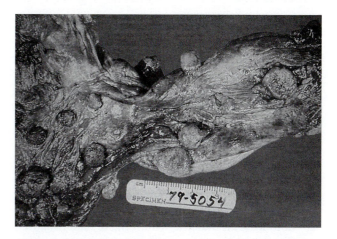

FIGURE 12–10 Bovine placenta. Caruncles are present on the maternal aspect of this bovine placenta. Think c**a**runcle = m**a**ternal; cotyl**e**don = f**e**tus. *Source:* Courtesy of Ron Fabrizius, DVM, Diplomat ACT

The postpartum period begins after delivery of the fetus. The newborn fetus is considered a **neonate** (nē-ō-nāt). The neonatal period varies from species to species, but usually indicataes under 4 weeks in length. The first stools of a newborn that consist of material collected in the intestine of the fetus is called the **meconium** (meh-kō-nē-uhm).

After delivery of the fetus the uterus of the mother will return to its normal size. The process of returning the muscular walled hollow organ that houses and nourishes the fetus is called **uterine involution** (yoo-tər-ihn ihn-vō-loo-shuhn). The mammary glands of the mother will secrete **colostrum** (kuh-lohs-truhm), which is a thick fluid that contains nutrients and antibodies needed by the neonate. **Lactation** (lahck-tā-shuhn) is the process of forming and secreting milk.

Presentation

Presentation is the orientation of the fetus prior to delivery. Presentation varies with species. For example, in cattle and sheep the fetus adopts an **anterior presentation,** in which the legs and head are directed towards the cervix. In swine anterior and posterior presentations are considered normal. **Posterior presentation** is when the pelvis is directed towards the cervix. **Transverse presentation** involves the fetus lying across the cervix and normal parturition is not achieved. **Breech presentation** is where the tail of the fetus is presented first and may or may not obstruct delivery.

▶ Test Me: Reproductive System ◀

Diagnostic procedures performed on the reproductive system include

- **radiography** (rā-dē-ohg-rah-fē) = procedure in which film is exposed as ionizing radiation passes through the patient and shows the internal body structures in profile.
- **ultrasound** (uhl-trah-sownd) = diagnostic test using high-frequency waves to evaluate internal structures. Ultrasonography works well in evaluating the uterus during pregnancy because the fluid present in the uterus helps define structures.
- **amniocentesis** (ahm-nē-ō-sehn-tē-sihs) = surgical puncture with a needle through the abdominal and uterine walls to obtain amniotic fluid to evaluate the fetus

▶ Path of Destruction: Reproductive System ◀

Pathologic conditions of the reproductive system include
- **ovarian cyst** (ō-vahr-ē-ahn sihst) = collection of fluid or solid material within the female gonad
- **fibroid** (fī-broyd) = benign tumor arising from the smooth muscle of the uterus; also called **leiomyoma** (lī-ō-mī-ō-mah)

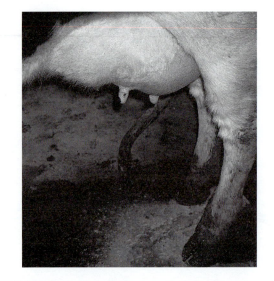

FIGURE 12–11 Mastitis in the rear quarter of a cow udder. *Source:* Courtesy of Ron Fabrizius, DVM, Diplomat ACT

- **metritis** (meh-trī-tihs) = inflammation of the uterus
- **vaginitis** (vahj-ih-nī-tihs) = inflammation of the tube that connects the cervix to outside the body
- **mastitis** (mahs-tī-tihs) = inflammation of the mammary gland(s) (Figure 12–11).
- **pyometra** (pī-ō-mē-trah) = pus in the uterus (Figure 12–12).
- **dystocia** (dihs-tōs-ah) = difficult birth; the female is having difficulty in expelling the fetus
- **transmissible venereal tumor** (trahnz-mihs-ih-buhl vehn-ēr-ē-ahl too-mər) = naturally occurring, sexually transmitted tumor of dogs that affects the external genitalia and other mucous membranes; abbreviated TVT
- **pseudocyesis** (soo-dō-sī-ē-sihs) = false pregnancy; also called **pseudopregnancy.** A behavioral and physical syndrome (most commonly seen in bitches 2–3 months after estrus) in which mammary glands develop, lactation occurs, and moth-

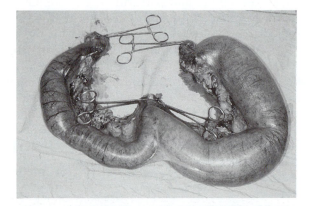

FIGURE 12–12 Pyometra in a bitch. *Source:* Courtesy of Ron Fabrizius, DVM, Diplomat ACT

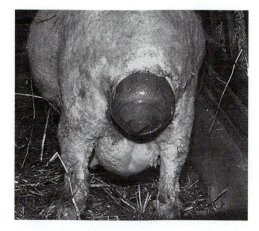

FIGURE 12–13 Vaginal prolapse in an ewe. *Source:* Courtesy of Ron Fabrizius, DVM, Diplomat ACT

ering behaviors occur. **Cyesis** means pregnancy; **pseudo-** is the prefix for false.

- **uterine prolapse** (yoo-tər-ihn prō-lahps) = falling forward or sinking down of the uterus through the vaginal wall
- **vaginal prolapse** (vah-jih-nahl prō-lahps) = falling forward or sinking down of the vagina through the vaginal wall (Figure 12–13).
- **cervicitis** (sihr-vih-sī-tihs) = inflammation of the neck of the uterus
- **hermaphroditism** (hər-mahf-rō-dih-tihzm) = condition of an individual having both ovarian and testicular tissue. **Pseudohermaphroditism** (soo-dō-hər-mahf-rō-dih-tihzm) is the condition of having gonads of one sex, but having the physical characteristics of both sexes.
- **supernumerary** (soo-pər-nū-mahr-ē) = more than the normal number. Supernumerary teats is a condition in which an animal has more than the normal number of nipples (commonly seen in ruminants).
- **abortion** (ah-bōr-shuhn) = termination of pregnancy
- **azoospermia** (ā-zō-ō-spər-mē-ah) = absence of sperm in the semen
- **oligospermia** (ohl-ih-gō-spər-mē-ah) = deficient amount of sperm in semen; **oligo-** is the prefix for scant or few
- **sterility** (stər-ihl-ih-tē) = inability to reproduce
- **orchitis** (ōr-kī-tihs) = inflammation of the gonads of the male; also called **testitis**
- **cryptorchidism** (krihp-tōr-kih-dihzm) = developmental defect in which one or both testis fails to descend into the scrotum; also termed **undescended testicle(s).** Animals may be unilaterally or bilaterally cryptorchid. Unilaterally cryptorchid is sometimes called **monorchid** (mohn-ōr-kihd).

- **epididymitis** (ehp-ih-dihd-ih-mī-tihs) = inflammation of the tube at the upper part of each testis that secretes semen, stores sperm before ejaculation, and provides a passageway for sperm out of the testes (epididymis).
- **scrotal hydrocele** (skrō-tahl hī-drō-sēl) = hernia of fluid in the testes or along the spermatic cord. **Hydro-** is the prefix for water; **-cele** is the suffix meaning hernia (Figure 12–14).
- **prostatitis** (prohs-tah-tī-tihs) = inflammation of the gland surrounding the urethra that secretes a thick fluid to aid motility of sperm
- **benign prostatic hypertrophy** (beh-nīn prohs-tah-tihck hī-pər-trō-fē) = abnormal noncancerous enlargement of the gland surrounding the urethra that secretes a thick fluid to aid motility of sperm; also called **prostatomegaly** or **enlarged prostate**
- **phimosis** (fih-mō-sihs) = narrowing of the skin of the prepuce so it cannot be retracted to expose the glans penis
- **paraphimosis** (pahr-ah-fih-mō-sihs) = retraction of the skin of the prepuce causing a painful swelling of the glans penis that prevents the penis from being retracted
- **pneumovagina** (nū-mō-vah-jī-nah) = conformational defect in the perineum of cows and mares that allows air to enter the vagina; also called **windsuckers**
- **priapism** (prī-ahp-ihzm) = persistent penile erection not associated with sexual excitement
- **fetal defects** may occur. **Teratogens** (tər-ah-tō-jehnz) are substances that produce defects in the fetus. **Mutagens** (mū-tah-jehnz) are substances that produce change or create genetic abnormalities.
- **ectopic pregnancy** (ehck-tohp-ihck prehg-nahn-sē) = fertilized ovum implanted outside the uterus

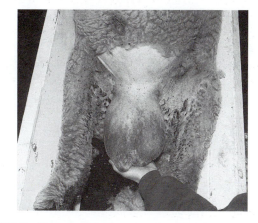

FIGURE 12–14 Scrotal hydrocele in a ram. *Source:* Courtesy of Ron Fabrizius, DVM, Diplomat ACT

▶ Procede With Caution: Reproductive System ◀

Procedures performed on the reproductive system include

- **mastectomy** (mahs-tehck-tō-mē) = surgical removal of the mammary gland or breast
- **cesarean section** (sē-sā-rē-ahn sehck-shuhn) = delivery of offspring through an incision in the maternal abdominal and uterine wall; also called a **C-section** (Figure 12–15).
- **episiotomy** (eh-pihz-ē-oht-ō-mē) = surgical incision of the perineum and vagina to facilitate delivery of the fetus and to prevent damage to maternal structures
- **ovariohysterectomy** (ō-vahr-ē-ō-hihs-tər-ehck-tō-mē) = surgical removal of the ovaries, oviducts, and uterus; also called a **spay;** abbreviated OHE or OVH
- **hysterectomy** (hihs-tər-ehck-tō-mē) = surgical removal of the uterus
- **oophorectomy** (ō-ohf-ō-rehck-tō-mē) = surgical removal of the ovary (ovaries)
- **electroejaculation** (ē-lehck-trō-ē-jahck-yoo-lā-shuhn) = method of collecting semen for artifi-

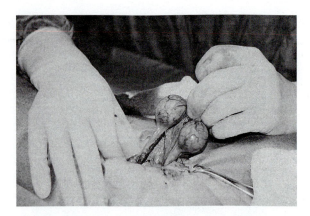

FIGURE 12–16 Canine castration. *Source:* Lodi Veterinary Hospital, S.C.

cial insemination or examination in which electrical stimulation is applied to the nerves to promote ejaculation. Electroejaculation is achieved by use of an **electroejaculator** (ē-lehck-trō-ē-jahck-yoo-lā-tər) which is a probe and power source to apply current to the nerves that promote ejaculation.

- **fetotomy** (fē-toh-tō-mē) = cutting apart of a fetus to enable removal from the uterus; also called **embryotomy** (ehm-brē-ah-tō-mē)
- **neuter** (nū-tər) = to sexually alter; usually used to describe sexually altering of males. An animal that is not neutered is **intact** (ihn-tahck) or has reproductive capability.
- **orchidectomy** (ōr-kih-dehck-tō-mē) = surgical removal of the testis (testes); also known as **orchectomy** (ōr-kehck-tō-mē), **orchiectomy** (ōr-kē-ehck-tō-mē), or **castration** (kahs-trā-shuhn) (Figure 12–16)
- **vasectomy** (vah-sehck-tō-mē) = sterilization of a male in which a portion of the vas deferens is surgically removed yet the animal may retain its libido. **Libido** (lih-bē-dō) is sexual desire.

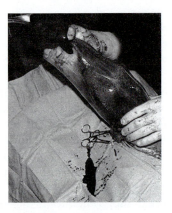

FIGURE 12–15 C-section in a bitch. *Source:* Courtesy of Ron Fabrizius, DVM, Diplomat ACT

REVIEW EXERCISES

Multiple Choice—Choose the correct answer.

1. The inner layer of the uterus is called the
   a. endohysteria
   b. myometrium
   c. perimetrium
   d. endometrium

2. The area between the vaginal orifice/scrotum and the anus is called the
   a. clitoris
   b. perineum
   c. vulva
   d. inguinal area

3. Copulation is also called
   a. coitus
   b. impotence
   c. sterility
   d. zygote

4. The act of giving birth is
   a. freshening
   b. calving
   c. gestation
   d. parturition

5. A difficult birth is known as
   a. dystocia
   b. dyshernia
   c. dyspartia
   d. dyslaboratum

6. A false pregnancy is also called
   a. pseudo
   b. pseudopara
   c. pseudocyesis
   d. pseudogestia

7. A condition of an individual having both ovarian and testicular tissue is termed
   a. hemisexual
   b. hermaphroditism
   c. supernumerary
   d. orchioovaris

8. Pyometra is
   a. pus in the uterus
   b. increased temperature of the uterus
   c. tumors in the uterus
   d. necrosis of the uterus

9. The innermost membrane enveloping the embryo in the uterus is the
   a. allantois
   b. umbilicus
   c. amnion
   d. chorion

10. Attachment and embedding of the zygote within the uterus is
    a. zygotion
    b. conception
    c. fertilization
    d. implantation

Word Building—Build the following terms using word parts.

pus in the uterus:

    word part for pus _____
    word part for uterus _____
    term for pus in the uterus _____

near the ovary:

    word part for near _____
    word part for ovary _____
    term for near the ovary _____

pertaining to the urinary and reproductive systems:

    word part for urinary _____
    word part for reproductive system _____
    word part for pertaining to _____
    term for pertaining to the urinary and reproductive systems _____

capable of stimulating milk production:

    word part for milk _____
    word part for producing _____
    word part for pertaining to _____
    term for capable of stimulating milk production _____

surgical removal of the uterus :

    word part for uterus _____
    word part for surgical removal _____
    term for surgical removal of the uterus _____

Case Studies—Define the terms underlined in each case study.

A 12 yr old F miniature poodle was presented with lethargy, anorexia, and PU/PD. The bitch had been in proestrus 1 mo prior to presentation. On PE, the dog was pyrexic and tachypnic. Abdominal palpation yielded an enlarged uterus, and a purulent vaginal discharge was noted. Blood was collected for a CBC and chem panel. The CBC results included leuko-cytosis with a left shift (a left shift is the increase in band neutrophils in peripheral blood due to many causes including infection). The dx of pyometra was made, antibiotics were started, and the dog was scheduled for an emergency OHE.

yr_____

F _____

lethargy_____

anorexia _____

PU/PD:

  abbreviations for_____

  meaning _____

bitch _____

proestrus _____

PE _____

pyrexic _____

tachypnic _____

abdominal _____

palpation _____

uterus _____

purulent _____

vaginal _____

CBC _____

chem panel _____

leukocytosis _____

dx _____

pyometra _____

OHE _____

A 10 yr old intact M German shepherd was presented with stranguria and hematuria. PE revealed pyrexia, anorexia, and a stiff gait. On rectal palpation, the veterinarian discovered that the prostate gland was bilaterally enlarged. Upon review of the record it was noted that this dog has had a hx of recurrent UTIs. Radiographs were taken, and prostatomegaly was noted. After a dx of prostatitis was made, the dog was scheduled to be neutered the following day.

yr _____

intact _____

M _____

stranguria _____

hematuria _____

PE _____

pyrexia _____

anorexia _____

gait _____

rectal palpation _____

prostate gland _____

bilaterally _____

hx _____

recurrent _____

UTI _____

radiographs _____

prostatomegaly _____

dx _____

prostatitis _____

neutered _____

another term for neutering in male dogs _____

A 2 yr old Holstein <u>cow</u> was examined because the farmer noted that she was <u>off feed</u>. PE revealed that the cow had a slightly elevated rectal temperature. The farmer told the veterinarian that this cow had stepped on her <u>teat</u> previously, but it had appeared to be healing. The <u>udder</u> was palpated and it was not warm to the touch or swollen. Milk was expressed from each <u>quarter</u>, and the milk appeared more watery than normal. A <u>CMT</u> paddle test was performed on the milk and moderate precipitation was noted (Figure 12–17). The <u>diagnosis</u> of <u>mastitis</u> was made, and milk samples were taken for <u>culture</u>. Antibiotic treatment was started pending culture results and milking hygiene was discussed with the farmer.

cow _____

off feed _____

  medical term for off feed _____

teat _____

udder_____

quarter _____

CMT _____

diagnosis _____

mastitis _____

culture_____

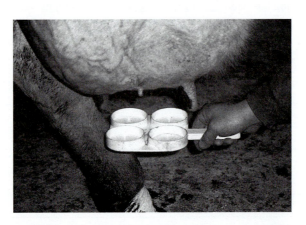

FIGURE 12–17 CMT paddle. The CMT (California Mastitis Test) paddle is a cow side test to detect mastitis. *Source:* Courtesy of Ron Fabrizius, DVM, Diplomat ACT

A 5 yr old quarter horse <u>mare</u> had <u>foaled</u> 10 hours ago. The foal appeared normal, and the mare was letting the foal suckle. The mare had not moved since foaling, and the owner was concerned because the mare seemed quieter than normal. PE revealed a mildly elevated rectal temperature, <u>vaginal discharge</u> was noted, milk production seemed adequate, and the milk appeared normal. The owner was questioned whether the mare had passed its <u>placenta</u>, and the owner was unaware if she had. The stall was examined for remnants of the placenta, and none were found. The veterinarian was concerned that the mare had a <u>retained placenta</u> and that an infection may be starting. The equine placenta is normally passed within a few hours of foaling and since it has been at least 10 hours <u>postpartum</u>, an injection of oxytocin was given slow <u>IV</u>. Blood was drawn for a <u>CBC</u> to assess <u>leukocyte</u> numbers. The owner was advised to watch for the passing of the placenta, and general hygiene was discussed with the client.

mare _____

foaled _____

vaginal discharge_____

placenta_____

retained placenta_____

postpartum _____

IV _____

CBC _____

leukocyte_____

 **Nerves of Steel**

► Identify and describe the structures and functions of the nervous system
► Identify the divisions of the nervous system and describe the structures and functions of each
► Recognize, define, spell, and pronounce terms related to the diagnosis, pathology, and treatment of the nervous system

FUNCTIONS OF THE NERVOUS SYSTEM

The **nervous** (nər-vuhs) **system** coordinates and controls body activity. It detects and processes internal and external information and formulates appropriate responses.

STRUCTURES OF THE NERVOUS SYSTEM

The major structures of the nervous system are the brain, spinal cord, peripheral nerves, and sensory organs. The two major divisions of the nervous system are

• the **central nervous system** (sehn-trahl nər-vuhs sihs-tehm) = portion of the nervous system that consists of the brain and spinal cord; abbreviated CNS
• the **peripheral nervous system** (pehr-ihf-ər-ahl nər-vuhs sihs-tehm) = portion of the nervous system that consists of the cranial and spinal nerves, autonomic nervous system, and ganglia; abbreviated PNS (Figure 13–1).

Back to Basics

The basic unit of the nervous system is the **neuron** (nū-rohn). There are three types of neurons based on their function.

1. **sensory neurons** (sehn-sōr-ē nū-rohnz) = nerves that carry sensory impulses towards the CNS; also called **afferent** (ahf-fər-ahnt) or **ascending tracts** because they carry information towards the CNS. Sensory information such as sound or light is converted into electrical impulses so that the nerves can transport them.
2. **associative neurons** (ahs-ō-shē-ah-tihv nū-rohnz) = nerves that carry impulse from one neuron to another; also called **connecting neurons**
3. **motor neurons** (mō-tər nū-rohnz) = nerves that carry impulses away from the CNS and toward the muscles and glands; also called **efferent** (ē-fər-ahnt) or **descending tracts** because they carry information away from the CNS

The parts of a neuron are the cell body, one or more dendrites, one axon, and terminal end fibers. The cell body or **soma** (sō-mah) has a nucleus and is responsible for maintaining the life of the neuron. The **dendrites** (dehn-drīts) are root-like structures that receive impulses and conduct them toward the cell body. The combining form for dendrite is **dendr/o**. The **axon** (ahcks-ohn) is a single process that extends away from the cell body and conducts impulses away from the cell body. The combining form **ax/o** means axis or mainstem. The terminal end fibers are the branching fibers that lead the impulse away from the axon and toward the synapse.

177

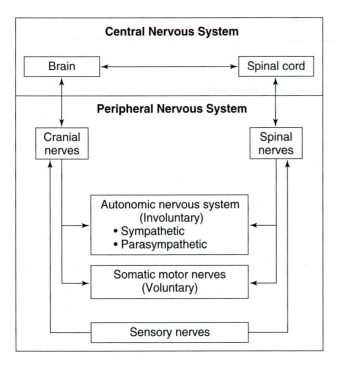

FIGURE 13–1 Divisions of the nervous system

The dendrites and axons are also called **nerve fibers.** Bundles of nerve fibers bound together by specialized tissues are called **nerves** or **nerve trunks.** Nerve fibers are covered with a tube like membrane called the **neurolemma** (nū-rō-lehm-ah) or neurilemma (nū-rih-lehm-ah). Neuron cell bodies grouped together within the CNS are called **nuclei** (nū-klē-ī) and those outside the CNS are called **ganglia** (gahng-glē-ah) (Figure 13–2).

The Gap

The space between two neurons or between a neuron and receptor is the **synapse** (sihn-ahps). The combining forms for this space or point of contact are **synaps/o** and **synapt/o.** A chemical substance called a **neurotransmitter** (nū-rō-trahnz-miht-ər) allows the impulse to move across the synapse from one neuron to another. There are different neurotransmitters each with specific functions.

Supporting Role

The **neuroglia** (nū-rohg-lē-ah) or **glial** (glē-ahl) **cells** are the supportive cells of the nervous system. Glial cells consist of **astrocytes** (ahs-trō-sītz), **microglia** (mī-krō-glē-ah), **oligodendrocytes** (ohl-ih-gō-dehn-drō-sītz), and **Schwann** (shwahn) **cells.** The combining form **gli/o** means glue, which helps explain the function of glial cells.

▶ **astr/o** means star. Astrocytes are star-shaped and cover the capillary surface of the brain and help form the blood-brain barrier in the CNS.

▶ **micro-** means small. Microglia are small cells that are phagocytic in nature to help fight infection in the CNS.
▶ **oligo-** means few, **dendr/o** means branching, and **-cyte** means cell. Oligodendrocytes are cells with few branches that hold the nerve fibers together and help form myelin in the CNS.
▶ Schwann cells (named after a German anatomist) help form myelin in the PNS.

Surrounding Some

Myelin (mī-e-lihn) is a protective covering over some nerve cells including parts of the spinal cord, white matter of the brain, and most peripheral nerves. Myelin is also referred to as the **myelin sheath.** Myelin serves as an electrical insulator. Nerves that are described as **myelinated** (mī-lihn-āt-ehd) have myelin surrounding them. Myelin gives nerve fibers a white color, and myelinated nerves are referred to as **white matter.** The **gray matter** does not contain myelinated fibers; hence it is darker in color. The **gray matter** is composed of cell bodies, branching dendrites, and neuroglia.

Myelin is interrupted at regular intervals along the length of a fiber by gaps called **nodes of Ranvier** (nōdz of rohn-vē-ā). Ionic exchange takes place at the nodes of Ranvier. Some nerve fibers have a very thin layer of

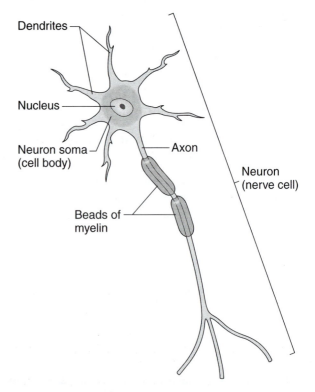

FIGURE 13–2 Structures of a neuron. Dendrites conduct impulses **t**owards the neuron cell body, while **a**xons conduct impulses **a**way from the cell body

myelin and are referred to as **nonmyelinated.** The autonomic nervous system contains nonmyelinated fibers.

Carry an Impulse

A **nerve** (nərv) is one or more bundles of impulse-carrying fibers that connect the CNS to the other parts of the body. **Neur/i** and **neur/o** are combining forms for nerve or nerve tissue. Terms used in reference to nerves include:

▶ A **tract** (trahck) is a group of nerve fibers located within the CNS. Ascending tracts (groups going *up*) carry nerve impulses toward the brain, while descending tracts (groups going *down*) carry nerve impulses away from the brain.

▶ A **ganglion** (gahng-glē-ohn) is a knot-like mass of neuron cell bodies located *outside* the CNS. More than one ganglion are termed **ganglia** or less frequently **ganglions.** The combining forms **gangli/o** and **ganglion/o** mean knot-like mass of nerve cell bodies located outside the CNS.

▶ A **plexus** (plehck-suhs) is a network of intersecting nerves or vessels. **Plexi** (plehck-sī) are networks of intersecting nerves or vessels. **Plex/o** is the combining form for a plexus or network.

▶ **Innervation** (ihn-nər-vā-shuhn) is the supply or stimulation of a body part through the action of nerves.

▶ **Receptors** (rē-cehp-tərz) are sensory organs that receive external stimulation and transmit that information to the sensory neurons. There are different types of receptors. For example, **nociceptive** (nō-sih-sehp-tihv) receptors are pain receptors, while **proprioceptive** (prō-prē-ō-sehp-tihv) receptors are spatial orientation or perception of movement receptors.

▶ A **stimulus** (stihm-yoo-luhs) is something that excites or activates. Multiple excitations or activations are termed **stimuli** (stihm-yoo-lī).

▶ An **impulse** (ihm-puhlz) is a wave of excitation transmitted through nervous tissue.

▶ A **reflex** (rē-flehcks) is an automatic, involuntary response to change. Reflex actions include heart and respiration rates, coughing, and sneezing.

CENTRAL NERVOUS SYSTEM

The central nervous system, cerebrospinal system, or CNS is made up of the brain and the spinal cord. The combining form for brain is **encephal/o** and the combining form for spinal cord is **myel/o** (remember myel/o also means bone marrow). The CNS contains both white and gray matter; white matter is caused by myelinated fibers and gray matter is caused by nerve cell bodies.

Membrane

The brain and spinal cord are encased in connective tissue called the **meninx** (meh-nihcks). Since this connective tissue has three layers, it more commonly is referred to by its plural form of **meninges** (meh-nihn-jēz). **Mening/o** and **meningi/o** are combining forms for the layers of connective tissue enclosing the CNS.

The three layers of the meninges are

1. **dura mater** (doo-rah mah-tər) = thick, tough, outermost layer of the meninges. **Dura** means tough and this layer of the meninges is tough and strong. **Dur/o** is the combining form for dura mater. The dura mater is also referred to as **pachymeninx** (pahck-ē-meh-nihcks); the prefix **pachy-** means thick.

2. **arachnoid** (ah-rahck-noyd) **membrane** = second layer of the meninges. **Arachn/o** means spider and the arachnoid membrane resembles a spider web. The arachnoid membrane is loosely attached to the other layers of the meninges to allow space between the layers.

3. **pia mater** (pē-ah mah-tər) = third and deepest layer of the meninges. **Pia** means soft or tender, and this layer of the meninges is soft with a rich supply of blood vessels. The pia is very adherent to the CNS. The pia mater and arachnoid membranes are collectively called the **leptomeninges** (lehp-tō-meh-nihn-jēz) (Figure 13–3).

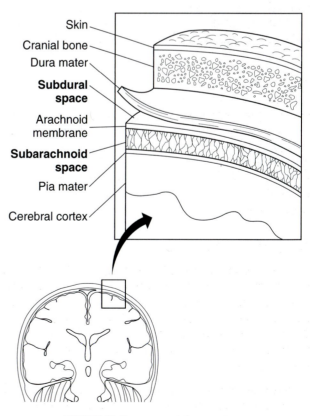

Skin
Cranial bone
Dura mater
Subdural space
Arachnoid membrane
Subarachnoid space
Pia mater
Cerebral cortex

FIGURE 13–3 Layers of the meninges

180 *Chapter 13*

There are terms used to describe the location of structures in reference to the meninges. **Epidural** (ehp-ih-doo-rahl) means located above the dura mater (**epi-** is the prefix meaning above or upon). The **subdural** (suhb-doo-rahl) space is the area located below the dura mater and above the arachnoid membrane (**sub-** is the prefix meaning below). The **subarachnoid** (suhb-ah-rahck-noyd) **space** is the area located below the arachnoid membrane and above the pia mater. The subarachnoid space contains cerebrospinal fluid.

Fluid

Cerebrospinal fluid (sər-ē-brō-spīn-ahl flū-ihd) is the clear, colorless ultrafiltrate that nourishes, cools, and cushions the CNS. Cerebrospinal fluid is abbreviated CSF. CSF is produced by special capillaries within the ventricles of the brain. The ventricles of the brain are cavities. Vascular folds of the pia mater in the ventricles called the **choroid plexus** (kōr-oyd plehck-suhs) secrete CSF.

Brain

The brain is the enlarged and highly developed portion of the CNS that lies in the skull and is the main site of nervous control. The portion of the skull that encloses the brain is called the **cranium** (krā-nē-uhm). **Crani/o** is the combining form for skull. **Intracranial** (ihn-trah-krā-nē-ahl) means within the cranium.

The brain is commonly divided into parts based either on functional group or on location. **Encephal/o** is the combining form for brain; however, other combining forms that refer to specific regions of the brain are also used.

The divisions of the brain (Figure 13–4) based on functional group include

▶ **cerebrum** (sər-ē-bruhm). The combining form for cerebrum is **cerebr/o**. The cerebrum is the largest part of the brain responsible for receiving and processing stimuli, initiating voluntary movement, and storing information. The **cerebral cortex** (outer region) is made up of gray matter and is arranged in folds (Figure 13–5). The elevated portions of the cerebral cortex are known as **gyri** (jī-rī). The combining form **convolut/o** means coiled and **gyr/o** means folding. The grooves of the cerebral cortex are called fissures or **sulci** (suhl-sī). The combining form **sulc/o** means groove. The medullary substance of the cerebrum is made up of white matter. The cerebrum also has small cavities called **ventricles** (vehn-trih-kuhlz) (Figure 13–6). There are four ventricles of the brain: two lateral, a third, and a fourth. The ventricles of the brain (and central canal of the spinal cord) are lined with a membrane called the **ependyma** (eh-pehn-dih-mah).

▶ **cerebellum** (sehr-eh-behl-uhm). The combining form for cerebellum is **cerebell/o**. The cerebellum is the second largest part of brain that coordinates muscle activity for smooth movement. The cerebellum has a inner portion called the **vermis** (vər-mihs) because it is worm-like and other portions divided into right and left cerebellar hemispheres.

▶ **brainstem** (brān-stehm). The brainstem is the stalk-like portion of the brain that connects the cerebral hemisphere with the spinal cord. The brainstem is

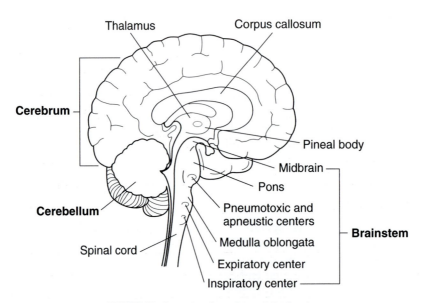

FIGURE 13–4 Sagittal section of the brain

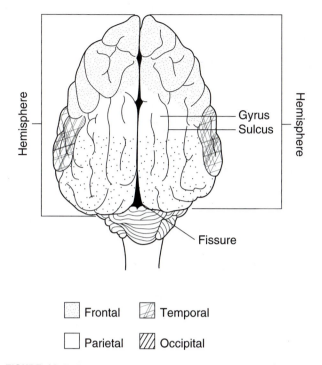

Frontal ▨ Temporal

Parietal ▥ Occipital

FIGURE 13–5 Dorsal view of the cerebral cortex. The cerebrum is divided into right and left hemispheres. Each hemisphere is further divided into lobes and each lobe is named for the bone plate covering it: **frontal** (frohn-tahl) lobe = most cranial lobe controls motor function; **parietal** (pahr-ī-ih-tahl) lobe = receives and interprets sensory nerve impulses; **occipital** (ohcks-ihp-ih-tahl) lobe = most caudal lobe that controls vision; **temporal** (tehmp-ruhl) lobe = laterally located lobe that controls hearing and smell.

comprised of the pons, medulla oblongata, midbrain, and interbrain. The **interbrain** contains structures like the pituitary gland, hypothalamus, and thalamus. These structures are responsible for endocrine activity, regulation of thirst and water balance, and regulation of body temperature. The **midbrain** contains structures responsible for visual and auditory reflexes, posture, and muscle control. The **pons** (pohnz) is the bridge at base of the brain that allows nerves to cross over so that one side of the brain controls the opposite side of the body. The **medulla oblongata** (meh-duhl-ah ohb-lohng-gah-tah) is the cranial continuation of the spinal cord that controls basic life functions. Divisions of the brain are also based on location (Figure 13–7 and Table 13–1).

Some parts of the brain are not names based on location or division, but rather based on how they look. The vermis of the cerebellum is one example. Another example is the hippocampus, a portion of the limbic system that involves memory. The **hippocampus** (hihp-ō-kahm-puhs) is shaped like a seahorse, which exists in mythology as a sea monster with the head of a horse and the tail of a fish and as an actual sea creature. The name hippocampus comes from the Greek hippos, meaning horse, and kampos, meaning sea monster.

Spinal Cord

The **spinal cord** is the continuation of the medulla oblongata. The spinal cord passes through an opening in the occipital bone called the **foramen magnum** (fōr-ā-mehn mahg-nuhm). **Foramen** means passage and **magnum** means great. The spinal cord carries all the tracts

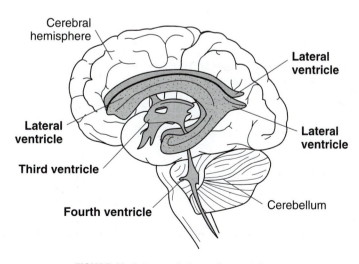

FIGURE 13–6 Lateral view of ventricles

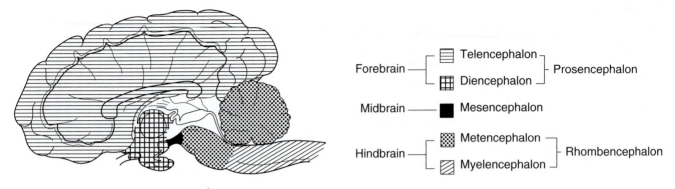

FIGURE 13–7 Brain divisions based on location

that influence the innervation of the limbs and lower part of the body. In addition, the spinal cord is the pathway for impulses going to and from the brain. **Myel/o** is the combining form for spinal cord (remember myel/o is also the combining form for bone marrow).

The spinal cord is part of the central nervous system, which means it is surrounded by the meninges and is protected by cerebrospinal fluid. The gray matter of the spinal cord is located in the internal portion and is not protected by myelin. The white matter of the spinal cord is located in the external portion and is myelinated.

The spinal cord contains two areas of swelling. The medical term for swelling (normal or abnormal) is **intumescence** (ihn-too-meh-sehns). The swelling is due to an increase in white matter and cells bodies that are associated with the innervation of the limbs. One swelling occurs in the area of C6-T2, which is known as the **cervical intumescence.** The other swelling occurs in the area of the L4-caudal segment, which is known

as the **lumbosacral intumescence.** The cranial parts of the spinal cord have tracts with fibers from the cranial and caudal portions, but the caudal parts of the spinal cord only have tracts with fibers from the caudal portions. Therefore, as the spinal cord proceeds caudally, its cross-sectional area decreases. At the level of the cranial lumbar vertebrae, the spinal cord becomes cone shaped. This cone-shaped segment is called the **conus medullaris. Conus** means cone. At the caudal end of the spinal cord (caudal lumbar, sacral, and coccygeal nerve segments), the tracts terminate and the spinal nerves fan outward and backward giving the appearance of a horse's tail. This collection of spinal roots at the caudal part of the spinal cord is termed the **cauda equina** (kaw-dah ē-kwī-nah). The cauda equina includes the conus medullaris to the caudal vertebrae. The thread-like tapering section of the cauda equina is known as the **filum terminale. Filum** means thread-like structure. The filum terminale attaches the conus medullaris to the caudal vertebrae.

TABLE 13–1

Brain Divisions Based on Location

Brain Part	Divisions	Some Components
Forebrain; prosencephalon (prōs-ehn-cehf-ah-lohn)	**telencephalon** (tē-lehn-cehf-ah-lohn)	cerebral cortex and olfactory brain, **limbic** (lihm-bihck) system (affects emotion and behavior); limbic means border
	diencephalon (dī-ehn-cehf-ah-lohn)	thalamus, epithalamus, and hypothalamus limbic system optic chiasm (the crossing of the vision nerves); **chiasm** means crossing
midbrain; mesencephalon (mēz-ehn-cehf-ah-lohn)	**mesencephalon**	vision and hearing bodies, posture, and muscle control; limbic system
hindbrain; rhombencephalon (rohmb-ehn-cehf-ah-lohn)	**metencephalon** (meht-ehn-cehf-ah-lohn)	cerebellum and pons
	myelencephalon (mī-lehn-cehf-ah-lohn)	medulla oblongata

Cushions

The spinal cord is housed within the vertebrae to protect it from injury. The vertebrae are protected from each other by **intervertebral** (ihn-tər-vər-tē-brahl) discs or disks that are located between each vertebra (Figure 13–8). Intervertebral discs are layers of fibrocartilage that form pads separating and cushioning the vertebrae from each other. The center of the intervertebral disc is gelatinous (**nucleus pulposus**) and the outer layer is fibrous (**annulus fibrosis**).

PERIPHERAL NERVOUS SYSTEM

The peripheral nervous system or PNS consists of the cranial nerves, the autonomic nervous system, and the spinal nerves.

Cranial Nerves

There are twelve pairs of **cranial nerves** that originate from the undersurface of the brain. The cranial nerves are generally named for the area or function they serve and are represented by Roman numerals (Table 13–2).

Spinal Nerves

The **spinal nerves** arise from the spinal cord. With the exception of some cervical and coccygeal nerves, the spinal nerves are paired and emerge from behind the vertebra of the same number and name. The first cervical vertebra (C1 or the atlas) has a pair of spinal nerves emerging

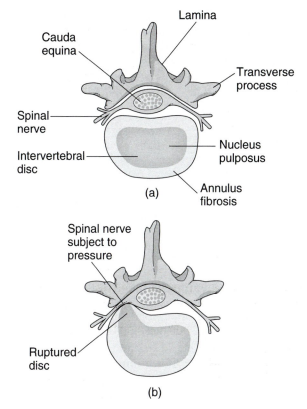

FIGURE 13–8 Intervertebral disc. (a) Normal intervertebral disc; (b) Ruptured disc

TABLE 13–2
Cranial Nerves

Cranial Nerve	Name	Function
I	olfactory	conduct sensory impulses from nose to the brain (smell)
II	optic	conduct sensory impulses from eyes to the brain (vision)
III	oculomotor	send motor impulses to the external eye muscles (dorsal, medial and ventral rectus, ventral oblique and levator superioris) as well as some internal eye muscles
IV	trochlear	send motor impulses to one external eye muscle (dorsal oblique)
V	trigeminal	three branches: ophthalmic = sensory to cornea; maxillary = motor to upper jaw; mandibular = motor to lower jaw
VI	abducens	innervate two muscles of the eye (retractor bulbi and lateral rectus)
VII	facial	motor to facial muscles, salivary glands, lacrimal glands, and taste sensation to anterior two-thirds of tongue
VIII	acoustic or cochleovestibular	two branches: cochlear = sense of hearing; vestibular = sense of balance
IX	glossopharyngeal	motor to the parotid glands, taste sensation to posterior third of tongue, sensory to the pharyngeal mucosa, and motor to the pharyngeal muscles
X	vagus	sensory to part of pharynx, larynx, and parts of thoracic and abdominal viscera; motor for swallowing and voice production
XI	spinal accessory	motor to shoulder muscles
XII	hypoglossal	motor to the muscles that control tongue movement

from in front of it and behind it; therefore, there are eight cervical spinal nerves. C1 exits the spinal canal in a foramen in the wing of the atlas and C2 emerges at the C1-C2 intervertebral foramen. C8 emerges caudal to vertebra C7, and T1 emerges caudal to T1 vertebra. The coccygeal vertebrae usually have fewer pairs of nerves than the number of vertebrae.

> Spinal nerves are named from where they arise from the spinal cord. C represents cervical, T represents thoracic, L represents lumbar, S represents sacral, and Co or Cy represents coccygeal (or Cd represents caudal). These letter abbreviations are followed by numbers to represent the vertebral area from where the nerve exits the spinal cord. C4 would represent cervical spinal nerve 4, T11 would represent thoracic spinal nerve 11, and so on.

Spinal nerves have dorsal and ventral roots. The **dorsal root** enters the dorsal portion of the spinal cord and carries afferent or sensory impulses from the periphery toward the spinal cord. The **ventral root** emerges from the ventral portion of the spinal cord and carries efferent or motor impulses from the spinal cord to muscle fibers or glands (Figure 13–9).

Spinal nerves supply sensory and motor fibers to the body region associated with their emergence from spinal cord. After spinal nerves exit the spinal cord, they branch to form the peripheral nerves of the trunk and limbs. Several spinal nerves may join together to form a single peripheral nerve. This braiding of branches is called a **plexus**. Each appendage is innervated by a plexus. Each forelimb is supplied from nerves that arise from the **brachial** (brā-kē-ahl) **plexus** (C6-T2), and each hindlimb is supplied from nerves that arise from the **lumbosacral plexus** (L4-S3). Brachial means the arm and lumbosacral means the loin and sacrum.

Autonomic Nervous System

The **autonomic nervous system** (aw-tō-nah-mihck nər-vuhs sihs-tehm) or ANS is that part of the peripheral nervous system that innervates smooth muscle, cardiac muscle, and glands. The two divisions of the autonomic nervous system are the **sympathetic nervous system** (sihm-pah-theh-tihck nər-vuhs sihs-tehm) and the **parasympathetic nervous system** (pahr-ah-sihm-pah-theh-tihck nər-vuhs sihs-tehm). The two divisions of the autonomic nervous system work together to maintain homeostasis within the body. **Homeostasis** (hō-mē-ō-stā-sihs) is the process of maintaining constant internal body environment.

sympathetic	emergency and stress response; "fight or flight"	↑ heart rate, respiratory rate and blood flow to muscles; ↓ gastrointestinal function; pupil dilation
parasympathetic	returns body to normal after stressful response; maintains normal body function	returns heart rate, respiratory rate, and blood flow to normal levels; returns normal gastrointestinal function; constricts pupil size to normal

▶ **Test Me: Nervous System** ◀

Diagnostic procedures performed on the nervous system include

- **pupillary light reflex** = response of pupil to a bright light source; abbreviated PLR. Light is

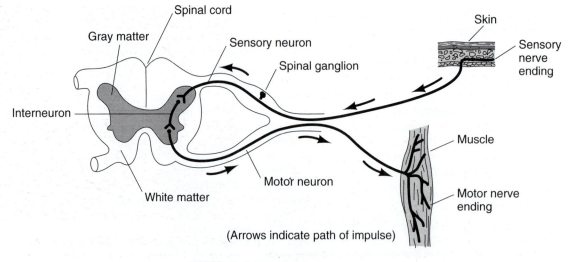

(Arrows indicate path of impulse)

FIGURE 13–9 Spinal nerve

Labels: Spinal cord, Gray matter, Sensory neuron, Spinal ganglion, Skin, Sensory nerve ending, Interneuron, Muscle, Motor neuron, Motor nerve ending, White matter

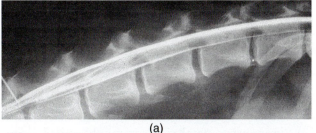

(a)

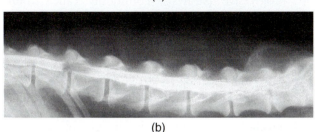

(b)

FIGURE 13–10 (a) Normal myelogram in a Labrador retriever; (b) Myelogram demonstrating disc disease (note that the dye does not flow continuously). *Source:* Photo by Anne E. Chauvet, DVM, Diplomate ACVIM–Neurology, University of Wisconsin School of Veterinary Medicine

shone in one eye and that eye (direct) and the opposite eye (consensual) should constrict. It is used to assess neurologic damage.

- **electroencephalography** (ē-lehck-trō-ehn-sehf-ah-lohg-rah-fē) = process of recording electrical activity of the brain; abbreviated EEG. An **electroencephalograph** (ē-lehck-trō-ehn-sehf-ah-lō-grahf) is the instrument used to record the electrical activity of the brain, and an **electroencephalogram** (ē-lehck-trō-ehn-sehf-ah-lō-grahm) is the record of the electrical activity of the brain.
- **myelography** (mī-eh-lohg-rah-fē) = diagnostic study of the spinal cord after injection of contrast material. A **myelogram** (mī-eh-lō-grahm) is the record of the spinal cord after injection of contrast material (Figure 13–10).
- **magnetic resonance imaging** and **computed axial tomography** will be covered in Chapter 16.
- **cerebrospinal fluid tap** (sər-ē-brō-spī-nahl flū-ihd tahp) = removal of cerebrospinal fluid; also called a CSF tap. CSF is obtained by inserting a needle or catheter into the **cisterna magna** (sihs-tər-nah mahg-nah) or lumbosacral area. The cisterna magna is the subarachnoid space located between the caudal surface of the cerebellum and the dorsal surface of the medulla oblongata. Intracranial pressures may also be measured before removal of CSF (Figure 13–11).
- **discography** (dihs-kō-grah-fē) = radiographic study of an intervertebral disc, after injection of contrast material into the disc; also spelled diskography. The record of this procedure is called a **discogram** (Figure 13–12).

Levels of Consciousness = descriptive terms used to describe mentation

- **conscious** (kohn-shuhs) = awake, aware, and responsive; also known as **alert**
- **BAR** = bright, alert, and responsive
- **lethargy** (lehth-ahr-jē) = drowsiness, indifference, and listlessness
- **obtunded** (ohb-tuhn-dehd) = depressed
- **disorientation** (dihs-ōr-ē-ehn-tā-shuhn) = condition in which the animal appears mentally confused
- **stupor** (stoo-pər) = impaired consciousness with unresponsiveness to stimuli
- **coma** (kō-mah) = deep state of unconsciousness

▶ The Path of Destruction: Nervous System ◀

Pathologic conditions of the nervous system include

- **syncope** (sihn-kō-pē) = fainting; sudden fall in blood pressure or cardiac systole resulting in cerebral anemia and more or less complete loss of consciousness
- **narcolepsy** (nahr-kō-lehp-sē) = syndrome of recurrent uncontrollable sleep episodes. The combining form **narc/o** means stupor and the suffix **-lepsy** means seizure (episode).
- **catalepsy** (kaht-ah-lehp-sē) = waxing rigidity of muscles accompanied with the animal having a trancelike state
- **cataplexy** (kaht-ah-plehck-sē) = sudden attacks of muscular weakness and hypotonia triggered by an emotional response
- **meningitis** (mehn-ihn-jī-tihs) = inflammation of the meninges
- **encephalopathy** (ehn-sehf-ah-lohp-ah-thē) = any disease of the brain
- **encephalitis** (ehn-sehf-ah-lī-tihs) = inflammation of the brain

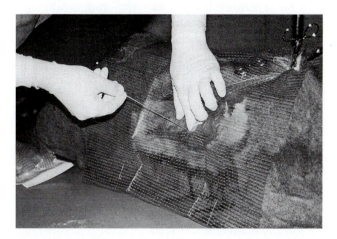

FIGURE 13–11 Lumbar CSF tap in a lion. *Source:* Photo by Anne E. Chauvet, DVM, Diplomate ACVIM–Neurology, University of Wisconsin School of Veterinary Medicine

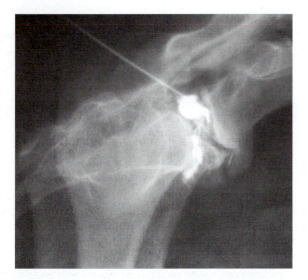

FIGURE 13–12 Discogram in a dog. *Source:* Photo by Anne E. Chauvet, DVM, Diplomate ACVIM–Neurology, University of Wisconsin School of Veterinary Medicine

- **hydrocephalus** (hī-drō-cehf-ah-luhs) = abnormally increased amount of cerebrospinal fluid within the ventricles of the brain; "water on the brain"
- **encephalocele** (ehn-sehf-ah-lō-sēl) = herniation of brain through a gap in the skull
- **meningocele** (meh-nihng-gō-sēl) = protrusion of the meninges through a defect in the skull or vertebrae
- **concussion** (kohn-kuhsh-uhn) = shaking of the brain caused by injury. The combining form **concuss/o** means shaken together.
- **contusion** (kohn-too-zhuhn) = bruising. The combining form **contus/o** means bruise.
- **hematoma** (hē-mah-tōh-mah) = mass or collection of blood. In the nervous system a hematoma is usually described by the area where it is found. An **epidural hematoma** (ehp-ih-doo-rahl hē-mah-toh-mah) is a collection of blood above or superficial to the dura mater. A **subdural hematoma** (suhb-doo-rahl hē-mah-toh-mah) is a collection of blood below the dura mater and above the arachnoid membrane.
- **amnesia** (ahm-nē-zē-ah) = memory loss
- **myelitis** (mī-eh-lī-tihs) = inflammation of the spinal cord (or bone marrow)
- **poliomyelitis** (pō-lē-ō-mī-eh-lī-tihs) = inflammation of the gray matter of the spinal cord. The combining form **poli/o** means gray.
- **polioencephalomyelitis** (pō-lē-ō-ehn-cehf-ah-lō-mī-eh-lī-tihs) = inflammation of the gray matter of the brain and spinal cord
- **polioencephalomalacia** (pō-lē-ō-ehn-cehf-ah-lō-mah-lā-shah) = abnormal softening of the gray matter of the brain; abbreviated PEM (Figure 13–13).

- **radiculitis** (rah-dihck-yoo-lī-tihs) = inflammation of the root of a spinal nerve. **Radicul/o** is the combining form for root.
- **polyneuritis** (poh-lē-nū-rī-tihs) = inflammation of many nerves
- **polyradiculoneuritis** (poh-lē-rah-dihck-yoo-lō-nū-rī-tihs) = inflammation of many peripheral nerves and spinal nerve roots that may lead to progressive paralysis; commonly called **coonhound paralysis** in coonhound dogs and **idiopathic polyradiculoneuropathy** in other dogs
- **demyelination** (dē-mī-eh-lih-nā-shuhn) = destruction or loss of myelin
- **neuralgia** (nū-rahl-jē-ah) = nerve pain
- **neuritis** (nū-rī-tihs) = inflammation of the nerve(s)
- **paralysis** (pahr-ahl-ih-sihs) = loss of voluntary movement or immobility. The suffix **-plegia** means paralysis. Paralysis is further described by the area(s) it involves.
- **tetraplegia** (teht-rah-plē-jē-ah) = paralysis of all four limbs; also called **quadriplegia** (kwohd-rih-plē-jē-ah)
- **hemiplegia** (hehm-ih-plē-jē-ah) = paralysis of one side of the body
- **paraplegia** (pahr-ah-plē-jē-ah) = paralysis of the lower body in bipeds/hind limbs in quadrupeds. In reference to the nervous system, the prefix **para-** means hind or lower portion.
- **monoplegia** (mohn-ō-plē-jē-ah) = paralysis of one limb
- **myoparesis** (mī-ō-pahr-ē-sihs) = weakness of muscles. The suffix **-paresis** means weakness. As is the case with **-plegia,** the suffix -paresis is modified to describe the area of weakness. **Hemiparesis** (hehm-ih-pahr-ē-sihs) is weakness on one side of the body; **paraparesis** (pahr-ah-pahr-ē-sihs) is weakness of the lower body in bipeds/hind limbs in quadrupeds.

FIGURE 13–13 Polioencephalomalacia in a sheep. Opisthotonos is one sign of PEM. *Source:* Courtesy of Ron Fabrizius, DVM, Diplomat ACT

- **ataxia** (ā-tahck-sē-ah) = without coordination; "stumbling"
- **paresthesia** (pahr-ehs-thē-zē-ah) = abnormal sensation. The suffix **-esthesia** means sensation or feeling. Abnormal sensations may include tingling, numbness, or burning and may be difficult to assess in animals. The combining forms for burning are **caus/o** and **caust/o.**
- **hyperesthesia** (hī-pər-ehs-thē-zē-ah) = excessive sensitivity
- **spasticity** (spahs-tihs-ih-tē) = a state of increased muscular tone
- **tremor** (treh-mər) = involuntary trembling
- **seizure** (sē-zhər) = sudden, involuntary contraction of some muscles caused by a brain disturbance; also called **convulsions.** The most common type of seizure is animals is **grand mal** (grahnd-mahl) in which the animal experiences loss of consciousness and muscle contractions. Other types of seizures include partial, which have a seizure focus that does not spread, and petit mal, which is a mild generalized seizure where loss of consciousness and generalized loss of muscle tone occur.
- **epilepsy** (ehp-ih-lehp-sē) = recurrent seizures of nonsystemic origin or of intracranial disease. Epilepsy may be described as **idiopathic** (ihd-ē-ō-pahth-ihck). Idiopathic means unknown cause or disease of an individual. **Idio-** is the prefix meaning individual.
- **hallucination** (hah-loo-sehn-ā-shuhn) = false sensory perception
- **hyperkinesis** (hī-pər-kihn-ē-sihs) = increased motor function or activity
- **hypnosis** (hihp-nō-sihs) = condition of altered awareness; "trancelike" state
- **astrocytoma** (ahs-trō-sī-tō-mah) = malignant intracranial tumor composed of astrocytes
- **chorea** (kōr-ē-ah) = repetitive, rhythmic contraction of limb or facial muscles; also called **myoclonus** (mī-ō-klō-nuhs); usually the result of distemper viral infection in dogs
- **choriomeningitis** (kōr-ē-ō-meh-nihn-jī-tihs) = inflammation of the choroid plexus and meninges
- **decerebration** (dē-sər-ē-brā-shuhn) = condition of loss of mental functions due to damage to the midbrain
- **discospondylitis** (dihs-kō-spohn-dih-lī-tihs) = destructive, inflammatory disorder that involves the intervertebral discs, vertebral end plates, and vertebral bodies.
- **encephalomalacia** (ehn-cehf-ah-lō-mah-lā-shē-ah) = abnormal softening of the brain
- **encephalomyelitis** (ehn-cehf-ah-lō-mī-ih-lī-tihs) = inflammation of the brain and spinal cord
- **encephalopathy** (ehn-cehf-ah-lah-pahth-ē) = disease of the brain

- **ptosis** (tō-sihs) = prolapse or drooping. The suffix **-ptosis** means prolapse, drooping, or falling downward; specifically refers to the upper eyelid
- **Horner's syndrome** (hōr-nərz sihn-drōm) = collection of signs relating to injury of the cervical sympathetic innervation to the eye; signs include sinking of the eyeball (enophthalmus), ptosis of the upper eyelid, pupil constriction, and prolapse of the third eyelid
- **intervertebral** (ihn-tər-vər-teh-brahl) **disc disease** = condition of pain and neurologic deficits resulting from the displacement of part or all of the material in the disc located between the vertebrae
- **macrocephaly** (mahck-rō-cehf-ah-lē) = abnormally large skull
- **microcephaly** (mī-krō-cehf-ah-lē) = abnormally small skull
- **meningioma** (meh-nihn-jē-ō-mah) = benign tumor of the meninges (Figure 13–14)
- **meningoencephalomyelitis** (meh-nihn-gō-ehn-cehf-ah-lō-mī-eh-lī-tihs) = inflammation of the meninges, brain, and spinal cord
- **meningoencephalitis** (meh-nihn-gō-ehn-cehf-ah-lī-tihs) = inflammation of the meninges and brain
- **myelopathy** (mī-eh-lah-pahth-ē) = disease of the spinal cord (or bone marrow)
- **opisthotonos** (ohp-ihs-thoht-ō-nohs) = tetanic spasm in which the head and tail are bent dorsally and the back is arched
- **conscious proprioceptive deficit** (kohn-shuhs prō-prih-ō-cehp-tihv dehf-ih-siht) = neurologic

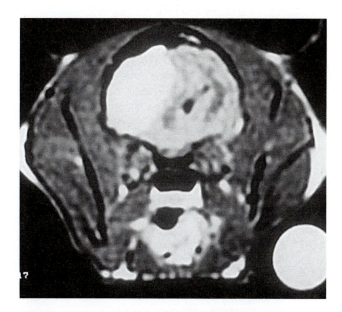

FIGURE 13–14 Computed tomogram of a cat with a meningioma. *Source:* Photo by Anne E. Chauvet, DVM, Diplomate ACVIM–Neurology, University of Wisconsin School of Veterinary Medicine

defect in which the animal appears to not know where its limbs are; abbreviated CP deficit, "knuckling" (Figure 13–15).

- **spina bifida** (spī-nah bihf-ih-dah) = congenital anomaly in which the spinal canal does not close over the spinal cord. The combining form **bifid/o** means split or cleft.
- **vestibular disease** (vehs-tihb-yoo-lahr dih-zēz) = neurologic disorder characterized by head tilt, nystagmus, rolling, falling, and circling. **Nystagmus** (nī-stahg-muhs) is involuntary, rhythmic movement of the eye and will be discussed in Chapter 14.
- **cerebellar hypoplasia** (sehr-eh-behl-ahr hī-pō-plā-zē-ah) = smaller than normal cerebellum; seen in cats secondary to feline panleukopenia virus, which leads to incoordination
- **cervical vertebral malformation** = abnormal formation or instability of the caudal cervical vertebrae that causes ataxia and incoordination; seen more frequently in horses and dogs; also called **wobbler's syndrome**
- **roaring** = noisy respiration caused by air passing through a narrowed larynx in horses; common term for equine laryngeal hemiplegia because of nerve fiber degeneration of the left recurrent laryngeal nerve

What side are you on?

Pathologic conditions of the nervous system may involve lesions that cause abnormal clinical signs on the same side or opposite side that the lesion occurs. In describing lesions of the nervous system the terms ipsilateral and contralateral are used. **Ipsi-** is the prefix meaning the same and **contra-** is the prefix meaning opposite. **Ipsilateral** (ihp-sē-laht-ər-ahl) means on the same side and **contralateral** (kohn-trah-laht-ər-ahl) means on the opposite side.

FIGURE 13–15 CP deficit in a dog

Seizure stages

Seizures are divided into stages to aide in identifying when they may start or end. These stages are:

pre-ictal (prē-ihck-tahl) = the period before a seizure; also called the **aura** (aw-rah). An animal may pace, excessively lick, fly bite, or seem anxious during this stage.

ictus (ihck-tuhs) = the attack or actual seizure. An animal may convulse, lose control of their excretory functions, shake, and appear confused during this stage.

post-ictal (pōst-ihck-tahl) = the period after a seizure. An animal may appear obtunded, tired, fearful, or anxious during this stage.

► Procede With Caution: Nervous System ◄

Procedures performed on the nervous system include

- **neurectomy** (nū-rehck-tō-mē) = surgical removal of a nerve
- **neuroplasty** (nū-rō-plahs-tē) = surgical repair of a nerve
- **neurorrhaphy** (nū-rōr-ah-fē) = suturing the ends of a severed nerve
- **neurotomy** (nū-roht-ō-mē) = surgical incision or dissection of a nerve
- **anesthesia** (ahn-ehs-thē-zē-ah) = absence of sensation. An **anesthetic** (ahn-ehs-theht-ihck) is a substance used to induce anesthesia. There are different types of anesthesia. Some types include: **topical anesthesia** = absence of sensation after a substance has been applied to the skin or external surface; **local anesthesia** = absence of sensation after chemical injection to an adjacent area; **epidural anesthesia** = absence of sensation to a region after injection of a chemical into the epidural space; **general anesthesia** = absence of sensation and consciousness
- **dysesthesia** (dihs-eh-stēsh-ah) = impaired sensation
- **analgesia** (ahn-ahl-jēz-ē-ah) = without pain. Analgesia is used to describe pain relief, which is different from anesthesia which is absence of sensation. **Endorphins** (ehn-dōr-fihnz) are natural, opioid-like chemicals that are produced in the brain that raise the pain threshold.
- **disc fenestration** (dihsk fehn-ih-strā-shuhn) = removal of intervertebral disc material by perforating and scraping out its contents
- **laminectomy** (lahm-ihn-ehck-tō-mē) = surgical removal of the posterior portion of the vertebral body to relieve pressure on the spinal cord
- **neuroanastomosis** (nū-rō-ahn-ahs-tō-mō-sihs) = connecting nerves together

REVIEW EXERCISES

Multiple Choice—Choose the correct answer.

1. The space between two neurons or between a neuron and a receptor is a/an
    a. synapse
    b. ganglion
    c. axon
    d. dendrite

2. Maintaining a constant internal environment is termed
    a. osmosis
    b. homeosmosis
    c. homeostasis
    d. ipsistasi

3. Inflammation of the root of a spinal nerve is
    a. myelitis
    b. radiculitis
    c. polyneuritis
    d. poliomyelitis

4. The three-layered membrane lining the CNS is called the
    a. hippocampus
    b. pons
    c. myelin
    d. meninges

5. The protective sheath that covers some nerve cells of the spinal cord, white matter of the brain, and most peripheral nerves is called
    a. pia
    b. dura
    c. glia
    d. myelin

6. The division of the autonomic nervous system that is concerned with body functions under emergency or stressful situations is the
    a. peripheral
    b. central
    c. sympathetic
    d. parasympathetic

7. A network of intersecting nerves is a
    a. bundle
    b. trunk
    c. tract
    d. plexus

8. What type of neuron carries impulses away from the CNS and toward the muscles?
    a. afferent (sensory)
    b. efferent (motor)
    c. sympathetic
    d. parasympathetic

9. An automatic, involuntary response to change is called a/an
    a. impulse
    b. stimulus
    c. reflex
    d. receptor

10. The largest portion of the brain that is involved with thought and memory is the
    a. cerebrum
    b. cerebellum
    c. brainstem
    d. spinal cord

Matching—Match the term in Column I with the definition in Column II.

Column I	Column II
_____ conscious	a. depressed
_____ BAR	b. impaired consciousness with unresponsiveness to stimuli
_____ coma	c. bright, alert, and responsive
_____ lethargy	d. deep state of unconsciousness
_____ obtunded	e. awake, aware, and responsive; also known as alert
_____ disorientation	f. condition in which the animal appears mentally confused
_____ stupor	g. drowsiness, indifference, and listlessness

Word Scramble—Use the definitions to unscramble the terms.

incision into a nerve	tronumoye
period before a seizure	uaar
disease of the spinal cord (or bone marrow)	pthayyelom
passage	fmnraoe
paralysis to the lower limbs in bipeds/hind limbs in quadrupeds	ppaaaliger
opposite	rcoatn
recurrent seizures of non-systemic origin	yspelipe

Case Studies—Define the terms underlined in each case study.

A 3 yr old F cocker spaniel was presented to the clinic for <u>convulsions</u>. A thorough history was taken including a description of what the convulsion episodes were like, possible toxin exposure, and eating history. PE revealed that the animal was <u>obtunded</u>, but otherwise in normal health. The <u>neurologic</u> examination was normal. Blood was drawn for CBC and chem panel. The CBC was normal, and the chem panel did not show any evidence of <u>renal</u> disease, <u>hepatic</u> disease, or <u>hypoglycemia</u>. Urine was collected for <u>UA</u> via <u>cystocentesis</u>. The results of the UA were normal. The dog's history was examined, and it was noted that the dog's vaccinations were current. The <u>signalment</u>, history, and <u>clinical</u> signs were used to <u>diagnose idiopathic epilepsy</u>. Diagnostic testing, including <u>CSF tap</u> and <u>EEG</u>, were recommended to the owners, but were declined at this time. The owners were advised to have the dog <u>spayed</u> (hormone levels may contribute to <u>seizure</u> activity) and to monitor the animal's activity for recurrence of seizures. If the seizures recur and persist, <u>anticonvulsant</u> medication will be used in an attempt to control them.

convulsions _____

obtunded _____

neurologic _____

renal _____

hepatic _____

hypoglycemia _____

UA_____

cystocentesis _____

signalment _____

clinical _____

diagnose _____

idiopathic _____

epilepsy _____

CSF tap _____

EEG _____

spayed _____

seizure _____

anticonvulsant _____

A 5 yr old <u>M/N</u> dachshund was presented to the clinic with a history of not being able to walk up stairs or jump on the bed. The owner states that the dog has decreased its eating and is more <u>lethargic</u> than normal. PE revealed an <u>obese</u> dog that had normal <u>vital signs</u>. The neurologic examination revealed normal <u>cranial nerves</u> and <u>CP deficit</u> present on both hind limbs. <u>Patellar reflexes</u> were <u>hyporeflexive</u> with the right side worse than the left. <u>Anal</u> tone was adequate. <u>Radiographs</u> were recommended to assess whether the dog had calcified or herniated <u>intervertebral discs</u>. Abnormal discs were noted in the <u>lumbar</u> region. The dog was referred for <u>myelography</u>. The <u>myelogram</u> confirmed that the dog had <u>herniated</u> discs in the lumbar region, and surgery was recommended. <u>Disc fenestration</u> surgery was performed on the following day. The dog went home with orders for strict cage rest and rechecks at the referring veterinarian.

M/N _____

lethargic _____

obese _____

vital signs _____

cranial nerves _____

CP deficit _____

patellar reflexes _____

hyporeflexive _____

anal _____

radiographs _____

intervertebral discs _____

lumbar _____

myelography _____

myelogram _____

herniated _____

disc fenestration _____

14 Seeing and Hearing

Objectives _In this chapter, you should learn to:_

▶ Identify and describe the structures and functions of the eyes and ears
▶ Recognize, define, spell, and pronounce terms related to the diagnosis, pathology, and treatment of eye and ear disorders

FUNCTIONS OF THE EYE

The **ocular** (ohk-yoo-lahr) **system** is responsible for vision. The **eyes** are the receptor organs for sight. Combining forms for the eye or sight include **opt/i, opt/o, optic/o, ocul/o,** and **ophthalm/o.** Extraocular (ehcks-trah-ohk-yoo-lahr) means outside the eyeball, while **intraocular** (ihn-trah-ohk-yoo-lahr) means within the eyeball. **Periocular** (pehr-ē-ohck-yoo-lahr) means around the eyeball.

STRUCTURES OF THE EYE

Accessories

The accessory structures of an organ are referred to as **adnexa** (ahd-nehck-sah). **Stroma** (strō-mah) is another term used to describe the supporting tissue of an organ. The adnexa of the eye include the orbit, eye muscles, eyelids, eyelashes, conjunctiva, and lacrimal apparatus (Figure 14–1).

▶ **orbit** (ohr-biht) = bony cavity of the skull that contains the eyeball. The term **periorbita** (pehr-ih-ōr-bih-tah) means eye socket.
▶ **eye muscles** = seven major muscles are attached to each eye making range of movement possible (two oblique muscles, four rectus muscles, and the retractor bulbi). The muscles of both eyes work together in coordinated movements to make normal

binocular vision possible. **Binocular** (bi-nohck-yoo-lahr) means both eyes. The **extrinsic muscles** (ehcks-trihn-sihck muhs-uhlz) are six muscles that attach the outside of the eyeball to the bones of the orbit. The **levator palpebrae muscles** (lē-vā-tər pahl-pē-brā muhs-uhlz) are muscles that raise the upper eyelid.

▶ **eyelids** = each eye has an upper and lower eyelid to protect the eye from injury, foreign material, and

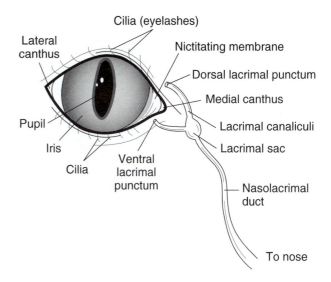

FIGURE 14–1 Adnexa of the eye

excessive light. The combining form **blephar/o** (blehf-ah-rō) means eyelid. **Palpebra** (pahl-pē-brah) is another term used for the eyelid; plural is **palpebrae** (pal-pē-brā). **Palpebral** (pahl-pē-brahl) means pertaining to the eyelid.

▶ The angle where the upper and lower eyelids meet is termed the **canthus** (kahn-thuhs). The combining form **canth/o** means corner of the eye. The medial canthus is the corner of the eye nearest the nose (also called the inner canthus). The lateral canthus is the corner of the eye farthest away from the nose (also called the outer canthus).

▶ The **tarsal** (tahr-sahl) **plate** (or **tarsus**) is the platelike framework within the upper and lower eyelides that provides stiffness and shape. The combining form **tars/o** means edge of the eyelid or "ankle" joint

▶ **Meibomian** (mī-bō-mē-ahn) **glands** are the sebaceous glands on the margins of each eyelid; also called **tarsal glands.**

▶ **eyelashes** = the edge of each eyelid has hair-like structures called **cilia** (sihl-ē-ah). The cilia or eyelashes protect the eye from foreign material.

▶ **conjunctiva** (kohn-juhnk-tī-vah) = mucous membrane that lines the underside of each eyelid. The conjunctiva forms a protective covering of the exposed surface of the eyeball when the eyelids are closed. The combining form **conjunctiv/o** means conjunctiva. The **nictitating membrane** (nihck-tih-tāt-ihng mehm-brān) is the conjunctival fold attached at the medial canthus that moves across the cornea when the eyelids close; it is also called the **third eyelid** or **nictitans** (nihck-tih-tahnz) or **haws** (hawz).

▶ **lacrimal apparatus** (lahck-rih-mahl ahp-ah-raht-tuhs) = structures that produce, store, and remove tears. **Lacrimation** (lahck-rih-mā-shuhn) is the condition of normal tear secretion. The combining forms **lacrim/o** and **dacry/o** mean teardrop, tear duct, or lacrimal duct. The lacrimal glands are glands that secrete tears. The **lacrimal canaliculi** (kahn-ah-lihck-yoo-lī) are the ducts at the medial canthus that collect tears and drain them into the lacrimal sac. The **lacrimal sac** or **dacryocyst** (dahck-rē-ō-sihst) is the enlargement that collects tears at the upper portion of the tear duct. The **nasolacrimal** (nā-sō-lahck-rih-mahl) **duct** is the passageway that drains tears into the nose. The **dorsal punctum** (dōr-sahl puhnk-tuhm) is the small spot near the upper medial canthus where the nasolacrimal duct begins; the **ventral punctum** (vehn-trahl puhnk-tuhm) is the small spot near the lower medial canthus where the nasolacrimal duct begins. A **punctum** (puhnk-tuhm) is a point or small spot.

Eyeball

The **eyeball** or **globe** is a sphere with multilayered walls. These walls are the sclera, choroid, and retina. Another term used for the eyeball is **orb** (ōrb) (Figure 14–2).

SCLERA

The **sclera** (sklehr-ah) is the fibrous outer layer of the eye that maintains the shape of the eye. It is sometimes referred to as the **white of the eye.** The combining form **scler/o** means sclera or hard.

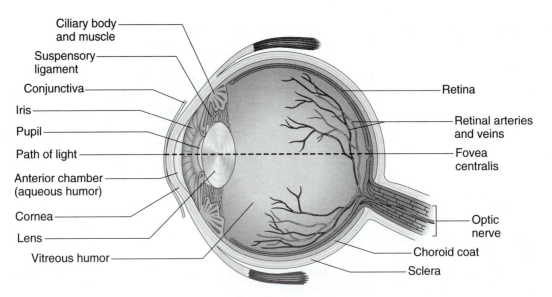

FIGURE 14–2 Cross-section of the eyeball

Labels: Ciliary body and muscle · Suspensory ligament · Conjunctiva · Iris · Pupil · Path of light · Anterior chamber (aqueous humor) · Cornea · Lens · Vitreous humor · Retina · Retinal arteries and veins · Fovea centralis · Optic nerve · Choroid coat · Sclera

The anterior portion of the sclera is transparent and is called the **cornea** (kōr-nē-ah). The cornea provides most of the optical power of the eye. The combining forms **corne/o** and **kerat/o** both mean cornea. **Descemet's membrane** (dehs-eh-māz mehm-brān) is the innermost or deepest layer of the cornea.

CHOROID

The **choroid** (kō-royd) is the opaque middle layer of the eyeball that contains blood vessels and supplies blood for the entire eye. **Opaque** (ō-pāk) means that light cannot pass thorough. The **tapetum lucidum** (tah-pē-duhm loo-sehd-uhm) is the brightly colored iridescent reflecting tissue layer of the choroid of most species. The tapetum lucidum is also called **choroid tapetum.** The **tapetum nigrum** (tah-pē-duhm nī-gruhm) is the black pigmented tissue layer of the choroid in some species. Tapetum is the medical term for a layer of cells. The combining form **choroid/o** means choroid. Associated with the choroid are the iris, pupil, lens, and ciliary muscles.

The **iris** (ī-rihs) is the pigmented muscular layer of the choroid that surrounds the pupil. The iris is composed of muscle fiber rings that contract to increase the size of the pupil and thus regulate the amount of light entering the lens. The **corpora nigra** (kōr-pōr-ah nī-grah) is the black pigmentation at the edge of the iris in equine and ruminants. The combining forms **ir/i, ir/o, irid/o,** and **irit/o** refer to the iris of the eye.

The **pupil** (pū-pihl) is the circular opening in the center of the iris. The combining forms **pupill/o** and **core/o** mean pupil. Muscles within the iris control the amount of light entering the pupil. To decrease the amount of light entering the eye, the iris muscles contract and make the opening smaller. Making the opening smaller is called **constriction;** when used in reference to pupillary constriction it is called **miosis** (mī-ō-sihs). To increase the amount of light entering the eye, the iris muscles relax and make the opening larger. Making the opening larger is called **dilation;** when used in reference to pupillary dilation it is called **mydriasis** (mih-drī-ah-sihs).

The **lens** is the clear, flexible, curved capsule located behind the iris and pupil. The shape of the lens is altered by the ciliary muscles. The varying shape of the lens affects the angle at which light rays enter the retina. The combining form **phac/o** means lens of the eye.

The **ciliary** (sihl-ē-ər-ē) **body** is the thickened extension of the choroid that assists in accommodation or adjustment of the lens. The **ciliary muscles,** located within the ciliary body, are muscles that adjust the shape and thickness of the lens. These adjustments make it possible for the lens to refine the focus of light rays on the retina.

Working Together Parts of the sclera and choroid are sometimes referred to together. Examples of these terms are:
iridocorneal (ihr-ihd-ō-kōr-nē-ahl) = pertaining to the iris and cornea
uvea (yoo-vē-ah) = term used to describe the iris, ciliary body, and choroid
limbus (lihm-buhs) = term used for the corneoscleral junction

---

RETINA

The **retina** (reht-ih-nah) is the nervous tissue layer of the eye that receives images. The retina is located in the vitreous chamber of the eye. The combining form **retin/o** means retina.

The retina contains specialized cells called rods and cones that convert nerve impulses from the eye to the brain via the optic nerve. **Rods** are specialized cells of the retina that react to light, and **cones** are specialized cells of the retina that react to color and fine detail.

The **optic** (ohp-tihck) **disk** is the region of the eye where nerve endings of the retina gather to form the optic nerve. It is also called the **blind spot** because it does not contain any rods or cones.

The **macula** (mahck-yoo-lah) **lutea** (lū-tē-ah) is a centrally depressed, clearly defined, yellow area in the center of the retina. The macula lutea surrounds a small depression called the fovea centralis. The **fovea centralis** (fō-vē-ah sehn-trah-lihs) contains the greatest concentration of cones in the retina. The combining form **macul/o** means spot; **lute/o** is the combining form for yellow; the combining form for pit is **fove/o.** The term macula is used in other parts of the body, such as the ear and kidney.

EYE CHAMBERS

The eye is divided into parts to make identification and location of structures easier. The **anterior segment** is the front one-third of the eyeball and is divided into anterior and posterior chambers. The **anterior chamber** is the eye cavity located behind the inner surface of the cornea and in front of the iris. The **posterior chamber** is the eye cavity located between the caudal surface of the iris and the cranial surface of the lens.

The anterior and posterior chambers of the eye are filled with a watery fluid called **aqueous humor** (ah-kwē-uhs hū-mər). Aqueous humor is the anterior segment fluid that nourishes the intraocular structures. The combining form **aque/o** means water. Humor is any clear body fluid.

The posterior two-thirds of the eyeball is called the **vitreous chamber.** The **vitreous humor** (viht-rē-uhs hū-mər) or vitreous is the soft, clear, jelly-like mass

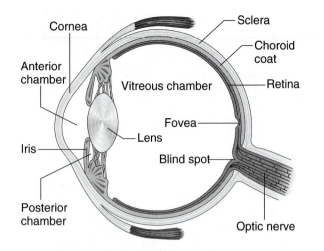

FIGURE 14–3 Chambers of the eye. The anterior and posterior chambers of the eye contain aqueous humor. The vitreous chamber of the eye contains vitreous humor.

that fills the vitreous chamber. The combining form **vitre/o** means glassy (Figure 14–3).

ACTION

Vision occurs when light rays enter the eye through the cornea, pass through the lens, and travel to the retina. The image is focused on the retina and is then transmitted to the optic nerve. Stimulations are transmitted from the optic nerve, to the optic chiasm, to the midbrain, and to the visual cortex of the occipital lobe of the cerebrum.

Accommodation (ah-kohm-ō-dā-shuhn) is the process of eye adjustments for seeing objects at various distances. These adjustments include dilation or constriction of the pupil, eye movement, and changes in lens shape.

Refraction (rē-frahck-shuhn) is the process of the lens bending the light rays to help them focus on the retina. Refraction is also referred to as **focusing.**

Convergence (kohn-vər-jehns) is simultaneous inward movement of both eyes. Convergence usually occurs in an effort to maintain single binocular vision as an object approaches.

Acuity (ah-kū-ih-tē) means sharpness or acuteness; usually used in reference to vision.

▶ **Test Me: Eyes** ◀

Diagnostic procedures performed on the eyes include

- **tonometry** (tō-nohm-eh-trē) = procedure using an instrument to indirectly measure intraocular pressure. **Intraocular pressure** is determined by the resistance of the eyeball to indentation by an applied force. Tonometry can be applanation,

which is directly placing the instrument (and weights) on the cornea and measuring resistance, or pneumatic, which is blowing a puff of air against the cornea to flatten it slightly to measure resistance. A **Schiotz** (shē-ohtz) **tonometer** is an example of an applanation tonometer (Figure 14–4).

- **slitlamp examination** = visual testing of the cornea, lens, fluids, and membranes of the interior of the eye using a narrow beam of light

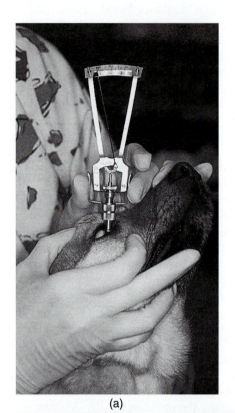

(a)

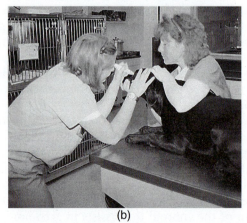

(b)

FIGURE 14–4 Detection of intraocular pressure by tonometry. (a) A Schiotz tonometer, an example of an applanation tonometer, measures corneal resistance to an applied force. (b) A pneumatic tonometer detects intraocular pressure by measuring corneal resistance to a puff of air. *Source (part b):* Lodi Veterinary Hospital, S.C.

- **pupillary light reflex** (pū-puhl-ār-ē līt rē-flehcks) = response of opening in the center of the iris to light; abbreviated PLR. When light is shown in pupil constriction should take place.
- **fluorescein** (fluhr-ō-sēn) **dye stain** = diagnostic test to detect corneal injury by placing dye onto the surface of the cornea
- **Schirmer** (shər-mər) **tear test** = diagnostic test using a graded paper strip to measure tear production
- **conjunctival** (kohn-junk-tī-vahl) **scrape** = diagnostic test using an instrument to peel cells from the conjunctiva so that the cells can be viewed microscopically.
- **electroretinography** (ē-lehck-trō-reh-tihn-ohg-rah-fē) = procedure of recording the electrical activity of the retina. An **electroretinogram** (ē-lehck-trō-reh-tihn-ō-grahm) is the record of electrical activity of the retina; abbreviated ERG.
- **goniometry** (gō-nē-ah-meh-trē) = procedure to measure the drainage angle of the eye. **Gon/i** is the combining form for angle or seed.
- **menace response** (mehn-ahs rē-spohnz) = diagnostic test to detect vision in which movement is made toward the animal to test if it will see movement and try to close its eyelids
- **palpebral** (pahl-pē-brahl) **reflex** = diagnostic test in which the eye should blink in response to touch to the medial canthus of the eye. This test is used in neurologic assessment of cranial nerves V and VII and to assess depth of anesthesia.
- **ophthalmoscope** (ohp-thahl-mō-skōp) is an instrument used for ophthalmoscopy.
- **ophthalmoscopy** (ohp-thahl-mō-skōp-ē) = procedure used to examine the interior eye structures; may be direct or indirect (Figure 14–5).

When describing the eyes, the abbreviations OD, OS, and OU are used. OD means right eye (oculus dexter), OS means left eye (oculus sinister), and OU means both eyes (oculus uterque). Similarly AD, AS, and AU are used to describe the right, left, and both ear(s). "A" stands for auris.

▶ Examination of the eyes may require treatment with chemicals to enhance the examination. Anesthetics may be used so that tonometers can be placed on the cornea and intraocular pressure can be measured. Retinal examination may be aided with the use of cycloplegics or mydriatics. **Cycloplegics** (sī-klō-plē-jihcks) cause paralysis of the ciliary muscle that may aide in dilation of the pupil and ease the pain of ciliary muscle spasms. **Mydriatics** (mihd-rē-ah-tihcks) are agents that dilate the pupil.

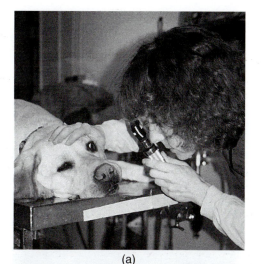

(a)

(b)

FIGURE 14–5 Ophthalmoscopy. Two methods of ophthalmoscopy are (a) direct and (b) indirect. *Source:* Lodi Veterinary Hospital, S.C.

▶ Path of Destruction: Eyes ◀

Pathologic conditions of the eyes include

- **blepharitis** (blehf-ah-rī-tihs) = inflammation of the eyelid
- **blepharoptosis** (blehf-ah-rō-tō-sihs) = drooping of the upper eyelid
- **ectropion** (ehck-trō-pē-ohn) = eversion or turning outward of the eyelid
- **entropion** (ehn-trō-pē-ohn) = inversion or turning inward of the eyelid (Figure 14–6)
- **hordeolum** (hōr-dē-ō-luhm) = infection of one or more glands of the eyelid; also called **stye** (stī)
- **microphthalmia** (mī-krohf-thahl-mē-ah) = abnormally small eyes
- **chalazion** (kah-lā-zē-ohn) = localized swelling of the eyelid resulting from the obstruction of a sebaceous gland of the eyelid

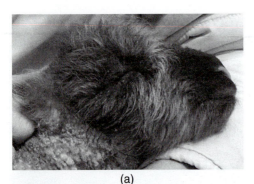

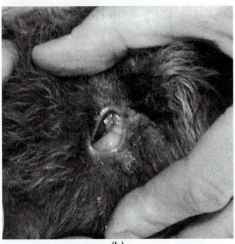

FIGURE 14–6 (a) Entropion in a lamb (b) Note severe conjunctivitis associated with entropion. *Source:* Courtesy of Ron Fabrizius, DVM, Diplomat ACT

- **dacryocystitis** (dahck-rē-ō-sihs-tī-tihs) = inflammation of the lacrimal sac and abnormal tear drainage
- **conjunctivitis** (kohn-juhnk-tih-vī-tihs) = inflammation of the conjunctiva
- **nystagmus** (nī-stahg-muhs) = involuntary, constant, rhythmic movement of the eye
- **strabismus** (strah-bihz-muhs) = disorder in which the eyes are not directed in a parallel manner; deviation of one or both eyes (Figure 14–7)
- **convergent strabismus** (kohn-vər-gehnt strah-bihz-muhs) = deviation of the eyes toward each other; **crossed eyes;** also called **esotropia** (ehs-ō-trō-pē-ah). The suffix **-tropia** means turning; the prefix **eso-** means inward.
- **divergent strabismus** (dī-vər-gehnt strah-bihz-muhs) = deviation of the eyes away from each other; also called **exotropia** (ehck-sō-trō-pē-ah). The prefix **exo-** means outward.
- **hypertropia** (hī-pər-trō-pē-ah) = deviation of one eye upward
- **hypotropia** (hī-pō-trō-pē-ah) = deviation of one eye downward
- **scleritis** (skleh-rī-tihs) = inflammation of the sclera

- **keratitis** (kehr-ah-tī-tihs) = inflammation of the cornea (Figure 14–8).
- **iritis** (ī-rī-tihs) = inflammation of the iris
- **synechia** (sī-nēk-ē-ah) = adhesion that binds the iris to an adjacent structure; plural is **synechiae** (sī-nēk-ē-ā)
- **anisocoria** (ahn-ih-sō-kō-rē-ah) = condition of unequal pupil size. **Anis/o** is the combining form meaning unequal (**an-** is not; **iso-** is equal) (Figure 14–9).
- **cataract** (kaht-ah-rahkt) = cloudiness or opacity of the lens (Figure 14–10).
- **glaucoma** (glaw-kō-mah) = group of disorders resulting from increased intraocular pressure
- **floaters** (flō-tərz) = particles that cast shadows on the retina suspended in the vitreous fluid; also called **vitreous floaters**
- **retinopathy** (reht-ih-nohp-ah-thē) = any disorder of the retina
- **retinal detachment** (reht-ih-nahl dē-tahch-mehnt) = separation of the nervous layer of the eye from the choroid; also called **detached retina**
- **papilledema** (pahp-ehl-eh-dē-mah) = swelling of the optic disc
- **macular degeneration** (mahck-yoo-lahr dē-jehn-ər-ā-shuhn) = condition of central vision loss
- **diplopia** (dih-plō-pē-ah) = double vision. **Dipl/o** is the combining form for double; **-opia** is the suffix meaning vision.
- **monochromatism** (mohn-ō-krō-mah-tihzm) = lack of ability to distinguish colors; also called **color blindness. Mono-** is the prefix meaning one; **chrom/o** is the combining form for color.
- **nyctalopia** (nihck-tah-lō-pē-ah) = condition of inability or having difficulty seeing at night; also called **night blindness**

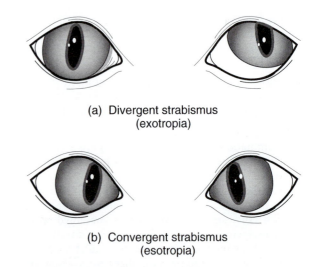

(a) Divergent strabismus (exotropia)

(b) Convergent strabismus (esotropia)

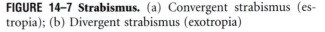

FIGURE 14–7 Strabismus. (a) Convergent strabismus (estropia); (b) Divergent strabismus (exotropia)

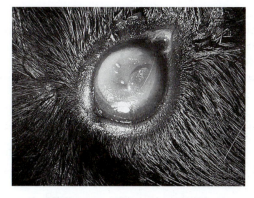

FIGURE 14–8 Ulcerative keratitis

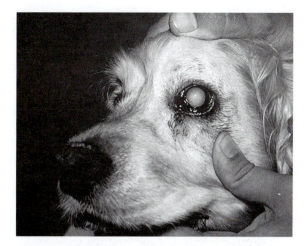

FIGURE 14–10 **Cataract in a dog.** *Source:* Courtesy of Mark Jackson, DVM

- **blindness** (blīnd-nehs) = inability to see
- **amblyopia** (ahm-blē-ō-pē-ah) = dimness or loss of sight without detectable eye disease. The combining form **ambly/o** means dim.
- **keratoconjunctivitis** (kehr-ah-tō-kohn-juhnk-ih-vī-tihs) = inflammation of the cornea and conjunctiva
- **uveitis** (yoo-vē-ī-tihs) = inflammation of the uvea
- **photophobia** (fō-tō-fō-bē-ah) = fear or intolerance of light
- **blepharospasm** (blehf-rō-spahzm) = rapid involuntary contractions of the eyelid
- **distichiasis** (dihs-tē-kē-ī-ah-sis) = abnormal condition of a double row of eyelashes that usually result in conjunctival injury. **Distichia** (dihs-tēck-ē-ah) is a double row of eyelashes.
- **buphthalmos** (boof-thahl-muhs) = abnormal enlargement of the eye
- **proptosis** (prohp-tō-sihs) = displacement of the eye from the orbit
- **epiphora** (ē-pihf-ōr-ah) = excessive tear production
- **corneal ulceration** (kōr-nē-ahl uhl-sihr-ah-shuhn) = surface depression on the cornea
- **dacryoadenitis** (dahck-rē-ō-ahd-ehn-ī-tihs) = inflammation of the lacrimal gland
- **scleral injection** (skleh-rahl ihn-jehck-shuhn) = dilation of blood vessels into the sclera
- **episcleritis** (ehp-ih-sklehr-ī-tihs) = inflammation of the tissue of the cornea
- **panophthalmitis** (pahn-ohp-thahl-mih-tī-tihs) = inflammation of all eye structures

- **anophthalmos** (ahn-ohp-thahl-mōs) = without development of one or both eyes
- **exophthalmos** (ehcks-ohp-thahl-mōs) = abnormal protrusion of the eyeball
- **microphthalmos** (mī-krohp-thahl-mōs) = abnormally small eye(s)
- **cyclopia** (sī-klō-pē-ah) = congenital anomaly characterized by single orbit
- **aphakia** (ah-fahk-ē-ah) = absence of lens
- **hypopyon** (hī-pō-pē-ohn) = pus in the anterior chamber of the eye
- **nuclear sclerosis** (nū-klē-ahr sklehr-ō-sihs) = drying out of the lens with age
- **ophthalmoplegia** (ohp-thahl-mō-plē-gē-ah) = paralysis of eye muscles

▶ Procede With Caution: Eyes ◀

Procedures performed on the eyes include
- **tarsectomy** (tahr-sehck-tō-mē) = surgical removal of all or part of the tarsal plate of the third eyelid
- **tarsorrhaphy** (tahr-sōr-ah-fē) = suturing together of the eyelids
- **conjunctivoplasty** (kohn-juhnk-tī-vō-plahs-tē) = surgical repair of the conjunctiva
- **iridectomy** (ihr-ih-dehck-tō-mē) = surgical removal of a portion of the iris
- **lensectomy** (lehn-sehck-tō-mē) = surgical removal of a lens (usually performed on cataracts)
- **extracapsular extraction** (ehcks-trah-kahp-soo-lahr ehck-trahck-shuhn) = removal of a cataract that leaves the posterior lens capsule intact
- **intracapsular extraction** (ihn-trah-kahp-soo-lahr ehck-trahck-shuhn) = cataract removal that includes the surrounding capsule
- **blepharectomy** (blehf-ār-ehck-tō-mē) = surgical removal of all or part of the eyelid

FIGURE 14–9 **Anisocoria**

- **blepharorrhaphy** (blehf-ār-ōr-rah-fē) = suturing together of the eyelids; also called **tarsorrhaphy** (tahr-soh- rah-fē)
- **blepharoplasty** (blehf-ār-rō-plahs-tē) = surgical repair of the eyelid
- **blepharotomy** (blehf-ahr-ah-tō-mē) = incision of the eyelid; also called **tarsotomy** (tahr-soh-tō-mē)
- **canthectomy** (kahn-thehck-tō-mē) = surgical removal of the corner of the eyelid
- **canthoplasty** (kahn-thō-plahs-tē) = surgical repair of the palpebral fissure
- **canthotomy** (kahn-thoh-tō-mē) = incision into the corner of the eyelid
- **dacryocystectomy** (dahck-rē-ō-sihs-tehck-tō-mē) = surgical removal of the lacrimal sac
- **dacryocystotomy** (dahck-rē-ō-sihs-tah-tō-mē) = incision into the lacrimal sac
- **enucleation** (ē-nū-klē-ā-shuhn) = removal of the eyeball
- **goniotomy** (gō-nē-ah-tō-mē) = incision into the anterior chamber angle for treatment of glaucoma
- **keratectomy** (kehr-ah-tehck-tō-mē) = surgical removal of part of the cornea
- **keratocentesis** (kehr-ah-tō-sehn-tē-sihs) = puncture of the cornea to allow for aspiration of aqueous humor

- **keratoplasty** (kehr-ah-tō-plahs-tē) = surgical repair of the cornea (may include corneal transplant)
- **keratotomy** (kehr-ah-toh-tō-mē) = incision into the cornea
- **lacromotomy** (lahck-rō-moh-tō-mē) = incision into the lacrimal gland or duct

FUNCTIONS OF THE EAR

The **ear** is the sensory organ than enables hearing and helps to maintain balance. The combining forms **audit/o, aud/i,** and **ot/o** mean ear. **Acoust/o** and **acous/o** are combining forms for sound or hearing. Therefore, the term **auditory** pertains to the ear while **acoustic** pertains to sound.

STRUCTURES OF THE EAR

The ear is divided into the outer, middle, and inner portions. Each part has its own unique structures (Figure 14–11).

Outer or External Ear

The **pinna** (pihn-ah) is the external portion of the ear that catches sound waves and transmits them into the external auditory canal. The pinna is also known as the

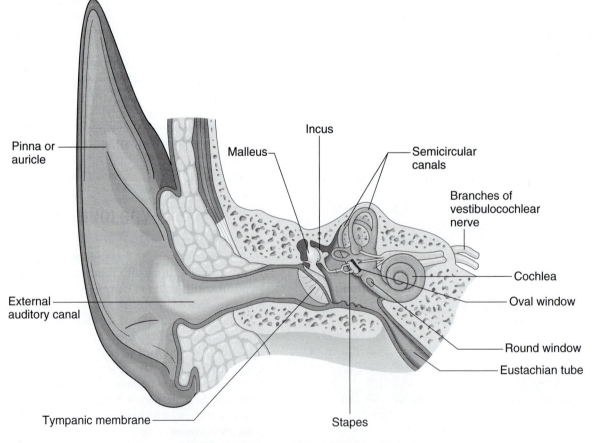

FIGURE 14–11 Cross-section of ear structures

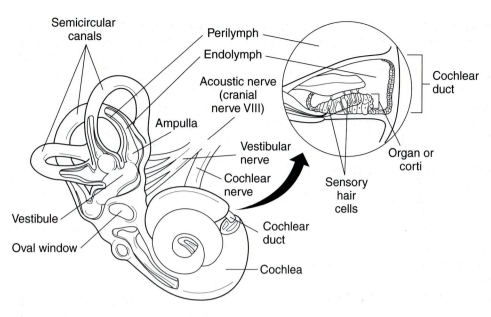

FIGURE 14–12 The inner ear. The bony labyrinth (semicircular canals, vestibule, and cochlea) is the hard outer wall of the entire inner ear. The membranous labyrinth is located within the bony labyrinth. The membranous labyrinth is surrounded by perilymph and is filled with endolymph.

auricle (awr-ihck-kuhl). The combining form **pinn/i** means external ear. The combining forms **aur/i** and **aur/o** mean external ear, but are commonly used to simply mean ear.

The **external auditory canal** is the tube that transmits sound from the pinna to the tympanic membrane. The external auditory canal is also known as the **external auditory meatus.** Glands that line the external auditory canal secrete **cerumen** (seh-roo-mehn), which is commonly known as **earwax.**

Middle Ear

The middle ear begins with the **eardrum.** The medical term for eardrum is **tympanic membrane** (tihm-pahn-ihck mehm-brān). The tympanic membrane is the tissue that separates the external ear from the middle ear. When sound waves reach the tympanic membrane it transmits sounds to the ossicles. The combining forms **tympan/o** and **myring/o** both mean eardrum.

The **auditory ossicles** (aw-dih-tōr-ē ohs-ih-kulz) are three little bones of the middle ear that transmit sound vibrations. The three bones are called

► **malleus** (mahl-ē-uhs) = auditory ossicle known as the hammer
► **incus** (ihng-kuhs) = auditory ossicle known as the anvil
► **stapes** (stā-pēz) = auditory ossicle known as the stirrup

The **eustachian tube** (yoo-stā-shuhn toob) follows the auditory ossicles. The eustachian tube or **auditory**

tube is the narrow duct that leads from the middle ear to the nasopharynx. It helps equalize air pressure in the middle ear with that of the atmosphere.

The **oval window** (ō-vahl wihn-dō), located at the base of the stapes, is the membrane that separates the middle and inner ear.

The **round window** (rownd wihn-dō) is the membrane that receives sound waves through fluid after it has passed through the cochlea.

The **tympanic bulla** (tihm-pahn-ihck buhl-ah) is the osseous chamber that houses the middle ear at the base of the skull. **Bulla** (buhl-ah) is the medical term for large vesicle.

Inner Ear

The inner ear contains sensory receptors for hearing and balance (Figure 14–12). The inner ear consists of three spaces in the temporal bone assembled into the **bony labyrinth** (lahb-ih-rihnth). The combining form **labyrinth/o** means maze, labyrinth, and the inner ear. The bony labyrinth is filled with a water-like fluid called **perilymph** (pehr-eh-lihmf). A membranous sac is suspended in the perilymph and follows the shape of the bony labyrinth. This membranous labyrinth is filled with a thicker fluid called **endolymph** (ehn-dō-lihmf).

The bony labyrinth is divided into three parts:

1. **vestibule** = located adjacent to the oval window and between the semicircular canals and cochlea. The vestibule (and semicircular canals) contain specialized mechanoreceptors for balance and equilibrium.

2. **semicircular canals** = located adjacent to the vestibule. The semicircular canals are oriented at right angles to each other. The three canals are the **vestibular, tympanic,** and **cochlear.** Each canal has a dilated area called the **ampulla** that contains sensory cells with hair like extensions. These sensory cells are suspended in endolymph and when the head moves, the hair-like extensions bend. This bending generates nerve impulses for the regulation of equilibrium.

3. **cochlea** (kōck-lē-ah) = spiral shaped passage that leads from the oval window to the inner ear. The combining form **cochle/o** means snail or spiral. Located within the cochlea are:

   - **cochlear duct** (kōck-lē-ahr duhckt) = membranous tube within the bony cochlea that is filled with endolymph. Endolymph vibrates when sound waves strike it.
   - **organ of Corti** (ōr-gahn of kōr-tē) = spiral organ of hearing located within the cochlea that receives and relays vibrations. Specialized cells in the organ of Corti generate nerve impulses when they are bent by the movement of endolymph. Nerve impulses are relayed to the auditory nerve fibers that transmit them to the cerebral cortex.

Do You Hear What I Hear?

Sound waves enter the pinna Travel through the auditory canal Strike the tympanic membrane	**air conduction**
Tympanic membrane vibrates Ossicles conduct sound waves 　through middle ear	**bone conduction**
Sound vibrations enter inner ear 　via the oval window Structures of inner ear receive 　sound waves Sound waves are relayed to the brain	**sensorineural conduction**

MECHANISM OF HEARING

Sound waves enter the ear through the pinna, travel through the auditory canal, and strike the tympanic membrane. This is called **air conduction.** As the tympanic membrane vibrates, it moves the ossicles. The ossicles conduct the sound waves through the middle ear. This is called **bone conduction.** Sound vibrations reach the inner ear via the round window. The structures of the inner ear receive the sound waves and relay them to the brain. This is called **sensorineural conduction.**

MECHANISM OF EQUILIBRIUM

In addition to the structures listed previously, the ear also has structures that function to maintain equilibrium. **Equilibrium** (ē-kwihl-ih-brē-uhm) is the state of balance.

There are three semicircular canals in the inner ear that lie in planes at right angles to each other. The inner ear also has a **saccule** (sahck-yoo-uhl) and **utricle** (yoo-trih-kuhl), which are small, hair-like sacs partially responsible for equilibrium. The saccule and utricle contain **otoliths** (ō-tō-lihthz), which are small stones.

In maintaining equilibrium, the otoliths press on the hair cells due to gravity. The otoliths initiate impulses from the hair cells to the brain. The saccule and utricle are responsible for equilibrium from positional change and linear movement. The semicircular canals maintain equilibrium from rotary movement. Inside the semicircular canals are hair cells that bend in response to rotational movement. Impulses are carried to the vestibular branch of cranial nerve VIII and then to the brain.

▶ **Test Me: Ears** ◀

Diagnostic procedures performed on the ear include

- **otoscopy** (ō-tō-skōpē) = procedure used to examine the ear for parasites, irritation to the ear lining, discharge, and the integrity of the tympanic membrane. **Otoscope** (ō-tō-skōp) is the instrument used for otoscopy (Figure 14–13).

FIGURE 14–13 Otoscope

► Path of Destruction: Ears ◄

Pathologic conditions of the ears include

- **otitis** (ō-tī-tihs) = inflammation of the ear; usually has second term that describes location
  otitis externa (ō-tī-tihs ehcks-tər-nah) = inflammation of the outer ear
  otitis media (ō-tī-tihs mē-dē-ah) = inflammation of the middle ear
  otitis interna (ō-tī-tihs ihn-tər-nah) = inflammation of the inner ear
- **panotitis** (pahn-ō-tī-tihs) = inflammation of all ear parts
- **aural hematoma** (awr-ahl hēm-ah-tō-mah) = collection or mass of blood on the outer ear. **Aural** means pertaining to the ear (external ear).
- **otopathy** (ō-tohp-ah-thē) = disease of the ear
- **otalgia** (ō-tahl-gē-ah) = ear pain
- **otorrhea** (ō-tō-rē-ah) = ear discharge
- **otopyorrhea** (ō-tō-pī-ō-rē-ah) = pus discharge from the ear
- **otomycosis** (ō-tō-mī-kō-sihs) = fungal infection of the ear
- **myringitis** (mihr-ihn-jī-tihs) = inflammation of the ear drum

- **vertigo** (vər-tih-gō) = sense of dizziness
- **deafness** (dehf-nehs) = complete or partial hearing loss
- **otoplasty** (ō-tō-plahs-tē) = surgical repair of the ear

What is in a name?

The names of parasites seem long and cumbersome at first glance; however, many times the name of something can tell a lot about what it does or where it lives. The scientific name of the common ear mite of dogs and cats is *Otodectes cynotis*. From the combining form **ot/o** in its name, it can be concluded that *Otodectes* are parasites that infiltrate the ear. **Dectes** means biter; therefore, *Otodectes* are biting parasites found in the ear.

► Procede With Caution: Ears ◄

Procedures performed on the ears include

- **ablation** (ah-blā-shuhn) = removal of a part
- **myringectomy** (mihr-ihn-jehck-tō-mē) = surgical removal of all or part of the eardrum; also called **tympanectomy** (tihm-pahn-ehck-tō-mē)

REVIEW EXERCISES

Multiple Choice—Choose the correct answer.

1. The state of balance is
   a. vertigo
   b. hemostasis
   c. vestibular
   d. equilibrium

2. The outer or external ear is separated from the middle ear by the
   a. oval window
   b. round window
   c. tympanic membrane
   d. pinna

3. Another term for ear wax is
   a. pinna
   b. auricle
   c. cerumen
   d. corti

4. The fibrous tissue that maintains the shape of the eye is the
   a. choroid
   b. sclera
   c. white of the eye
   d. b and c

5. The term for corner of the eye is
   a. canthus
   b. cilia
   c. cerumen
   d. cornus

6. The colored muscular layer of the eye that surrounds the pupil is the
   a. cornea
   b. choroid
   c. lens
   d. iris

7. Involuntary, constant, rhythmic movement of the eyeball is termed
   a. ectropion
   b. nystagmus
   c. strabismus
   d. entropion

8. A group of eye disorders resulting from increased intraocular pressure is
   a. ophthalmopathy
   b. glaucoma
   c. floaters
   d. hypertension

9. Opacity of the lens is termed
   a. opague
   b. turgid
   c. cataract
   d. diplopia

10. The process of the lens bending the light ray to help focus the rays on the retina is called
    a. convergence
    b. refraction
    c. humor
    d. fovea

Matching—Match the term in Column I with the definition in Column II.

Column I
_____ palpebra

_____ orbit

_____ cilia

_____ cornea

_____ conjunctiva

_____ tarsus

_____ uvea

Column II
a. iris, ciliary body, and choroid

b. platelike frame within the upper and lower eyelids

c. eyelid

d. eyelashes

e. bony cavity of the skull that contains the eyeball

f. transparent anterior portion of the sclera

g. mucous membrane that lines the underside of each eyelid

Case Studies—Define the terms underlined in each case study.

A 6 yr old DSH M/N cat was presented for inappetence and blepharospasm. PE revealed T = 103.4 °F, HR = 200 bpm, RR = 40 breaths/min, mm = pink and dry, CRT = 1.5 sec. The conjunctiva was reddened and the sclera was infected. Heart and lungs ausculted normally. Abdominal palpation was normal. Oral examination was normal. Ocular exam: anterior chamber was cloudy. Iris appeared normal. Tonometer readings were OS 7 mm Hg OD 10 mm Hg. Dx was uveitis—anterior chamber. Possible causes include infections (FIP, FeLV, toxoplasmosis), immune mediated, systemic disease, and trauma. Further diagnostic tests are being pursued by the veterinarian.

blepharospasm _____

conjunctiva _____

sclera _____

ocular _____

anterior chamber _____

iris _____

tonometer _____

OS _____

mm Hg _____

OD _____

uveitis _____

An 8 mo old M black Labrador retriever was presented to the clinic for ocular discharge and rubbing at the eyes. Upon presentation to the clinic the vital signs were normal, attitude was normal for a puppy, and bilateral mucopurulent ocular discharge was noted. Ophthalmic examination revealed normal Schirmer tear test values, a normal looking retina, and no stain retention via fluorescence staining. Upon examination of the eyelids it was noted that the dog had blepharospasm, and entropion was noted. The entropion was most likely the cause of the eye infection because the eyelashes were brushing against the cornea. Blepharoplasty was recommended to this owner. Topical antibiotics were dispensed pending surgery.

ocular _____

bilateral mucopurulent ocular discharge _____

ophthalmic _____

Schirmer tear test _____

retina _____

fluorescence staining_____

blepharospasm _____

entropion _____

cornea_____

blepharoplasty _____

topical_____

Buddy, a 3 yr old M/N springer spaniel, was presented to the clinic with <u>recurrent</u> ear problems. The dog has had <u>bilat-eral</u> ear problems in the past that have responded well to treatment. Buddy had been seen 2 wk earlier for ear problems and <u>otitis externa</u> was diagnosed. Ear <u>cytology</u> revealed that Buddy had a severe yeast infection of both ears and an anti-fungal drug was prescribed. On today's examination <u>mucopurulent</u> discharge was noted <u>AU</u>. <u>Otoscopic</u> examination re-vealed <u>AD</u> was <u>hyperemic</u> and <u>hyperkeratotic</u>, and the ear was so swollen that the <u>tympanic membrane</u> could not be vi-sualized. <u>AS</u> was hyperemic, hyperkeratotic, and the tympanic membrane was intact. Ear cytology revealed yeast, and the dog was getting worse. Oral antifungal drugs as well as anti-inflammatory drugs were prescribed. The owner was advised that if the problem did not resolve <u>ear ablation</u> surgery may be warranted.

recurrent_____

bilateral_____

otitis externa_____

cytology _____

mucopurulent_____

AU_____

Otoscopic _____

AD_____

hyperemic_____

hyperkeratotic_____

tympanic membrane _____

AS _____

ear ablation_____

15 Feed and Protect Me

Objectives *In this chapter, you should learn to:*

▶ Identify and describe the major structures and functions of the hematologic and immune systems
▶ Recognize, define, spell, and pronounce the major terms related to diagnosis, pathology, and treatment of the hematologic, immune, and lymphatic systems
▶ Recognize, define, spell, and pronounce terms related to oncology

HEMATOLOGIC SYSTEM

Functions of Blood

Blood supplies body tissues with oxygen, nutrients, and various chemicals. Blood transports waste products to various organs for removal from the body. Blood cells also play important roles in the immune and endocrine systems.

Structures of Blood

Blood is composed of 55% liquid plasma and 45% formed elements. Formed elements include red blood cells, white blood cells, and clotting cells. The combining forms for blood are **hem/o** and **hemat/o.**

Blood is formed in the bone marrow. **Hematopoiesis** (hē-mah-tō-poy-ē-sihs) is the medical term for formation of blood. The suffix **-poiesis** means formation.

The components of blood can be separated clinically and examined microscopically. A blood sample is collected with a needle/catheter and syringe. **Drawing blood** is a common term for collecting a blood sample. Blood can be collected in a tube that has an **anticoagulant** (ahn-tih-kō-ahg-yoo-lahnt). An anticoagulant is a substance that prevents clotting of blood. **EDTA** (ethylenediamine tetraacetic acid) and **heparin** (hehp-ahr-ihn) are types of anti-

coagulants found in blood tubes and used clinically as drugs. **Coagulation** (kō-ahg-yoo-lā-shuhn) is the process of clotting. Sometimes coagulation of blood is desired after blood is placed in a collection tube. When blood coagulates, a layer of leukocytes and thrombocytes forms, which is located between erythrocytes and plasma. This layer is called the **buffy coat** (buhf-ē kōt) (Figure 15–1).

LIQUID

The liquid portion of blood consists of

- **plasma** (plahz-mah) is the straw-colored fluid portion of blood that transports nutrients, hormones, and waste products.
- **serum** (sē-ruhm) is the liquid portion of blood with clotting proteins removed
- Clotting proteins are found in plasma. Examples include **fibrinogen** (fih-brihn-ō-jehn) and **prothrombin** (prō-throhm-bihn). The combining form **fibrin/o** means fibrin or threads of a clot, the prefix **pro-** means before, and the combining form **thromb/o** means clot. **Albumin** (ahl-byoo-mihn) is another example of a plasma protein.
- **fats** also circulate in plasma. **Cholesterol** (kō-lehs-tər-ohl) and **triglyceride** (trī-glihs-ər-īd) are types of lipids that circulate in blood.

207

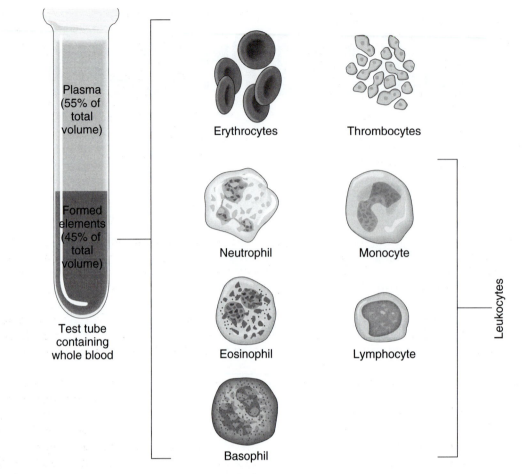

Plasma
(55% of
total
volume)

Formed
elements
(45% of
total
volume)

Test tube
containing
whole blood

Erythrocytes

Thrombocytes

Neutrophil

Monocyte

Eosinophil

Lymphocyte

Basophil

Leukocytes

FIGURE 15–1 Blood components

FORMED ELEMENTS

ERYTHROCYTES

An **erythrocyte** (eh-rihth-rō-sīt) is a mature red blood cell (oxygen-carrying cell) and is abbreviated RBC. The combining form **erythr/o** means red and the suffix **-cyte** means cell. Erythrocytes contain **hemoglobin** (hē-mō-glō-bihn), a blood protein that transports oxygen. **Heme** (hēm) is the nonprotein, iron-containing portion of hemoglobin.

Erythrocytes are produced in the bone marrow. The combining form for bone marrow (and spinal cord) is **myel/o.** Erythrocytes vary in appearance from species to species.

A **reticulocyte** (reh-tihck-yoo-lō-sīt) is an immature erythrocyte characterized by polychromasia (Wright's stain) or a meshlike pattern of threads (new methylene blue stain).

When RBCs are no longer useful, they are destroyed by macrophages. A **macrophage** (mahck-rō-

TABLE 15–1
Terms Used to Describe Erythrocytes

Term	Pronunciation	Description
poikilocytosis	(poy-kē-loh-sī-tō-sihs)	condition of irregular cells; clinically means varied shapes of erythrocytes
polychromasia	(poh-lē-krō-mah-zē-ah)	condition of many colors; clinically means erythrocytes with varied or multicolored staining qualities
anisocytosis	(ahn-eh-sō-sī-tō-sihs)	condition of unequal cell size
hypochromic	(hī-pō-krō-mihck)	less than normal color
hyperchromic	(hī-pər-krō-mihck)	more than normal color
macrocytic	(mahck-rō-siht-ihck)	large cell size
microcytic	(mī-krō-siht-ihck)	small cell size

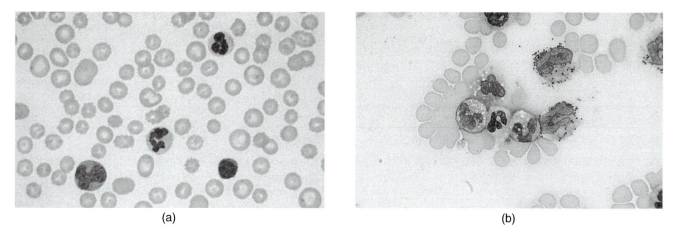

(a) (b)

FIGURE 15–2 Photomicrographs of canine blood cells. (a) Peripheral blood smear of canine blood demonstrating RBCs, platelets, and WBCs (2 PMNs (segmented neutrophils), a monocyte, and a lymphocyte); magnification 63× (b) Canine granulocytic WBCs at the feathered edge of a peripheral blood smear. Three basophils are on the right and a PMN is in the center of three eosinophils that are on the left; magnification 63×

fahj or mahck-rō-fāj) is a large cell that destroys by eating. The combining form **macr/o** means large and the suffix **-phage** means eating. A **phagocyte** (fā-gō-sīt) is "a cell that eats." The formal definition of a phagocyte is a leukocyte that ingests foreign material.

Hematology (hē-mah-tah-lō-gē) is the study of blood. When blood cells are studied, it is important to note the cell's morphology. **Morphology** (mōr-fah-lō-jē) is the study of form (Table 15–1 and Figure 15–2).

LEUKOCYTES

A **leukocyte** (loo-kō-sīt) is a white blood cell and is abbreviated WBC. The combining form **leuk/o** means white. Leukocytes are produced in the bone marrow (and other places) and function primarily in fighting disease in the body. **Leukocytopoiesis** (loo-koo-sīt-ō-poy-ē-sihs) is the production of white blood cells. The production of leukocytes is also called **leukopoiesis** (loo-koo-poy-ē-sihs) (Table 15–2).

A **granulocyte** (grahn-yoo-lō-sīt) is a cell that contains prominent grain-like structures in its cytoplasm; an **agranulocyte** (ā-grahn-yoo-lō-sīt) is a cell that does not contain prominent grain-like structures in its cytoplasm.

A **band cell** (bahnd sehl) is an immature polymorphonuclear leukocyte.

Some cells are described as basophilic or eosinophilic. **Basophilic** (bā-sō-fihl-ihck) means things that stain readily with basic, or blue, dyes in many commonly used stains such as hematoxylin and eosin (H&E), Giemsa, and Wright's. **Eosinophilic** (ē-ō-sihn-ō-fihl-ihck) means things that stain readily with acidic, or pink, dyes in many commonly used stains such as H&E, Giemsa, and Wright's.

CLOTTING CELLS

Clotting cells are also produced in the bone marrow and play a part in the clotting of blood. A **thrombocyte** (throhm-bō-sīt) is a nucleated clotting cell and a **platelet** (plāt-leht) is an anucleated clotting cell. Occasionally these terms are used interchangeably; however, they do have different meanings.

A **megakaryocyte** (mehg-ah-kahr-ē-ō-sīt) is a large nucleated cell found in the bone marrow from which platelets are formed.

TABLE 15–2
Major Groups of Leukocytes

Type of Leukocyte	Pronunciation	Description
lymphocyte	(lihm-fō-sīt)	class of *agranulocytic* leukocyte that has phagocytic and antibody formation functions
monocyte	(mohn-ō-sīt)	class of *agranulocytic* leukocyte that has phagocytic function
eosinophil	(ē-ō-sihn-ō-fihl)	class of *granulocytic* leukocyte that detoxifies allergens or parasitic infections
neutrophil	(nū-trō-fihl)	class of *granulocytic* leukocyte that has phagocytic function; also called **polymorphonuclear leukocyte** (poh-lē-mōr-fō-nū-klē-ahr) or PMN, polymorphonuclear means multishaped nucleus
basophil	(bā-sō-fihl)	class of *granulocytic* leukocyte that promotes the inflammatory response

► Test Me: Hematologic System ◄

Diagnostic tests performed on the hematologic system will be covered in Chapter 16.

► Path of Destruction: Hematologic System ◄

Pathologic conditions of the hematologic system include

- **phagocytosis** (fahg-ō-sī-tō-sihs) = condition of engulfing or eating cells
- **hemorrhage** (hehm-ōr-ihdj) = loss of blood (usually in a short period of time). **Hemostasis** (hē-mō-stā-sihs) is the act of controlling blood or bleeding. A **hemostat** (hē-mō-staht) is an instrument to control bleeding.
- **septicemia** (sehp-tih-sē-mē-ah) = blood condition in which pathogenic microorganisms (bacteria) and/or their toxins are present. The suffix **-emia** means blood condition. **Pathogenic** (pahth-ō-jehn-ihck) means producing disease. **Bacteremia** (bahk-tər-ē-mē-ah) is the blood condition in which bacteria are present.
- **hyperalbuminemia** (hī-pər-ahl-byoo-mih-nē-mē-ah) = blood condition of abnormally high albumin levels
- **hyperlipidemia** (hī-pər-lihp-ih-dē-mē-ah) = blood condition of abnormally high fat levels; more accurately means abnormally high fat levels due to fat metabolism. **Lip/o** is the combining form for fat.
- **lipemia** (lī-pē-mē-ah) = excessive amount of fats in the blood
- **lipemic serum** (lī-pē-mihck sē-ruhm) = fats from blood that have settled in the serum. Clinically the serum will appear cloudy and white.
- **hyperemia** (hī-pər-ē-mē-ah) = excess blood in a part; engorgement
- **erythrocytosis** (eh-rihth-rō-sī-tō-sihs) = abnormal increase in red blood cells . The suffix **-cytosis** means condition of cell, but implies increased cell numbers.
- **thrombocytopenia** (throhm-bō-sīt-ō-pē-nē-ah) = abnormal decrease in the number of clotting cells. The suffix **-penia** means less than normal or deficiency.
- **anemia** (ah-nē-mē-ah) = blood condition of less than normal levels of red blood cells or hemoglobin
- **thrombocytosis** (throhm-bō-sī-tō-sihs) = abnormal increase in the number of clotting cells
- **leukopenia** (loo-kō-pē-nē-ah) = abnormal decrease in the number of white blood cells
- **leukocytosis** (loo-kō-sī-tō-sihs) = abnormal increase in the number of white blood cells

- **leukemia** (loo-kē-mē-ah) = abnormal increase in the number of malignant white blood cells
- **hemolytic** (hē-mō-liht-ihck) = removing and destroying red blood cells. Hemolytic anemia is excessive RBC destruction, resulting in lower than normal levels of RBCs. **Hemolysis** (hē-mohl-eh-sihs) is the breaking down of blood cells. **Lysis** (lī-sihs) is the medical term for destruction or breakdown
- **dyscrasia** (dihs-krā-zē-ah) = any abnormal condition of the blood
- **left shift** (lehft shihft) = common term for an alteration in the distribution of leukocytes in which there are increases in band forms usually in response to infection
- **polycythemia** (poh-lē-sī-thē-mē-ah) = condition of many cells; clinically means excessive erythrocytes
- **pancytopenia** (pahn-sīt-ō-pē-nē-ah) = deficiency of all types of blood cells
- **neutropenia** (nū-trō-pē-nē-ah) = decreased number of neutrophilic leukocytes in the blood
- **lymphocytosis** (lihm-fō-sī-tō-sihs) = increased numbers of lymphocytic leukocytes in the blood
- **lymphopenia** (lihm-fō-pē-nē-ah) = decreased numbers of lymphocytic leukocytes in the blood
- **monocytosis** (moh-nō-sī-tō-sihs) = increased numbers of monocytic leukocytes in the blood
- **hemophilia** (hē-mō-fihl-ē-ah) = hereditary condition of deficient blood coagulation
- **monocytopenia** (moh-nō-sīt-ō-pē-nē-ah) = decrease in the number of monocytic leukocytes in the blood
- **edema** (eh-dē-mah) = accumulation of fluid in the intercellular space
- **exudate** (ehcks-yoo-dāt) = material that has escaped from blood vessels and is high in protein, cells, or solid materials derived from cells
- **transudate** (trahnz-yoo-dāt) = material that has passed through a membrane and is high in fluidity and low in protein, cells, or solid materials derived from cells

THE IMMUNE SYSTEM

The immune system functions to protect the body from harmful substances. The term **immunity** comes from the Latin term immunitas, which means exemption. Immunity was used to imply an animal was exempt from or protected against foreign substances. The combining form **immun/o** means protected. **Immunology** (ihm-yoo-nohl-ō-jē) is the study of the immune system.

The immune system is not contained within one set of organs or within one area. Many structures from different body systems aide in protecting the body. The lymphatic system, respiratory tract, gastrointestinal tract, integumentary system, and others all work together to prevent the body from being harmed from foreign invaders.

> **Anti- Means Against:** To understand immunology, the terms antigen and antibody must be distinguished. **Antigen** (ahn-tih-jehn) is a substance that the body regards as foreign (such as a virus, bacterium, or toxin). **Antibody** (ahn-tih-boh-dē) is a disease-fighting protein produced by the body in response to the presence of a specific antigen.

Lymphatic System

FUNCTIONS OF THE LYMPHATIC SYSTEM

The lymphatic system functions as part of the immune system, returns excess lymph to the blood, and absorbs fats and fat-soluble vitamins from the digestive system and transports them to cells. The combining form **lymph/o** means lymph fluid, lymph vessels, and lymph nodes. **Lymphoid** (lihm-foyd) pertains to lymph or tissue of the lymphatic system

STRUCTURES OF THE LYMPHATIC SYSTEM

The major structures of the lymphatic system include lymph vessels, lymph nodes, lymph fluid, tonsils, spleen, thymus, and lymphocytes.

LYMPH FLUID

Lymph (lihmf) or **interstitial fluid** (ihn-tər-stihsh-ahl flū-ihd) is the clear, colorless tissue fluid that leaves the capillaries and flows in the spaces between the cells of a tissue or organ. Interstitial pertains to the spaces within a tissue or organ. Lymph brings nutrients and hormones to cells and carries waste products from tissue back to the bloodstream (Figure 15–3).

LYMPH VESSELS

Lymph is carried from the tissue space via thin-walled tubes called **lymph capillaries.** Lymph capillaries take the lymph to the lymphatic vessels. **Lymphatic vessels** (lihm-fah-tihck vehs-uhlz) are like veins in that they have valves to prevent the backflow of lymph. Lymph always travels toward the thoracic cavity. In the thoracic cavity, the right lymphatic duct and thoracic duct empty lymph into veins. The **cisterna chyli** (sihs-tər-nah kī-lē) is the origin of the thoracic duct and saclike structure for the lymph collection.

Lacteals (lahck-tē-ahls) located in the small intestine are specialized lymph vessels that transport fats and fat-soluble vitamins.

LYMPH NODES

Lymph nodes (lihmf nōdz) are small bean-shaped structures that filter lymph and store B and T lymphocytes. The primary function of lymph nodes is to filter lymph to remove harmful substances such as bacteria and viruses. Because cells are destroyed in lymph nodes, swollen lymph nodes are often an indication of disease. Lymph nodes are described according to their location: mandibular lymph nodes are located near the mandible, parotid lymph nodes are located near the ear (**para-** means near and **otos** is Greek for ear), and so on.

TONSILS

The **tonsils** (tohn-sahlz) are masses of lymphatic tissue that protect the nose and upper throat. Tonsils are described according to their location: lingual tonsils are located near the tongue, palatine tonsils are located near the palate or roof of the mouth, and pharyngeal tonsils are located near the throat. **Tonsill/o** is the combining form for tonsil.

SPLEEN

The **spleen** (splēn) is a mass of lymphatic tissue located in the cranial abdomen that produces lymphocytes and monocytes, filters foreign material from the blood, stores red blood cells, and maintains an appropriate

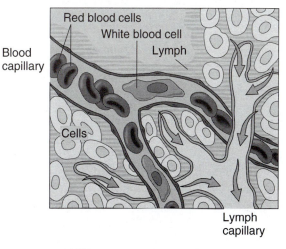

FIGURE 15-3 Lymph circulation

balance of cells and plasma in the blood. The combining form **splen/o** means spleen.

THYMUS

The **thymus** (thī-muhs) is a gland that has an immunologic function and is found predominantly in young animals. The thymus is located near midline in the cranioventral portion of the thoracic cavity. The immunologic role of the thymus is development of T cells. Some of the lymphocytes formed in the bone marrow migrate to the thymus where they multiply and mature into T cells. T cells play an important role in the immune response. The thymus also has endocrine function. The combining form for thymus is **thym/o.**

SPECIALIZED CELLS

There are some cells that are specialized for immune reactions. The **lymphocyte** is a type of white blood cell that attacks specific organisms. Lymphocytes are formed in lymphatic tissue throughout the body, such as lymph nodes, spleen, thymus, tonsils, and bone marrow. **Monocytes** are a type of leukocyte formed in the bone marrow and transported to other parts of the body. Monocytes may travel to the thymus, where they become macrophages. A **macrophage** is a phagocytic cell that protects the body by eating invading cells and by interacting with other cells of the immune system. **Histiocytes** (hihs-tē-ō-sītz) are large macrophages found in loose connective tissue.

T lymphocytes are small circulating lymphocytes produced in the bone marrow. These cells mature in the thymus; hence, they are called **T cells.** The primary function of T lymphocytes is to coordinate immune defenses and kill organisms on contact. T cells are involved in cell-mediated immunity. There are different types of T lymphocytes. **T helper cells** secrete substances, like lymphokines, that stimulate the production of B lymphocytes. **Suppressor T cells** stop B lymphocyte activity when this activity is no longer needed. **Memory T cells** remember a specific antigen and stimulate a faster and more intense response if that same antigen is presented to the body.

B lymphocytes are produced and mature in the bone marrow. B lymphocytes are responsible for antibody-mediated immunity or humoral immunity. Each B lymphocyte makes its own specific antibody against a specific antigen. In the presence of a specific antigen, B lymphocytes are transformed into plasma cells. A **plasma cell** (plahz-mah sehl) is an immune cell that produces and secretes a specific antibody for a specific antigen. Plasma cells are also called **plasmocytes** (plahz-mō-sītz) (Figure 15–4).

The antibodies made by plasma cells are called **immunoglobulins** (ihm-yoo-nō-glohb-yoo-lihnz). Immunoglobulin is abbreviated Ig. There are five distinct immunoglobulins (Table 15–3 and Figure 15–5).

Complement is a series of enzymatic proteins that occur in normal serum. Complement aids phagocytes in destruction of antigens and causes lysis.

How Does the Immune System Work?

The body has many ways to protect itself from invading organisms and substances. The first line of defense is intact skin. **Intact** (ihn-tahckt) means having no cuts, scrapes, openings, or alterations. Intact skin makes it more difficult for invading organisms and substances to obtain access to an animal. Oil secreted by sebaceous glands discourages bacteria growth on the skin.

The respiratory system has its own line of defense. Foreign material breathed in is trapped in the cilia of the nares and the moist mucous membrane that lines the respiratory tract. Mucus continually flushes away trapped debris. Coughing and sneezing also remove foreign material.

Cell-Mediated and Antibody-Mediated Immunity Compared	
T cells are responsible for cell-mediated immunity.	**B cells** are responsible for antibody-mediated immunity.
T cells directly attack the invading antigen.	B cells produce antibodies that react with the antigen or substances produced by the antigen.
Cell-mediated immunity is most effective against viruses that infect body cells, cancer cells, and foreign tissue cells.	Antibody-mediated immunity is most effective against bacteria, viruses that are outside body cells, and toxins. It is also involved in allergic reactions.

FIGURE 15–4 Comparison of cell-mediated and antibody-mediated immunity

TABLE 15–3
Types of Immunoglobulin and Their Functions

IgA	found in the mucous membrane lining of intestines, bronchi, saliva, and tears; protects these areas
IgD	found in small amounts in serum tissue; unknown function
IgE	found in lungs, skin, and cells of mucous membranes; provides defense against the environment
IgG	synthesized in response to invading germs like bacteria, fungi, and viruses
IgM	found in circulating fluids; first immunoglobulin produced when antigens invade

The digestive system destroys invading organisms that are swallowed by the acidic nature of the stomach.

An animal's health, age, and heredity also play roles in protecting the body. Some animals may be immunodeficient or hypersensitive.

The immune system is activated when the previously listed defenses fail. Activation of the immune system is outlined below:

Organisms invade the body. Macrophages ingest the invading organisms. T helper cells are activated.

↓

T helper cells multiply. Complement goes to the affected area. B cells sensitive to the organism multiply. B cells make antibody.

↓

Complement proteins lyse affected cells. Antibodies bind to organisms.

↓

If infection is contained, suppressor T cells stop the immune response. B cells remain ready.

Immunity (ihm-yoo-nih-tē) is the state of being resistant to a specific disease. Different forms of immunity are obtained during life. The types of immunity are

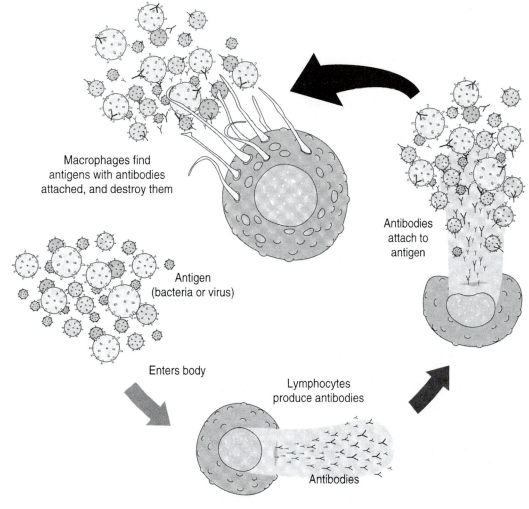

Macrophages find antigens with antibodies attached, and destroy them

Antigen (bacteria or virus)

Antibodies attach to antigen

Enters body

Lymphocytes produce antibodies

Antibodies

FIGURE 15–5 Antibody formation

- **natural passive immunity** = resistance to a specific disease by the passing of protection from mother to offspring before birth or through colostrum
- **natural active immunity** = resistance to a specific disease after the development of antibodies during the actual disease
- **artificial passive immunity** = resistance to a specific disease by receiving antiserum containing antibodies from another host
- **artificial active immunity** = resistance to a specific disease through vaccination

Words used when discussing the immune system
resistant (rē-zihs-tahnt) = not susceptible
heredity (hər-eh-dih-tē) = genetic transmission of characteristics from parent to offspring
vaccination (vahck-sihn-ā-shuhn) = administration of antigen (vaccine) to stimulate a protective immune response against a specific infectious agent; also called **immunization** (ihm-yoo-nih-zā-shuhn)
vaccine (vahck-sēn) = a preparation of pathogen (live, weakened, or killed) or portion of pathogen that is administered to stimulate a protective immune response against that pathogen
multiplication (muhl-tih-plih-kā-shuhn) = reproducing
inhibit (ihn-hihb-iht) = slows or stops
opportunistic (ohp-ər-too-nihs-tihck) = ability to cause disease due to debilitation when disease would not normally be produced
debilitated (dē-bihl-ih-tāt-ehd) = weakened or loss of strength

▶ **Test Me: Immune/Lymphatic Systems** ◀

Diagnostic tests performed on the immune/lymphatic systems will be covered in Chapter 16.

▶ **Path of Destruction: Immune/Lymphatic Systems** ◀

Pathologic conditions of the immune/lymphatic systems include
- **lymphadenitis** (lihm-fahd-eh-nī-tihs) = inflammation of the lymph nodes; also called **swollen glands**
- **tonsillitis** (tohn-sih-lī-tihs) = inflammation of the tonsils
- **lymphangioma** (lihm-fahn-jē-ō-mah) = abnormal collection of lymphatic vessels forming a mass (usually benign)
- **splenomegaly** (splehn-ō-mehg-ah-lē) = enlargement of the spleen
- **allergy** (ahl-ər-gē) = overreaction by the body to a particular antigen; also called **hypersensitivity** (hī-pər-sehn-sih-tihv-ih-tē). An **allergen** (ahl-ər-gehn) is a substance capable of inducing an allergic reaction.
- **anaphylaxis** (ahn-ah-fih-lahck-sihs) = severe response to a foreign substance. Signs develop acutely and may include swelling, blockage of airways, tachycardia, and ptylism.
- **autoimmune disease** (aw-tō-ihm-yoon dih-zēz) = disorder in which the body makes antibodies directed against itself
- **immunosuppression** (ihm-yoo-nō-suhp-prehsh-uhn) = reduction or decrease in the state of resistance to disease. An **immunosuppressant** (ihm-yoo-nō-soo-prehs-ahnt) is a chemical that prevents or reduces the body's normal reaction to disease.
- **lymphadenopathy** (lihm-fahd-eh-nohp-ah-thē) = disease of the lymph nodes; an example of incorrect use of "aden/o" because nodes are not glands

▶ **Procede With Caution: Immune/Lymphatic Systems** ◀

Procedures performed on the immune/lymphatic systems include
- **splenectomy** (splehn-ehck-tō-mē) = surgical removal of the spleen
- **thymectomy** (thī-mehck-tō-mē) = surgical removal of the thymus
- **tonsillectomy** (tohn-sih-lehck-tō-mē) = surgical removal of the tonsils

ONCOLOGY

Oncology (ohng-kohl-ō-jē) is the study of tumors. The combining form **onc/o** means tumor. The term **tumor** does not mean cancerous. If a tumor is cancerous the term **malignant** is used to describe it. Nonmalignant tumors are called **benign.** The term neoplasm is another word used to describe abnormal growths. A **neoplasm** (nē-ō-plahz-uhm) is any abnormal new growth of tissue in which the multiplication of cells is uncontrolled, more rapid than normal, and progressive.

Tumors are often described by their appearance. **Pedunculated** (peh-duhnk-yoo-lā-tehd) means having a peduncle or stalk. **Well-circumscribed** means that the mass has well-defined borders. **Invasive** means that the mass does not have well-defined borders and is spreading.

Malignant growths tend to spread to distant body sites. **Metastasis** (meh-tahs-tah-sihs) is a pathogenic growth distant from the primary disease site. Metastasis literally means "beyond control." The plural form of metastasis is **metastases** (meh-tahs-tah-sēz). The term **metastasize** (meh-tahs-tah-sīz) is used to describe invasion by the pathogenic growth to a point distant from the primary disease site.

Malignant growths may also be described by their tissue of origin. A **carcinoma** (kahr-sih-nō-mah) is a malignant growth of epithelial cells, whereas a **sarcoma** (sahr-kō-mah) is a malignant neoplasm arising from any type of connective tissue. The combining form **carcin/o** means cancer and the combining form **sarc/o** means flesh. Connective tissue is sometimes referred to as fleshy.

The combining form carcin/o is used to form the term carcinogen. A **carcinogen** is a substance that produces cancer. The suffix **-gen** means producing.

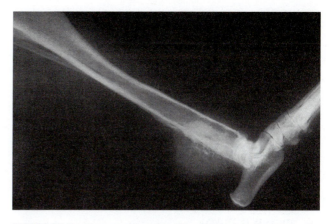

FIGURE 15–7 Osteosarcoma in a dog. Lateral radiograph of a dog with osteosarcoma of the distal tibia.

▶ Test Me: Oncology ◀

Diagnostic tests performed on tumors include
- **touch preps** = collection of cells via pressing a glass slide against a part of the mass. The slide is then examined under a microscope.
- **biopsies** = biopsies are covered in Chapter 10
- **radiographs** = record of ionizing radiation used to visualize internal body structures. Radiographs are taken in oncology patients to assess the extent of some tumors or to check for metastases (Figure 15–6).

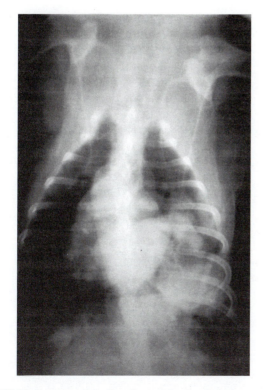

FIGURE 15–6 Ventrodorsal radiograph of pulmonary metastases. *Source:* Photo by Anne E. Chauvet, DVM, Diplomate ACVIM–Neurology Department, University of Wisconsin School of Medicine

▶ Path of Destruction: Oncology ◀

Tumors are named for the tissues involved. Some examples include
- **melanoma** (mehl-ah-nō-mah) = neoplasm composed of melanin-pigmented cells
- **adenocarcinoma** (ahd-eh-nō-kahr-sih-nō-mah) = malignant growth of epithelial glandular tissue
- **osteosarcoma** (ohs-tē-ō-sahr-kō-mah) = malignant neoplasm composed of bone (Figure 15–7)
- **myosarcoma** (mī-ō-sahr-kō-mah) = malignant neoplasm composed of muscle
- **myeloma** (mī-eh-lō-mah) = malignant neoplasm composed of bone marrow
- **lymphoma** (lihm-fō-mah) = general term for neoplasm composed of lymphoid tissue (usually malignant); also called **lymphosarcoma;** abbreviated LSA
- **blastoma** (blahs-tō-mah) = neoplasm composed of immature undifferentiated cells
- **neuroblastoma** (nū-rō-blahs-tō-mah) = malignant neoplasm of nervous tissue origin
- **myxoma** (mihx-ō-mah) = tumor of connective tissue
- **hemangioma** (hē-mahn-jē-ō-mah) = benign neoplasm composed of newly formed blood vessels
- **hemangiosarcoma** (hē-mahn-jē-ō-sahr-kō-mah) = malignant tumor of vascular tissue
- **mast cell tumor** = malignant growth of tissue mast cells (cells that release histamine); abbreviated MCT. Mast cell tumors are associated with

vomiting, anorexia, and various signs depending on the tissue involved.

- **squamous cell carcinoma** (skwā-mohs sehl kahr-sih-nō-mah) = malignant tumor developed from squamous epithelial tissue; abbreviated SCC (Figure 15–8).

▶ Procede With Caution: Oncology ◀

Procedures performed on tumors include

- **surgical excision** = removal of the entire mass plus some normal tissue to ensure that the entire mass is removed
- **radiation therapy** (rā-dē-ah-shuhn thehr-ah-pē) = treatment of neoplasm through the use of X-rays
- **chemotherapy** (kē-mō-thehr-ah-pē) = treatment of neoplasm through the use of chemicals.
- **lymphadenectomy** (lihm-fahd-ehn-ehck-tō-mē) = surgical removal of lymph node. The name for re-

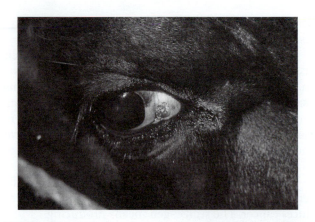

FIGURE 15–8 SCC (cancer eye) of a bovine eye. *Source:* Courtesy of Ron Fabrizius, DVM, Diplomat ACT

moval of any tissue that may have a mass or tumor can be derived by adding the combining form for the area being removed.

REVIEW EXERCISES

Multiple Choice—Choose the correct answer.

1. Red blood cells are called

   a. erythrocytes
   b. leukocytes
   c. thrombocytes
   d. platelets

2. White blood cells are called

   a. erythrocytes
   b. leukocytes
   c. thrombocytes
   d. platelets

3. An abnormal increase in white blood cells is called

   a. leukopenia
   b. leukocytosis
   c. leukemia
   d. leukosis

4. An immature erythrocyte that is characterized by polychromasia (Wright's stain) or a meshlike pattern of threads (new methylene blue stain) is called a

   a. macrophage
   b. prothrombocyte
   c. reticulocyte
   d. phagocyte

5. An immature polymorphonuclear leukocyte is called a

   a. baby PMN
   b. band cell
   c. left shift
   d. micro PMN

6. Lacteals are located in the

   a. groin
   b. loin
   c. small intestine
   d. stomach

7. Lymph is also known as

   a. serum
   b. plasma
   c. interstitial fluid
   d. a and b

8. T cells are responsible for

   a. allergies
   b. humoral immunity
   c. cell-mediated immunity
   d. thymus production

9. B cells are responsible for

   a. allergies
   b. humoral immunity
   c. cell-mediated immunity
   d. thymus production

10. The spleen is

    a. hemolytic
    b. a producer of lymphocytes
    c. a storage area for RBC's
    d. all of the above

Matching—Match the term in Column I with the definition in Column II.

Column I

_____ resistant

_____ heredity

_____ vaccination

_____ immunization

_____ vaccine

_____ multiplication

_____ inhibit

_____ opportunistic

_____ debilitated

Column II

a. weakened or loss of strength

b. ability to cause disease due to debilitation when normally disease would not be produced

c. administration of antigen (vaccine) to stimulate a protective immune response against a specific infectious agent

d. reproducing

e. not susceptible

f. genetic transmission of characteristics from parent to offspring

g. slows or stops

h. a preparation of pathogen (live, weakened, or killed) or portion of pathogen that is administered to stimulate a protective immune response against that pathogen

Fill in the Blanks

1. _____ means no cuts, scrapes, openings, or alterations.

2. The state of being resistant to a specific disease is _____.

3. A substance the body regards as foreign is a/an _____.

4. Less than normal levels of red blood cells or hemoglobin is termed _____.

5. Excessive blood in a part is called _____.

6. _____ is a severe response to a foreign substance.

7. A substance capable of inducing an allergic reaction is called a/an _____.

8. The study of tumors is _____.

9. A general term for a malignant neoplastic disorder of lymphoid tissue is _____.

10. A blood condition in which pathogenic microorganisms or their toxins are present is called _____.

 **Testing, Testing**

Objectives *In this chapter, you should learn to:*

▶ Describe terms and equipment for the basic physical examination
▶ Recognize, define, spell, and pronounce terms associated with physical examinations
▶ Recognize, define, spell, and pronounce terms associated with laboratory analysis
▶ Describe positioning for radiographic and imaging procedures
▶ Recognize, define, spell, and pronounce terms associated with radiographic and imaging procedures

BASIC PHYSICAL EXAMINATION

Physical examinations are performed to assess a patient's condition. **Assessment** is the term used to describe the evaluation of a condition. After a patient is assessed, the information is written into a medical record. The animal's signalment should always be included. A **signalment** is a description of the animal and includes information about an animal including the species, breed, age, and sexual status (intact or neutered).

Vital Signs

Vital signs are parameters taken from the animal to assess its health.

Temperature is a vital sign that tells about the degree of heat or cold. An animals' temperature is recorded in degrees Fahrenheit or Celsius. Different species of animals have different normal temperature ranges. An elevated body temperature is referred to as a **fever**. **Febrile** (fē-brī-ahl) is the medical term for fever; **afebrile** (ā-fē-brī-ahl) means without a fever. **Pyrexia** is another medical term for fever. **Pyr/o** means fire. A decrease in body temperature is known as **hypothermia.**

Pulse is another vital sign that tells the number of times the heart beats per minute. The pulse is also called the **pulse rate.** Pulse is taken by palpation of an artery. Heart rate may also be considered a vital sign and is taken by ausculting the heart with a stethoscope (Figure 16–1). **Heart rate** is the number of times the heart contracts and relaxes per minute. Heart rate is abbreviated HR.

Respiration is a vital sign that tells the number of respirations per minute. Respiration is one total inhale and one total exhale. Respiration is also called the **respiration rate** and is abbreviated RR (Figure 16–2).

Blood pressure is another vital sign that may be taken on veterinary patients. A **sphygmomanometer** (sfihg-mō-mah-nohm-eh-tər) is an instrument used to measure blood pressure (Figure 16–3). A Doppler is used to listen to blood sounds during the measurement of blood pressure in animals.

Listening

Auscultation (ahws-kuhl-tā-shuhn) is listening to body sounds, which usually involves the use of a stethoscope. Auscultation can be used to assess the condition of the heart, lungs, pleura, and abdomen.

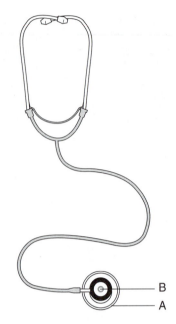

FIGURE 16–1 Parts of a stethoscope. (a) Diaphragm. The diaphragm is the flat, circular portion of the chestpiece covered with a thin membrane. The diaphragm transmits high-pitched sounds, such as those produced by the bowel, lungs, and heart. (b) Bell. The bell is not covered by a membrane. The bell facilitates auscultation of lower-frequency sounds, such as the third and fourth sounds of the heart.

Touching

Palpation (pahl-pā-shuhn) is examination by feeling. During palpation one can feel the texture, size, consistency, and location of body parts or masses.

Percussion (pər-kuhsh-uhn) is examination by tapping the surface to determine density of a body area. Sound may be produced by tapping the surface with a

	Heart rate (beats/min)	Respiratory rate (breaths/min)	Rectal temperature
Dogs	70–160	8–20	37.5°–39° C
Cats	150–210	8–30	38°–39° C
Hamsters	250–500	35–135	37°–38° C
Guinea pigs	230–280	42–104	37°–39.5° C
Rabbits	130–325	30–60	38.5°–40° C
Horses	28–50	8–16	37.5°–38.5° C
Cattle	40–80	12–36	38°–39° C
Sheep	60–120	12–72	39°–40° C
Pigs	58–100	8–18	38°–40° C
Llama	60–90	10–30	37°–39° C
Ferret	230–250	33–36	38°–40° C

FIGURE 16–2 Normal vital sign ranges

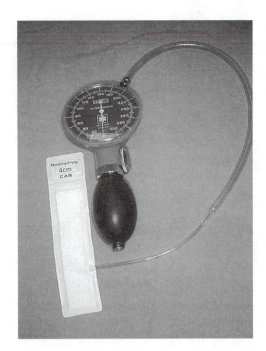

FIGURE 16–3 Sphygmomanometer. *Source:* Teri Raffel, CVT

finger or instrument. The sound produced by percussion will vary depending upon the amount of fluid, solid, or air present in the area being examined.

Looking

Various instruments can be used during the physical examination to obtain a better view of a body system. Examples include an ophthalmoscope and otoscope. A **speculum** (spehck-yoo-luhm) is an instrument to enlarge the opening of a canal or cavity. A speculum is attached to an otoscope (or other scope) to provide a better view of a canal or cavity. A mouth speculum is used to better visualize the oral cavity (Figure 16–4).

LABORATORY TERMINOLOGY

Specialized terminology has been developed to describe tests and results of laboratory tests. Blood for laboratory tests is usually collected via venipuncture. **Venipuncture** (vehn-ih-puhnk-tər) is withdrawing blood from a vein (usually with a needle and syringe).

 Records are kept under many different methods. One method is the SOAP method. **SOAP** is an acronym for subjective, objective, assessment, and plan analysis (Figure 16–5). An acronym is a word formed by the initial letter of the major part of the name. Some diseases, structures, and procedures are derived from a person's name, and the name is known as an **eponym.**

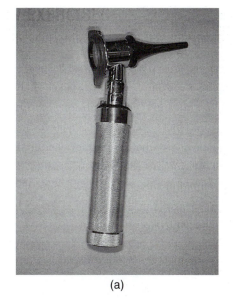
(a)

(b)

FIGURE 16–4 Specula. (a) A speculum is attached to an otoscope to allow better visualization of a canal or cavity. (b) A mouth speculum is used to enhance visualization of the oral cavity. *Source:* (b) Courtesy of Ron Fabrizius, DVM, Diplomat ACT

Examples of this terminology are given in Table 16–1. Laboratory tests often require special equipment to prepare the samples for analysis. A **refractometer** (rē-frahck-tah-mē-tər) is an instrument for determining the deviation of light through objects. Refractometers are used to measure solute (particle) concentration of serum, urine, and other body fluids (Figure 16–7).

A **centrifuge** (sehn-trih-fūj) is a machine that spins samples very rapidly to separate elements based on weight. A centrifuge is used to separate the formed

TABLE 16–1
Blood Test Terminology

Term	Pronunciation	Definition
profile	(prō-fīl)	group of laboratory tests performed on serum; also called **screen** or **panel;** includes tests like glucose, liver enzymes, and kidney enzymes
hemogram	(hē-mō-grahm)	record of the findings in examination of blood
serology	(sē-rohl-ō-jē)	laboratory study of serum and the reactions of antigens and antibodies
differential	(dihf-ər-ehn-shahl)	diagnostic evaluation of the number of blood cell types per cubic millimeter of blood
complete blood count		diagnostic evaluation of blood to determine the number of erythrocytes, leukocytes, and thrombocytes per cubic millimeter of blood; abbreviated CBC
hematocrit	(hē-maht-ō-kriht)	percentage of erythrocytes in blood; "to separate blood"; also called **crit, PCV,** or **packed cell volume** (Figure 16–6)
white cell count		number of leukocytes per cubic millimeter of blood
red cell count		number of erythrocytes per cubic millimeter of blood
prothrombin time	(prō-throhm-bihn)	diagnostic evaluation of the number of seconds required for thromboplastin to coagulate plasma
assay	(ahs-ā)	assessment or test to determine the number of organisms, cells, or amount of a chemical substance found in a sample
radio-immunoassay	(rā-dē-ō-ihm-yoo-nō-ahs-ā)	laboratory technique in which a radioactively labeled substance is mixed with a blood specimen to determine the amount of a particular substance in the mixture; also called **radioassay**
immunofluorescence	(ihm-yoo-nō-floo-rehs-ehns)	method of tagging antibodies with a luminating dye to detect antigen-antibody complexes
agglutination	(ah-gloo-tih-nā-shuhn)	clumping together of cells or particles

Animal Medical Hospital
(S) Dog presented c̄ hx of persistent cough, anorexia, and wt loss.
(O) T = 103.4 °F HR = 120 bpm   RR = 30 breaths/min CRT = 1 sec   mm = pink. Lungs ausculted harshly, Heart sounds appear (N). Rest of PE - WNL
(A) Respiratory Disease — R/O Kennel cough, pneumonia
(P) Radiograph chest CBC medication pending results.

FIGURE 16–5 SOAP method of record keeping

FIGURE 16–6 Microhematocrit tubes. Used to determine a patient's hematocrit

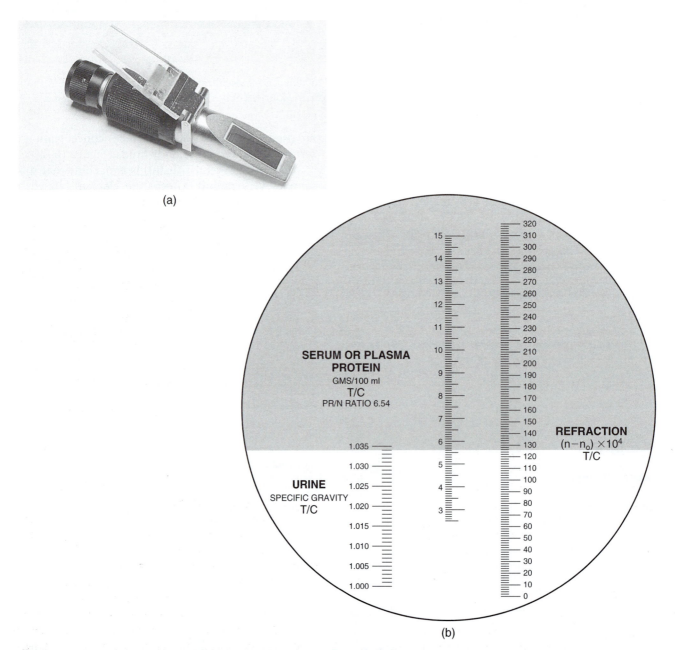

FIGURE 16–7 Refractometer. (a) A refractometer measures specific gravity optically. (b) This refractometer scale shows a urine specific gravity of 1.034 (lower left scale) or plasma protein of 5.6 (middle scale).

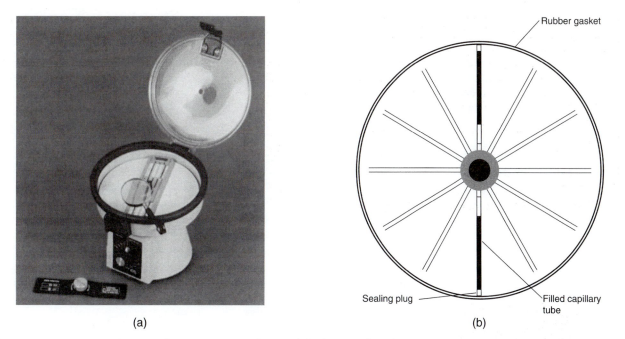

FIGURE 16–8 Centrifuge. (a) Microhematocrit centrifuge with built-in reader; (b) Proper placement of sealed capillary tubes in a microhematocrit centrifuge

elements of blood from the liquid portion of blood. A centrifuge also separates the liquid portion of urine from the heavier solids (Figure 16–8).

The pH of a sample may also provide information about a patient's status. The terms acid and alkaline are used to describe a patient's pH status. **Acid** (ah-sihd) is the property of low pH or increased hydrogen ions. **Alkaline** (ahl-kah-lihn) is the property of high pH or decreased hydrogen ions. Alkaline is also called **basic** (Figure 16–9).

Urinalysis terms are covered in Chapter 7.

BASIC MEDICAL TERMS

Basic medical terms are listed in Table 16–2.

PATHOGENIC ORGANISMS

A **pathogen** (pahth-ō-jehn) is a microorganism that produces disease. A **microorganism** (mī-krō-ōr-gahn-ihzm) is a living organism of microscopic dimensions. Not all microorganisms are pathogens. The term **virulence** (vihr-yoo-lehns) is used to describe the strength of an organism to cause disease (Table 16–3).

TYPES OF DISEASES

Disease is deviation from normal. There are different types of diseases, depending upon how they are spread, what causes them, etc. Types of diseases include

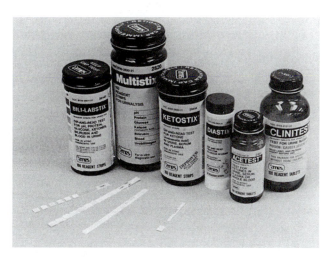

FIGURE 16–9 Reagent strips (dipsticks) for chemical testing of urine

► **infectious disease** (ihn-fehck-shuhs dih-zēz) = disorder caused by pathogenic organisms
► **noninfectious disease** (nohn-ihn-fehck-shuhs dih-zēz) = disorder not caused by organisms (examples include genetic, traumatic, or iatrogenic)
► **contagious** (kohn-tā-juhs) = disease that can be spread from one animal to another by direct or indirect contact. Direct contact is spread from animal to animal, while indirect contact is spread by contact with contaminated objects. Contagious disease may also be referred to as **communicable** (kuh-mū-nih-kuh-buhl).

TABLE 16–2
Basic Medical Terms

diagnosis (dī-ahg-nō-sihs) is the determination of the cause of disease—to "know completely"; plural is **diagnoses** (dī-ahg-nō-sēz). To **diagnose** (dī-ahg-nōs) is the process of reaching the cause of disease.	**differential diagnosis** (dihf-ər-ehn-shahl dī-ahg-nō-sihs) is the determination of possible causes of diseases; a list of possible disease causes	**prognosis** (prohg-nō-sihs) is the prediction of the outcome of disease—to "know before"
sign (sīn) is a characteristic of disease that can be observed by others	**symptom** (sihmp-tuhm) is a characteristic of disease that can only be sensed by the patient; incorrect term in veterinary medicine	A **syndrome** (sihn-drōm) is a set of signs that occur together
acute (ah-kūt) means having a short course with a sudden onset; implies severe. **Peracute** (pər-ah-kyūt) means having an excessively acute onset.	**chronic** (krohn-ihck) means having a long course with a progressive onset; persisting for a long time	**remission** (rē-mihs-shuhn) is partial or complete disappearance of disease signs and/or symptoms
endemic (ehn-dehm-ihck) is the ongoing presence of disease within a group also called **enzootic** (ehn-zō-oh-tihck) if the disease is always present within an *animal* community	**epidemic** (ehp-ih-dehm-ihck) is the sudden and widespread outbreak of disease within a group; also called **epizootic** (ehp-ih-zō-oh-tihck) if the outbreak attacks many *animals* within a group	**pandemic** (pahn-dehm-ihck) means occurring over a large geographic area

- **noncontagious** (nohn-kohn-tā-juhs) = disease that cannot be spread to another animal by contact or contact with an infected object
- **iatrogenic disease** (ī-aht-rō-jehn-ihck dih-zēz) = disorder caused by treatment
- **nosocomial infection** (nōs-ō-kō-mē-ahl ihn-fehck-shuhn) = disorder caused by pathogenic organisms contracted in a facility or clinic
- **idiopathic disease** (ihd-ē-ō-pahth-ihck) = disorder of unknown cause. Idiopathic disease is a disease peculiar to an individual and not likely to be seen in others.

DISEASE TERMINOLOGY

- **disease** (dih-zēz) = deviation from normal health or production
- **transmission** (trahnz-mihs-shuhn) = transfer from one animal to the next . There are different types of disease transmission including **bloodborne transmission** (spread of disease via blood or body fluids), **sexual transmission** (spread of disease via contract with reproductive areas or through copulation), **airborne transmission** (spread of disease via respiratory droplets), and **fecal/oral transmission** (spread of disease via eating, drinking, or licking contaminated food, water, or object).
- **transmissible** (trahnz-mihs-ih-buhl) = ability to transfer from one animal to the next

- **labile** (lā-bīl) = unstable
- **phobia** (fō-bē-ah) = extreme fear
- **asymptomatic** (ā-sihmp-tō-mah-tihck) = without signs of disease
- **clinical** (klihn-ih-kahl) = visible, readily observed, pertaining to treatment
- **subclinical** (suhb-klihn-ih-kahl) = without showing signs of disease
- **etiology** (ē-tē-ohl-ō-jē) = study of disease causes
- **focus** (fō-kuhs) = localized region
- **incidence** (ihn-sih-dehns) = number of new cases of disease occurring during a given time
- **lethal** (lē-thahl) = causing death
- **morbid** (mōr-bihd) = afflicted with disease
- **morbidity** (mōr-bihd-ih-tē) = ratio of diseased animals to well animals in a population
- **mortality** (mōr-tahl-ih-tē) = ratio of diseased animals to diseased animals that die
- **moribund** (mōr-ih-buhnd) = near death
- **palliative** (pah-lē-ah-tihv) = relief of condition, but not cure
- **prevalence** (preh-vah-lehns) = number of cases of disease in a population at a certain time
- **prophylaxis** (prō-fihl-ahcks-sihs) = prevention
- **sequela** (sē-kwehl-ah) = consequence of disease
- **zoonosis** (zō-ō-nō-sihs) = disease that is able to be transmitted between animals and humans
- **carrier** (kahr-ē-ər) = animal that harbors disease without clinical signs and serves as a distributor of infection

TABLE 16–3
Types of Organisms

Organism	Examples	
bacterium (bahck-tē-rē-uhm) is a microscopic, unicellular organism; plural is bacteria (bahck-tē-rē-ah)	**staphylococci** (stahf-ih-lō-kohck-sī) are grapelike clusters of round bacteria; **coccus** (kohck-uhs) means round.	Staphylococcus
	streptococci (strehp-tō-kohck-sī) are round bacteria that form twisted chains	Streptococcus
	bacilli (bah-sihl-ī) are rod-shaped bacteria	Bacillus
	spirochetes (spī-rō-kētz) are spiral-shaped bacteria	Spirochete
	spore (spōr) is a resistant, oval body formed within some bacteria	Spore
	rickettsia (rih-keht-sē-ah) is a bacterium transmitted by lice, fleas, ticks, or mites	
fungus (fuhng-guhs) is a simple parasitic plant; plural is **fungi** (fuh-jī)	**yeast** (yēst) is a budding form of fungus	Yeast
parasite (pahr-ah-sīt) is an organisms that lives on or within another living organism		
virus (vī-ruhs) is a small organism that is not visualized via microscopy		Virus

▶ **excessive** (ehck-sehs-ihv) = more than normal
▶ **susceptible** (sah-sehp-tih-buhl) = lacking resistance
▶ **contract** (kohn-trahckt) = common term for catching a disease
▶ **traumatic** (traw-mah-tihck) = pertaining to, resulting from, or causing injury
▶ **atraumatic** (ā-traw-mah-tihck) = pertaining to, resulting from, or caused by a noninjurious route
▶ **swollen** (swohl-ehn) = condition of enlargement of a body part due to fluid retention
▶ **germ** (jərm) = common term for any pathogenic microorganism, but especially bacterial or viral organisms
▶ **epidemiology** (ehp-ih-dē-mē-ohl-ō-gē) = study of relationships determining frequency and distribution of diseases

ENDOSCOPY

Endoscopy is the visual examination of the interior of any cavity of the body by means of an endoscope. The procedures and instruments are named for the body parts involved. Specific endoscopic procedures are covered in the chapter on the body system chapter they are used on.

Endoscopic surgery is a procedure using an endoscope to aide in surgical procedures so that only very small incisions are made. Some instruments used in endoscopic surgery include a trocar and cannula. A **trocar** (trō-kahr) is a sharp, needle-like instrument that has a cannula (tube), which is used to puncture the wall of a body cavity and withdraw fluid or gas. A **cannula** (kahn-yoo-lah) is a hollow tube.

CENTESIS

Centesis is the surgical puncture to remove fluid for diagnostic purposes or to remove fluid or gas. Specific centesis procedures are covered in the chapter on the body system they are used in.

IMAGING TECHNIQUES

Imaging techniques are used to visualize and examine internal structures of the body.

Radiology

The first imaging technique involves the use of ionizing radiation or X-rays to produce an image. **Radiography** (rā-dē-ohg-rah-fē) is the procedure in which film is exposed as ionizing radiation passes through the patient and shows the internal body structures in profile. A **radiograph** (rā-dē-ō-grahf) or X-ray is the record of ionizing radiation used to visualize internal body structures. Note that **graph** (as opposed to gram) is used to mean record in this case.

Radiographs are composed of shades of gray. Hard tissues such as bone are called radiopaque. **Radiopaque** (rā-dē-ō-pāg) is the quality of appearing white or light gray on a radiograph. Air and soft tissues are called radiolucent. **Radiolucent** (rā-dē-ō-loo-sehnt) is the quality of appearing black or dark gray on a radiograph.

Radiology (rā-dē-ohl-ō-jē) is the study of internal body structures after exposure to ionizing radiation. A radiologist (rā-dē-ohl-ō-jihst) is a specialist who studies internal body structures after exposure to ionizing radiation (Figure 16–10).

> X-rays were discovered in 1895 by Wilhelm Konrad Roentgen, a German physicist. Radiology comes from the Latin **radius** meaning a rod and the suffix **-logy** meaning "the science of" or "the study of." Radiology is also known as roentgenology as a tribute to its discoverer. A **roentgen** (rehnt-kihn) is the international unit of radiation. Another term, **rad** (an acronym for *r*adiation *a*bsorbed *d*ose) is a unit by which absorption of ionizing radiation is measured.

Two abbreviations commonly used in radiography are kVp and MAS. **kVp** stands for kilovolts peak and represents the strength of the X-ray beam. **MAS** stands for milliamperes per second and represents the number of X-ray beams (because it is based on time).

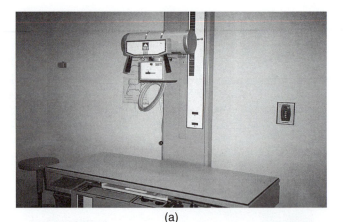

(a)

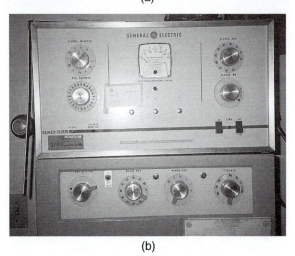

(b)

FIGURE 16–10 Radiographic machine. (a) Radiographic table and X-ray tube; (b) Radiographic control panel

There are different types of radiographs depending upon how they are taken or if any additional diagnostic tool is used. A **scout film** is a plain radiograph without the use of contrast medium. A scout radiograph "scouts" out whether or not there is an abnormality to test further.

Some body structures, such as the intestinal lumen, are difficult to assess using X-rays alone. In these cases, a radiographic contrast medium may be used. **Radiographic contrast** (rā-dē-ō-grahf-ihck kohn-trahst) medium or material is a substance used to show structures on X-ray that are otherwise difficult to see. **Barium sulfate** (bār-ē-uhm suhl-fāt) or **barium** is one example of contrast material.

The type of contrast radiograph taken depends upon what structures are to be visualized. The route of administration of contrast medium also varies depending upon what structures are to be visualized.

A **lower GI** is a type of contrast radiograph used to visualize the structures of the lower gastrointestinal tract. In a lower GI an enema is used to introduce contrast material into the colon; hence, it is also called a **barium enema** (bār-ē-uhm ehn-ah-mah). An **upper GI** is a type of contrast radiograph used to visualize the structures of the upper gastrointestinal tract. In an upper GI contrast material is swallowed; hence, it is also called a **barium swallow** (bār-ē-uhm swahl-ō).

Contrast material can also be injected intravenously. An intravenous contrast medium is injected into the vein to make visible the flow of blood through the blood vessels and organs.

Radiographic techniques are also named for the vessels or organs involved. An example is a **lymphangiography** (limh-fahn-jē-ohg-rah-fē), which is a radiographic examination of the lymphatic vessels after injection of contrast material.

The Two P's

Two terms frequently used in radiography are projection and positioning. **Projection** (prō-jehck-shuhn) is the path of the X-ray beam. **Positioning** (pō-sih-shuhn-ihng) is the specified body position and the part of the body closest to the film. **Recumbency** is used in reference to positioning (refer to Chapter 2 for positioning terms). **Anatomic position** (ahn-ah-tohm-ihck pō-sih-shuhn) refers to the animal in its normal standing position (Table 16–4).

Computed Tomography

Computed tomography (kohm-puh-tehd tō-moh-grah-fē) is the procedure in which ionizing radiation with computer assistance passes through the patient and shows the internal body structures in cross-sectional views. It is also called **CT scan** or **CAT** (computed axial tomography) **scan. Tomography** (tō-moh-grahf-ē) is a recording of the internal body structures at predetermined planes. Information obtained by radiation detectors is downloaded to a computer, analyzed, and converted into gray-scale images corresponding to anatomic body slices. These images are viewed on a television monitor or as a printed hard copy.

Magnetic Resonance Imaging

Magnetic resonance imaging (mahg-neh-tihck reh-sohn-ahns ih-mah-gihng) is the procedure in which radio waves and a strong magnetic field pass through the patient and show the internal body structures in three-dimensional views. Magnetic resonance imaging is abbreviated MRI. MRI is used for imaging the brain, spine, and joints (Figure 16–11).

Fluoroscopy

Fluoroscopy (floor-ohs-kō-pē) is the procedure to visually examine internal body structures in motion using radiation to project images on a fluorescent screen. The combining form **fluor/o** means luminous. **Luminous** (loo-mih-nuhs) means giving off a soft glowing light.

Ultrasound

Ultrasound (uhl-trah-sownd) or **ultrasonography** (uhl-trah-soh-noh-grah-fē) is the imaging of internal body structures by recording echoes of high-frequency

TABLE 16–4
Types of Projection

Projection	Pronunciation	Definition
craniocaudal projection	(krā-nē-ō-kaw-dahl prō-jehck-shuhn)	X-ray beam passes from cranial to caudal; used to describe extremity radiographs; also called **anteroposterior projection** (ahn-tēr-ō-poh-stēr-ē-ər prō-jehck-shuhn)
caudocranial projection	(kaw-dō-krā-nē-ahl prō-jehck-shuhn)	X-ray beam passes from caudal to cranial; used to describe extremity radiographs; also called **posteroanterior projection** (poh-stēr-ō-ahn-tēr-ih-ər prō-jehck-shuhn)
dorsoventral projection	(dōr-sō-vehn-trahl prō-jehck-shuhn)	X-ray beam passes from the back to the belly; abbreviated D/V
ventrodorsal projection	(vehn-trō-dōr-sahl prō-jehck-shuhn)	X-ray beam passes from the belly to the back; abbreviated V/D
lateral projection	(lah-tər-ahl prō-jehck-shuhn)	X-ray beam passes from side to side with the patient at right angles to the film
oblique projection	(ō-blēk prō-jehck-shuhn)	X-ray beam passes through the body on an angle

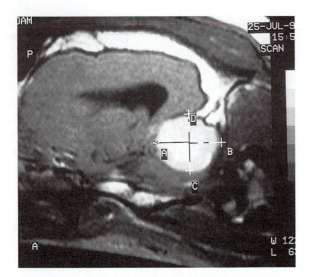

FIGURE 16–11 MRI with contrast material in a dog with a cerebellar tumor. *Source:* Photo by Anne E. Chauvet, DVM, Diplomate ACVIM–Neurology, University of Wisconsin School of Veterinary Medicine

waves. Ultrasound is most effective for viewing solid organs or soft tissues not blocked by bone or air. Ultrasound is also effective for viewing body parts through fluid, as in an ultrasound of a gravid uterus. A **sonogram** (soh-nō-grahm) is the record of the internal body structures by recording echoes of pulses of sound waves above the range of human hearing (Figure 16–12).

Terms related to ultrasound techniques are

▶ **amplitude** (ahm-plih-tood) = intensity of an ultra-sound wave
▶ **attenuation** (ah-tehn-yoo-ā-shuhn) = loss of intensity of the ultrasound beam as it travels through tissue
▶ **echoic** (eh-kō-ihck) = ultrasound property of producing adequate levels of reflections (echoes) when sound waves are returned to the transducer and displayed
▶ **anechoic** (ahn-eh-kō-ihck) = ultrasonic term for when waves are transmitted to deeper tissue and none are reflected back
▶ **hyperechoic** (hī-pər-eh-kō-ihck) = tissue that reflects more sound back to the transducer than the surrounding tissues; appears bright

▶ **hypoechoic** (hī-pō-eh-kō-ihck) = tissue that reflects less sound back to the transducer than the surrounding tissues; appears dark
▶ **isoechoic** (ī-sō-eh-kō-ihck) = tissue that has the same ultrasonic appearance as that of the surrounding tissue
▶ **frequency** (frē-kwehn-sē) = number of cycles per unit of time
▶ **velocity** (vehl-oh-sih-tē) = speed at which something travels through an object
▶ **wavelength** (wāv-lehngth) = length that a wave must travel in one cycle
▶ **resolution** (rehs-ō-loo-shuhn) = ability to separately identify different structures on radiograph or ultrasound

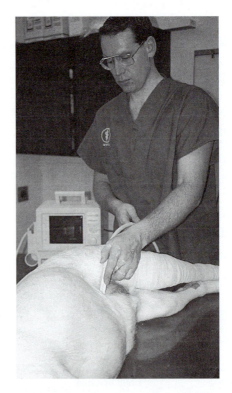

FIGURE 16–12 Ultrasound of a canine abdomen. *Source:* Lodi Veterinary Hospital, S.C.

REVIEW EXERCISES

Multiple Choice—Choose the correct answer.

1. Examination by tapping the surface to determine density of a body area is termed
   a. palpation
   b. auscultation
   c. percussion
   d. tapping

2. The percentage of RBCs in blood is called a
   a. hematocrit
   b. PCV
   c. crit
   d. all of the above

3. A machine that spins samples very rapidly to separate elements based on weight is a/an
   a. counter
   b. centrifuge
   c. refractometer
   d. cannula

4. The quality of appearing white or light gray on a radiograph is called
   a. radiopaque
   b. radiolucent
   c. radiodense
   d. radiopale

5. The determination of the cause of disease is the
   a. prognosis
   b. diagnosis
   c. symptom
   d. sign

6. A set of signs that occur together are referred to as a/an
   a. sign
   b. symptom
   c. endemic
   d. syndrome

7. The medical term for round is
   a. bacilli
   b. spirochete
   c. coccus
   d. strept/o

8. The term for relief of condition, but not a cure is
   a. prognosis
   b. diagnosis
   c. prevalence
   d. palliative

9. Pertaining to fever is
   a. febrile
   b. friable
   c. lethal
   d. morbid

10. A disease that is able to be transmitted between animals and humans is said to be
    a. animalistic
    b. zoonotic
    c. humanistic
    d. sequela

Matching—Match the term in Column I with the definition in Column II.

Column I

_____ infectious disease

_____ contagious disease

_____ noncontagious disease

_____ noninfectious disease

_____ communicable disease

_____ iatrogenic disease

_____ nosocomial infection

Column II

a. disorder caused by treatment

b. disorder caused by pathogenic organisms contracted in a facility or clinic

c. disease spread from one animal to another by direct or indirect contact

d. disorder caused by pathogenic organisms

e. disorder transmitted from animal to animal or by contact with contaminated objects.

f. disorder not caused by organisms (examples include genetic, traumatic, or iatrogenic)

g. disease that cannot be pread to another animal by contact or contact with an infected object

17 Drugs, Disease, and Dissection

Objectives *In this chapter, you should learn to:*

► Recognize, define, spell, and pronounce the terms associated with pharmacology and drugs used in various treatments
► Recognize, define, spell, and pronounce the terms associated with pathological procedures and processes
► Recognize, define, spell, and pronounce the terms associated with different types of surgery and the instruments used in surgery

PHARMACOLOGIC TERMS

Pharmacology (fahrm-ah-kohl-lō-jē) is the study of the nature, uses, and effects of drugs. Some drugs need to be dispensed by a licensed professional and other drugs do not. A **prescription** (per-skrihp-shuhn) **drug** is a medication that may be purchased by prescription or from a licensed professional. An **over-the-counter** (ō-vər theh kount-ər) **drug** is a medication that may be purchased without a prescription. A **generic** (jehn-ār-ihck) **drug** is a medication not protected by a brand name or trademark (it is also called a **nonproprietary drug**).

Terminology Related to Pharmacology

► **placebo** (plah-sē-bō) = inactive substance that is given for its suggestive effects or substance used as a control in experimental setting
► **regimen** (reh-geh-mehn) = directions
► **contraindication** (kohn-trah-ihn-dih-kā-shuhn) = recommended not to be used
► **dosage** (dō-sahj) = amount of medication based on units per weight of animal (i.e., 10 mg/lb, 2 mg/kg, etc.)
► **dose** (dōs) = amount of medication measured (i.e., milligrams, milliliters, units, grams, etc)

► **pharmacokinetics** (fahrm-ah-kō-kihn-eht-ihcks) = movement of drugs or chemicals; consists of absorption, distribution, biotransformation, and elimination
► **ionized** (ī-ohn-īzd) = electrically charged
► **nonionized** (nohn-ī-ohn-īzd) = not charged electrically
► **hydrophilic** (hī-drō-fihl-ihck) = water-loving; ionized form
► **lipophilic** (lihp-ō-fihl-ihck) = fat-loving; nonionized form
► **agonist** (ā-gohn-ihst) = substance that produces effect by binding to an appropriate receptor
► **antagonist** (ahn-tā-gohn-ihst) = substance that inhibits a specific action by binding with a particular receptor instead of allowing the agonist to bind to the receptor
► **dosage interval** (dō-sahj ihn-tər-vahl) = time between administration of drug
► **exogenous** (ehcks-ah-jehn-uhs) = originates from outside the body
► **endogenous** (ehn-dah-jehn-uhs) = originates from within the body
► **chelated** (kē-lā-tehd) = bound to and precipitated out of solution
► **drug** (druhg) = agent used to diagnose, prevent, or treat a disease

▶ **efficacy** (ehf-ih-kah-sē) = extent to which a drug causes the intended effects; "effectiveness"

▶ **bacterin** (bahck-tər-ihn) = killed bacterial vaccine

▶ **monovalent** (mohn-ō-vā-lehnt) = vaccine, antiserum, or antitoxin developed specifically for a single antigen or organism

▶ **polyvalent** (poh-lē-vā-lehnt) = vaccine, antiserum, or antitoxin active against multiple antigens or organisms; "mixed vaccine"

▶ **antitoxin** (ahn-tih-tohcks-sihn) = specific antiserum aimed against a poison that contains a concentration of antibodies extracted from the serum or plasma of a healthy animal

▶ **antiserum** (ahn-tih-sēr-uhm) = serum containing specific antibodies extracted form a hyperimmunized animal or an animal that has been infected with the microorganisms containing antigen

▶ **diffusion** (dih-fū-shuhn) = movement of solutes from an area of high concentration to low concentration.

▶ **hypertonic** (hī-pər-tohn-ihck) **solution** = solution that has fewer particles inside the cell than outside. Hypertonic solutions are usually compared with blood

▶ **hypotonic** (hī-pō-tohn-ihck) **solution** = solution that has more particles inside the cell than outside

▶ **isotonic** (ī-sō-tohn-ihck) **solution** = solution that has equal particles inside and outside the cell (Figure 17–1).

▶ **prevention** (prē-vehn-shuhn) = to avoid; also called **prophylaxis** (prō-fih-lahck-sihs)

▶ **hyperkalemia** (hī-pər-kā-lē-mē-ah) = excessive level of blood potassium

▶ **hypernatremia** (hī-pər-nā-trē-mē-ah) = excessive level of blood sodium

▶ **hypokalemia** (hī-pō-kā-lē-mē-ah) = deficient level of blood potassium

▶ **hyponatremia** (hī-pō-nā-trē-mē-ah) = deficient level of blood sodium

▶ **hypovolemia** (hī-pō-vō-lē-mē-ah) = decreased circulating blood volume

▶ **turgor** (tər-gər) = degree of fullness or rigidity due to fluid content

▶ **osmosis** (ohz-mō-sihs) = movement of water across a cell membrane

Routes of Administration

▶ **percutaneous** (pehr-kyoo-tā-nē-uhs) = through the skin

▶ **sublingual** (suhb-lihng-yoo-ahl or suhb-lihng-wahl) = under the tongue

▶ **oral** (ōr-ahl) = by mouth; abbreviated PO or p.o. Nothing orally is NPO or n.p.o.

▶ **rectal** (rehck-tahl) = by rectum

▶ **parenteral** (pah-rehn-tər-ahl) = administration through routes other than the gastrointestinal tract (Figure 17–2)

▶ **nonparenteral** (non-pah-rehn-tər-ahl) = administration via the gastrointestinal tract

▶ **intradermal** (ihn-trah-dər-mahl) = within the skin; abbreviated ID

▶ **transdermal** (trahnz-dər-mahl) = across the skin. Medication is stored in a patch placed on the skin and the medication is absorbed through the skin.

▶ **intramuscular** (ihn-trah-muhs-kyū-lahr) = within the muscle; abbreviated IM

▶ **intravenous** (ihn-trah-vehn-uhs) = within the vein; abbreviated IV

▶ **intraarterial** (ihn-trah-ahr-tehr-ē-ahl) = within the artery; abbreviated IA

▶ **intraperitoneal** (ihn-trah-pehr-ih-tohn-ē-ahl) = within the peritoneal cavity; abbreviated IP

▶ **subcutaneous** (suhb-kyoo-tā-nē-uhs) = under the skin or dermal layer; abbreviated SQ, SC, SubC, or SubQ

Hypertonic solution
(more particles
outside cell)

Hypotonic solution
(fewer particles
outside cell)

Isotonic solution
(equal particles inside
and outside cell)

FIGURE 17–1 Solution tonicity

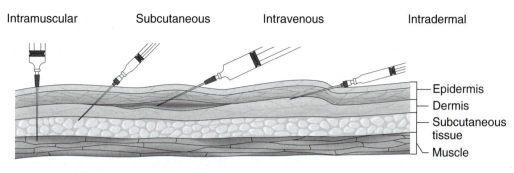

Intramuscular   Subcutaneous   Intravenous   Intradermal

— Epidermis
— Dermis
— Subcutaneous tissue
— Muscle

FIGURE 17–2 Some examples of parenteral routes of drug administration

▶ **intrathecal** (ihn-trah-thē-kahl) = within a shealth; injection of a substance through the spinal cord and into the subarachnoid space

▶ **intratracheal** (ihn-trah-trā-kē-ahl) = within the trachea or windpipe

▶ **intraosseous** (ihn-trah-ohs-ē-uhs) = within the bone (medullary cavity of a long bone)

▶ **nebulization** (nehb-yoo-lih-zā-shuhn) = process of making a fine mist; a method of drug administration

▶ **inhalation** (ihn-hah-lā-shuhn) = vapors and gases taken in through the nose and mouth and absorbed into the bloodstream through the lungs

Drug Categories

▶ **analgesic** (ahn-ahl-jē-zihck) = substance that relieves pain without affecting consciousness

▶ **anesthetic** (ahn-ehs-theht-ihck) = substance that produces a lack of sensation (Figure 17–3)

▶ **antibiotic** (ahn-tih-bī-ah-tihck) = substance that inhibits the growth of or kills bacteria. Antibiotics can be **bacteriostatic** (bahck-tē-rē-ō-stah-tihck), which means controlling bacterial growth , or **bacteriocidal** (bahck-tē-rē-ō-sī-dahl), which means killing bacteria

▶ **antipyretic** (ahn-tih-pī-reh-tihck) = substance that reduces fever

▶ **anticonvulsant** (ahn-tih-kohn-vuhl-sahnt) = substance that prevents seizures

▶ **anthelmintic** (ahn-thehl-mihn-tihck) = substance that works against intestinal worms

▶ **asepsis** (ā-sehp-sihs) = state without infection

▶ **antineoplastic agent** (ahn-tih-nē-ō-plah-stihck ā-jehnt) = substance that treats neoplasms; usually used against malignancies

▶ **emetic** (ē-meh-tihck) = substance that induces vomiting

▶ **antiemetic** (ahn-tih-ē-meh-tihck) = substance the prevents vomiting

▶ **antidiarrheal** (ahn-tih-dī-ər-rē-ahl) = substance that prevents watery, frequent bowel movements

▶ **inotrope** (ihn-ō-trōp) = substance affecting muscle contraction

▶ **antitussive** (ahn-tih-tuhs-ihv) = substance that reduces coughing

▶ **mucolytic** (mū-kō-lih-tihck) = substance that breaks up mucus and reduces its viscosity

▶ **antiseptic** (ahn-tih-sehp-tihck) = chemical agent that kills or prevents the growth of microorganisms on living tissue

▶ **disinfectant** (dihs-ehn-fehck-tahnt) = chemical agent that kills or prevents the growth of microorganisms on inanimate objects

▶ **sterilize** (stehr-ih-līz) = to destroy all organisms

▶ **endectocide** (ehnd-ehck-tō-sīd) = agent that kills both internal and external parasites

▶ **mydriatic agent** (mihd-rē-ah-tihck ā-jehnt) = substance used to dilate the pupils

▶ **miotic agent** (mī-ah-tihck ā-jehnt) = substance used to constrict the pupils

▶ **anticoagulant** (ahn-tih-kō-ahg-yoo-lahnt) = substance that inhibits clot formation

▶ **antipruritic agent** (ahn-tih-pər-ih-tihck ā-jehnt) = substance that controls itching

▶ **immunosuppressant** (ihm-yoo-nō-suhp-prehs-ahnt) = substance that prevents or decreases the body's reaction to invasion by disease or foreign material

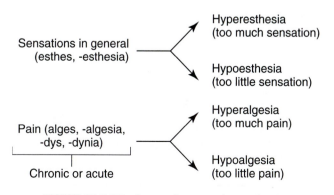

Sensations in general (esthes, -esthesia)
→ Hyperesthesia (too much sensation)
→ Hypoesthesia (too little sensation)

Pain (alges, -algesia, -dys, -dynia)
Chronic or acute
→ Hyperalgesia (too much pain)
→ Hypoalgesia (too little pain)

FIGURE 17–3 Word parts for sensation and pain

▶ **cytotoxic agent** (sī-tō-tohcks-ihck ā-jehnt) = substance that kills or damages cells

Weights and Measures

▶ **kilogram** (kihl-ō-grahm) = unit of weight that is 1000 grams; 1 kilogram is approximately 2.2 pounds; abbreviated kg
▶ **gram** (grahm) = metric base unit of weight, equal to 0.035 ounce; abbreviated g
▶ **milligram** (mihl-ih-grahm) = metric unit of weight, equal to one one-thousandth of a gram; abbreviated mg
▶ **liter** (lē-tər) = metric base unit of volume, equal to 0.2642 gallons ; abbreviated l
▶ **milliliter** (mihl-ih-lē-tər) = metric unit of volume, equal to 0.034 of an ounce or one one-thousandth of a liter; abbreviated ml; equivalent to 1 cubic centimeter (cc)
▶ **meter** (mē-tər) = metric base unit of length, equal to 1.09 yards; abbreviated m
▶ **millimeter** (mihl-ih-mē-tər) = metric unit of length, equal to one one-thousandth of a meter; abbreviated mm
▶ **percent** (pər-sehnt) = part per 100 parts; represented by % (Table 17–1)

TABLE 17–1
Frequently Used Drug Abbreviations

Abbreviation	Definition
bid	twice daily (bis in die)
c̄	with
cc	cubic centimeter (same as ml)
gt	drop (gutta); drops is gtt (guttae)
ml	milliliter
NPO/n.p.o.	nothing orally (non per os)
PO/p.o.	orally (per os)
q	every
qd	every day (same as sid)
qh	every hour
qid	four times daily
sid	once daily
tid	three times daily (ter in die)
tab	tablet (also abbreviated T)
T	tablespoon or tablet
prn	as needed
p̄	after
s̄	without
q8h	every 8 hours
q12h	every 12 hours
q4h	every 4 hours
q6h	every 6 hours
qn	every night
qod or eod	every other day

SURGICAL TERMINOLOGY

Surgery is that branch of science that treats diseases, injuries, and deformities by manual or operative methods. Surgical terms were developed to describe concisely many of the surgical procedures performed by surgeons. Some surgical terms include

▶ **excise** (ehck-sīz) = to surgically remove
▶ **incise** (ihn-sīz) = to surgically cut into
▶ **ligate** (lī-gāt) = to tie or strangulate. A **ligature** (lihg-ah-chūr) is any substance used to tie or strangulate. Ligatures are usually suture material.
▶ **fenestration** (fehn-ih-strā-shuhn) = perforation.
▶ **pinning** (pihn-ihng) = insertion of a metal rod into the medullary cavity of a long bone
▶ **friable** (frī-ah-buhl) = easily crumbled
▶ **lavage** (lah-vahj) = irrigation of tissue with copious amount of fluid
▶ **suction** (suhck-shuhn) = aspiration of gas or fluid by mechanical means
▶ **wicking** (wihck-ihng) = act of providing material that moves liquid from one area to another (thus provides a potential infection source)
▶ **appositional** (ahp-ō-sih-shuhn-ahl) = placing side to side
▶ **inversion** (ihn-vər-shuhn) = turning inward
▶ **eversion** (ē-vər-shuhn) = turning outward (Figure 17–4)
▶ **imbrication** (ihm-brih-kā-shuhn) = overlapping of apposing surfaces
▶ **fulguration** (fuhl-gər-ā-shuhn) = destruction of living tissue by electric sparks generated by a high-frequency current
▶ **debridement** (dē-brīd-mehnt) = removal of foreign material and devitalized or contaminated tissue
▶ **avulsion** (ā-vuhl-shuhn) = tearing away of a part
▶ **exteriorize** (ehcks-tēr-ē-ōr-īz) = to move an internal organ to the outside of the body
▶ **laceration** (lah-sihr-ā-shuhn) = act of tearing
▶ **transfix** (trahnz-fihcks) = pierce through and through. A transfixion suture pierces through an organ before ligation (Figure 17–5).
▶ **implant** (ihm-plahnt) = material inserted or grafted into the body
▶ **fracture** (frahck-chər) = breaking of a part, especially a bone
▶ **epithelialization** (ehp-ih-thē-lē-ahl-ih-zā-shuhn) = healing by growth of epithelium over an incomplete surface
▶ **enucleation** (ē-nū-klē-ā-shuhn) = removal of an organ in whole; usually used for removal of the eyeball
▶ **curettage** (kwoo-reh-tahj) = removal of material or growths from the surface of a cavity
▶ **sacculectomy** (sahk-yoo-lehck-tō-mē) = surgical

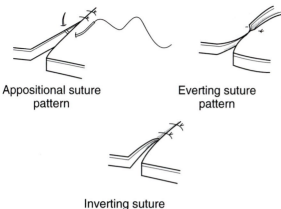

Appositional suture pattern

Everting suture pattern

Inverting suture pattern

FIGURE 17–4 Appositional versus everting versus inverting suture patterns

removal of a sac-like part; usually refers to surgical removal of the anal sacs

► **suture** (soo-chuhr) = to stitch or close an area; also refers to the material used in closing a surgical or traumatic wound with stitches. Suture material may be absorbable or nonabsorbable. (Suture is also a type of joint.)

► **transect** (trahnz-sehckt) = "to cut across"; a cross-section or section made across a long axis. **Sect** means to cut.

► **dissect** (dī-sehckt) = separation or cutting apart; **dissecare** is Latin for "to cut up"

► **resect** (rē-sehckt) = to remove an organ or tissue. Resect is used in reference to holding a tissue or organ out of the surgical field.

► **coaptation** (kō-ahp-tā-shuhn) = act of approximating

► **eviscerate** (ē-vihs-ər-āt) = removal or exposure of internal organs

► **postop** (pōst ohp) = common term for after surgery; postoperatively

► **preop** (prē ohp) = common term for before surgery; preoperatively

► **intraop** (ihn-trah ohp) = common term for during or within surgery; intraoperatively

► **flap** (flahp) = mass of tissue for grafting in which part of the tissue is still adhered to the body; used to repair defects adjacent to the mass sight

► **transplant** (trahnz-plahnt) = to transfer tissue from one part to another part

► **graft** (grahft) = tissue or organ for transplantation/implantation. There are different types of grafts. An **allograft** (ah-lō-grahft) is a graft from another individual of the same species. An **autograft** (aw-tō-grahft) is a graft from the same individual. An **isograft** (ī-sō-grahft) is a graft from

genetically identical animals, such as twins or inbred strains.

► **rupture** (ruhp-chuhr) = forcible tearing

► **involucrum** (ihn-voh-loo-kruhm) = covering or sheath that contains a sequestrum of bone

► **imbricate** (ihm-brih-kāt) = to tighten with sutures

► **lumpectomy** (luhmp-ehck-tō-mē) = general term for surgical removal of a mass

► **dehiscence** (dē-hihs-ehns) = disruption or opening of the surgical wound

► **seroma** (sehr-ō-mah) = accumulation of serum beneath the surgical incision

► **aseptic technique** (ā-sehp-tihck tehck-nēk) = precautions taken to prevent contamination of a surgical wound

Surgical Equipment

► **autoclave** (aw-tō-klāv) = apparatus for sterilizing by steam under pressure

► **drape** (drāp) = cloth arranged over a patient's body to provide a sterile field around the area to be examined, treated, or incised

► **scalpel** (skahl-puhl) = small, straight knife with a thin, sharp blade used for surgery and dissection

► **tissue forceps** (tihs-yoo fōr-sehps) = tweezer-like, nonlocking instruments used to grasp tissue

► **retractor** (rē-trahck-tər) = instrument used to hold back tissue (Figure 17–6)

► **boxlock** (bohcks-lohck) = movable joint of any ringed instrument (Figure 17–7)

► **rongeurs** (rohn-jūrz) = forceps with cupped jaws used to break large bone pieces into smaller ones

► **currette** (kwoor-reht) = instrument with cupped head to scrape material from cavity walls

► **elevator** (ehl-eh-vā-tər) = instrument used to reflect tissue from bone

► **emasculator** (ē-mahs-kwoo-lā-tər) = instrument used in closed castrations to crush and sever the spermatic cord

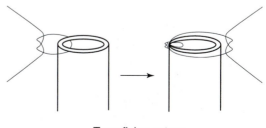

Transfixion suture

FIGURE 17–5 Transfixion suture. Transfixion sutures are used for large isolated vessels and organs to prevent slippage of the ligature.

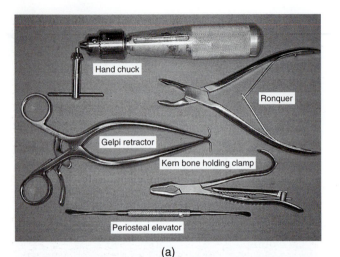

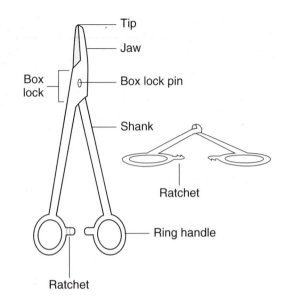

FIGURE 17–7 Parts of surgical instruments

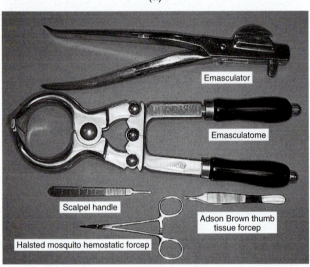

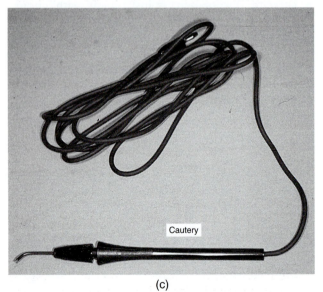

FIGURE 17–6 (a) and (b) Examples of surgical instruments; (c) Cautery unit for hemostasis. *Source:* Teri Raffel, CVT

▶ **emasculatome** (ē-mahs-kwoo-lah-tohm) = instrument used to crush and sever the spermatic cord through intact skin

▶ **elastrator** (ē-lahs-trā-tər) = bloodless castration device using small elastic bands

▶ **hemostatic forceps** (hē-mō-stah-tihck fōr-sehps) = locking instrument used for grasping and ligating vessels and tissues to control/stop bleeding; also called **hemostat**

▶ **ratchet** (rah-cheht) = graded locking portion of an instrument located near the finger rings

▶ **serration** (sihr-ā-shuhn) = sawlike edge or border

▶ **cautery** (caw-tər-ē) = application of a burning substance, hot instrument, electric current, or other agent to destroy tissue

▶ **chuck** (chuhck) = clamping device for holding a drill bit

▶ **prosthesis** (prohs-thē-sihs) = artificial substitute for a diseased or missing part of the body

▶ **belly band** (behl-ē bahnd) = common term for abdominal wrap; circumferentially wrapping the abdomen with bandages to apply pressure to this area

▶ **cerclage** (sihr-klahj) **wire** = band of metal that completely (cerclage) or partially (hemicerclage) goes around the circumference of bone that is used in conjunction with other stabilization techniques to repair bone fractures

▶ **dressing** (drehs-sihng) = various materials used to cover and protect a wound

▶ **drain** (drān) = device by which a channel may be established for the exit of fluids from a wound (Figure 17–8)

▶ **clamp** (klahmp) = instrument used to secure or occlude things

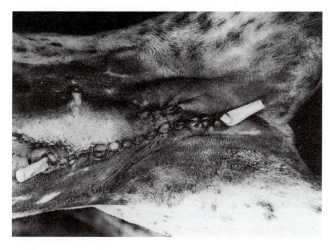

FIGURE 17-8 **Surgical drain.** Dog with a surgical drain placed on the ventral abdomen. *Source:* Photo by Linda Kratochwill, DVM

- **splint** (splihnt) = rigid or flexible appliance for fixation of movable or displaced parts
- **bandage** (bahn-dahj) = to cover by wrapping or the material to cover by wrapping (Figure 17–9)
- **sling** (slihng) = bandage for supporting part of the body
- **cast** (kahst) = stiff dressing used for immobilization of various body parts
- **intramedullary pins** (ihn-trah-mehd-yoo-lahr-ē pihnz) = metal rods that are inserted into the medullary cavity of long bones to repair stable fractures
- **bone screw** = screw that compresses bone fragments together to repair bone fractures
- **bone plate** = flat metal bar with screw holes that is used in bone fracture repair (Figure 17–10)

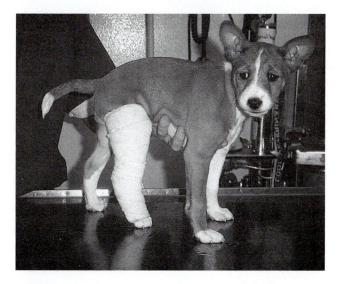

FIGURE 17-9 **Puppy with a Robert Jones bandage.** *Source:* Lodi Veterinary Hospital, S.C.

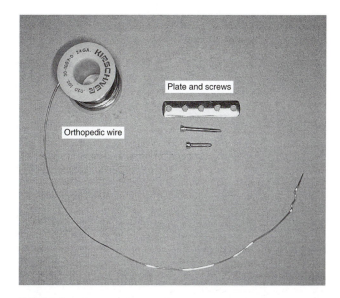

FIGURE 17-10 **Examples of orthopedic instruments.** *Source:* Teri Raffel, CVT

Cut on the Dotted Line

In surgery, the specific procedure by which an organ or part is exposed is called the **approach** (ah-prōch). There are different approaches to allow the best exposure to different parts of the body. The following are examples of different surgical approaches (Figure 17–11):

- **ventral midline incision** (vehn-trahl mihd-līn ihn-sihz-shuhn) = surgical cut along the midsagittal plane of the abdomen along the linea alba
- **paramedian incision** (pahr-ah-mē-dē-ahn ihn-sihz-shuhn) = surgical cut lateral and parallel to the ventral midline, but not on the midline
- **flank incision** (flahnk ihn-sihz-shuhn) = surgical cut perpendicular to the long axis of the body, caudal to the last rib
- **paracostal incision** (pahr-ah-kah-stahl ihn-sihz-shuhn) = surgical cut oriented parallel to the last rib

Procedure or Specimen?

The term **biopsy** (bī-ohp-sē) means removing living tissue to examine. Biopsy is also used to describe the actual specimen removed during the procedure. The first definition is the most correct; however, the term biopsy is commonly used both ways.

Types of biopsies include

- **excisional biopsy** (ehcks-sih-shuhn-ahl bī-ohp-sē) = removing entire mass, tissue, or organ to examine
- **incisional biopsy** (ihn-sih-shuhn-ahl bī-ohp-sē) = cutting into and removing part of a mass, tissue, or organ to examine

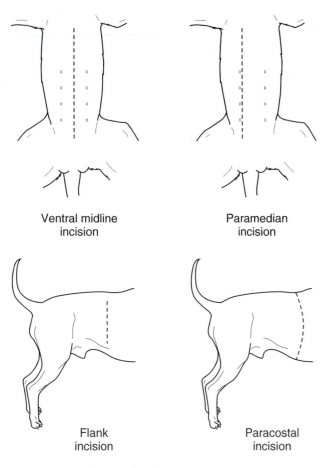

Ventral midline
incision

Paramedian
incision

Flank
incision

Paracostal
incision

FIGURE 17–11 Incision types

▶ **monofilament** (mohn-ō-fihl-ah-mehnt) = single strand of material; used to describe suture
▶ **multifilament** (muhl-tī-fihl-ah-mehnt) = several strands that are twisted together; used to describe suture
▶ **swaged** (swehgd) **needle** = needle in which the needle and suture material are joined in a continuous unit; "eyeless" needle
▶ **taper** (tā-pər) **needle** = needle that has a rounded tip which is sharp to allow piercing of, but not cutting of, tissue
▶ **cutting** (kuht-ihng) **needle** = needle that has two or three opposing cutting edges
▶ **blunt** (bluhnt) = dull, not sharp; used to describe needles or instrument ends
▶ **stapling** (stā-plihng) = method of suturing that involves the use of stainless steel staples to close a wound
▶ **surgical clip** (sihr-gih-kahl klihp) = metal staple-like device used for vessel ligation
▶ **ligation** (lī-gā-shuhn) = act of tying
▶ **ligature** (lihg-ah-chūr) = substance used to tie a vessel or strangulate a part

▶ **needle biopsy** (nē-dahl bī-ohp-sē) = insertion of a sharp instrument (needle) into a tissue for extraction of tissue to be examined

Needle and Thread

Suture material and needles are used by surgeons to close wounds or to tie things (Figure 17–12). Terminology used in reference to suture material and needles include

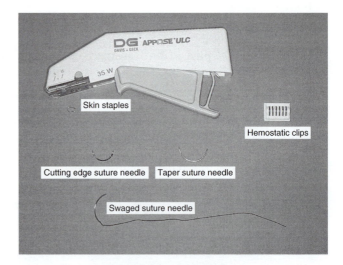

Skin staples

Hemostatic clips

Cutting edge suture needle   Taper suture needle

Swaged suture needle

FIGURE 17–12 Needles and staples

REVIEW EXERCISES

Multiple Choice—Choose the correct answer.

1. A monovalent vaccine, antiserum, or antitoxin is one that is developed for
   a. many organisms
   b. one organism
   c. monocyte activation
   d. subcutaneous injection

2. Hydrophilic substances are
   a. solid substances
   b. liquid substances
   c. water-loving substances
   d. fat loving substances

3. Movement of water across a cell membrane is
   a. diffusion
   b. efficacy
   c. chelation
   d. osmosis

4. Substances that control itching are termed
   a. antipruritic
   b. antimiotic
   c. antitussive
   d. antiseptic

5. The term used to describe surgically cutting out is
   a. incise
   b. excise
   c. friable
   d. eversion

6. The abbreviation for nothing orally is
   a. PO
   b. qid
   c. prn
   d. NPO

7. The surgical term for placing side by side is
   a. inversion
   b. eversion
   c. fulguration
   d. apposition

8. The term for a device by which a channel may be established for the exit of fluids from a wound is a/an
   a. sling
   b. cast
   c. drain
   d. dressing

9. A sawlike edge or border is a/an
   a. serration
   b. cautery
   c. chuck
   d. cerclage

10. Another name for an "eyeless" needle is
    a. monofilament
    b. multifilament
    c. swaged
    d. taper

Word Scramble—Use the definitions to unscramble the terms.

act of tying   ntoiaigl _____

blunt point   rtpae _____

cutting into   eiincs _____

stiff dressing   satc _____

bandage for supporting body part   nligs _____

graded locking portion of instrument   tthecar _____

movable joint of any ringed instrument   ooxkclb _____

forcible tearing   uutprer _____

removal of an organ in whole   eeucnlaiont _____

act of approximating   aattoinpco _____

exposure of internal organs   eevtiarsce _____

pierce through and through   xtriansf _____

18 Dogs and Cats

Objectives *In this chapter, you should learn to:*

▶ Recognize, define, spell, and pronounce terms related to dogs and cats
▶ Understand case studies and medical terminology in a practical setting

DOGS AND CATS

For many years, dogs and cats have been used by people for different purposes. Originally dogs and cats were domesticated for the work they provided, such as herding and rodent control. Although dogs and cats may still be used for work, they are more commonly kept as pets.

Many of the anatomy and physiology concepts, as well as medical terms, related to dogs and cats have been covered in previous chapters. The lists below contain terms that apply more specifically to the care and treatment of dogs and cats.

ANATOMY AND PHYSIOLOGY TERMS

anal sacs (ā-nahl sahcks)

pair of pouches that store an oily, foul-smelling fluid secreted by the anal glands located in the skin between the internal and external anal sphincters (located at the four o'clock and eight o'clock position); each sac has a duct that opens to the skin at the anal orifice, and fluid is expressed during defecation, excitement, or social interaction (Figures 18–1 and 18–2).

anal glands (ā-nahl)

secretory tissues that are composed of aprocine and sebaceous glands located within the anal sac; secretion of the anal glands is stored in the anal sacs and may play a role in marking terri-

tory, as a defense mechanism, or as a pheromone for sexual behavior

carnassial tooth (kahr-nā-zē-ahl)

large, shearing cheek tooth; the upper fourth premolar and lower first molar in dogs (Figure 18–3) and the upper third premolar and lower first molar in cats

constitution (kohn-stih-too-shuhn)

physical makeup of an animal

coprophagy (kōp-rohf-ah-jē)

ingestion of feces

or **debarking** (dē-bahrk-ihng)

surgical procedure that cuts vocal folds to soften a dog's bark; also called **devocalization** (dē-vō-kahl-ih-zā-shuhn)

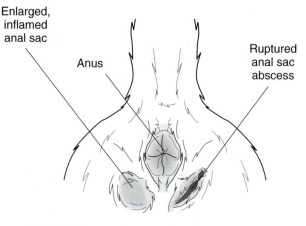

Enlarged, inflamed anal sac

Anus

Ruptured anal sac abscess

FIGURE 18–1 Line drawing of anal sac location

241

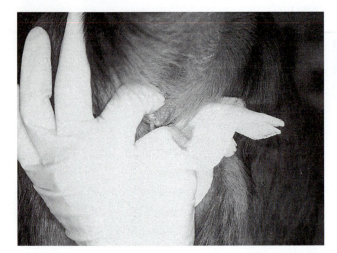

FIGURE 18–2 **Digital expression of fluid from anal sac**

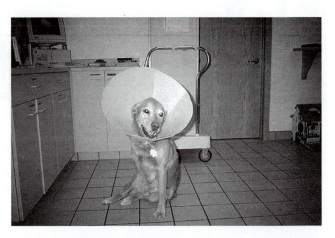

FIGURE 18–4 **Elizabethan collar on a dog**

steatitis (stē-aht-ī-tihs)
> inflammation of fat, usually caused by feeding cats too much oily fish; also called yellow fat disease

BREED-RELATED TERMINOLOGY

angora (ahn-gōr-ah)
> type of long fur on cats (and other species)

calico (kahl-ih-kō)
> cat having three colors of fur (black, orange, and white); usually female; if male they have a genotype of XXY

domestic short hair (dō-mehs-tihck)
> cat breed that has short guard hairs; abbreviated DSH

domestic long hair (dō-mehs-tihck)
> cat breed that has long guard hairs; abbreviated DLH

mackeral tabby (mahck-ər-ahl tah-bē)
> two-toned feline fur with stripes

mongrel (mohn-grehl)
> mixed breed of any animal

purebred (pər-brehd)
> member of a recognized breed

ruddy (ruhd-dē)
> orange-brown color with ticking of dark brown or black

self (sehlf)
> one color fur

tabby (tahb-bē)
> feline fur with two colors that may be either in stripes or spots

ticked coat (tihckd kōt)
> fur color where darker colors are found on the tips of each guard hair

Elizabethan collar (ē-lihz-ah-bēth-ahn)
> device placed around neck and head of dogs to prevent them from traumatizing an area; commonly called an E-collar (Figure 18–4)

induced ovulator (ihn-doozd ohv-yoo-lā-tər)
> species that ovulates only as a result of sexual activity (cats, rabbits, ferrets, llamas, camels, mink)

polydactyly (poh-lē-dahck-tih-lē)
> more than the normal number of digits (Figure 18–5)

spraying (sprā-ihng)
> urination on objects to mark territory

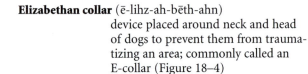
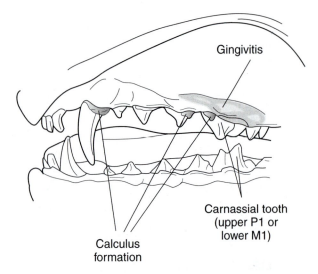

Gingivitis

Carnassial tooth
(upper P1 or
lower M1)

Calculus
formation

FIGURE 18–3 **Carnassial teeth of dogs**

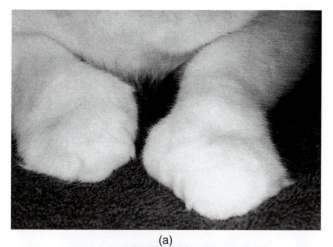

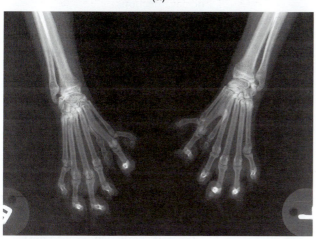

FIGURE 18–5 Polydactyly in a cat. (a) Paw of a polydactyl cat; (b) Radiograph of a polydactyl cat. *Source:* (a) Photo by Linda L. Kratochwill, DVM; (b) University of Wisconsin Veterinary Hospital–Radiology

euthanasia (yoo-thehn-ā-zhah)
 inducing death of an animal quickly and painlessly; "putting an animal to sleep"

feral (fēr-ahl)
 wild; not domesticated

gait (gāt)
 the way an animal moves

gloves (gluhvz)
 white paws

luster (luhs-tər)
 shine

obesity (ō-bē-siht-ē)
 excessive fat accumulation in the body

points (poyntz)
 color of nose (mask), ears, tail, and feet of an animal (Figure 18–6)

quarantine (kwahr-ehn-tēn)
 isolation of animals to determine if they have or carry a disease

retractile (rē-trahck-tīl)
 ability to draw back; feline claws can be drawn back

sheen (shēn)
 shininess or luster

staunchness (stawnch-nehs)
 strong and steady while on point

stud (stuhd)
 male animal used for breeding purposes

temperament (tehm-pər-mehnt)
 emotional and mental qualities of an individual

ticking (tihck-ihng)
 fur coat that has guard hairs with darker tips mixed in

tortoise shell (tōr-tihs shehl)
 feline fur with two colors (orange and black) producing a spotted or blotched pattern

DESCRIPTIVE TERMINOLOGY

docile (doh-sī-uhl)
 tame and easygoing

dull (duhl)
 lack of shine to haircoat; also used to describe a more lethargic than normal behavior

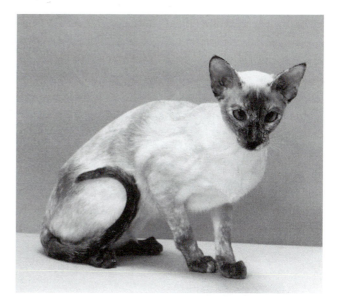

FIGURE 18–6 Points in a colorpoint shorthair cat

thorough (thər-ō)
dog that works every bit of ground and cover

throwbacks (thrō-bahcks)
offspring that shows an ancestor's characteristic that has not appeared in previous generations

timid (tihm-ihd)
showing lack of confidence or shy

underfur (uhn-dər-fər)
very dense, soft, short hair that is found beneath the longer, coarser guard hairs

unthrifty (uhn-thrihf-tē)
not thriving

vigor (vihg-ər)
healthy growth; also means having high energy

VACCINATIONS OF DOGS AND CATS

Bordetella (bōr-dih-tehl-ah)
upper respiratory infection caused by the bacteria *Bordetella bronchiseptica* that produces a severe hacking cough; also called kennel cough

canine adenovirus 2 (ahd-nō-vī-ruhs)
adenovirus infection in canines that causes signs of upper respiratory disease; abbreviated CAV-2

canine distemper (dihs-tehm-pər)
highly contagious paramyxovirus infection in canines that is associated with respiratory, digestive, and/or neurologic signs; abbreviated CDV

canine hepatitis virus (hehp-ah-tī-tihs)
highly contagious adenovirus I infection in canines that is associated with ocular ("blue eye"), abdominal, and liver signs; abbreviated ICH or CAV-I

canine herpesvirus (hər-pēz-vī-ruhs)
herpesvirus infection in canines that primarily affects newborn puppies and is associated with listlessness, nasal discharge, rash, neurologic signs, and death; abbreviated CHV

canine parainfluenza (pahr-ah-ihn-flū-ehn-zah)
paramyxovirus infection of canines that contributes to upper respiratory infections and causes subclinical bronchitis

canine parvovirus (pahr-vō-vī-ruhs)
highly contagious parvovirus infection in canines that is associated with severe diarrhea, vomiting, dehydration, and leukopenia

coronavirus (kō-rō-nah-vī-ruhs)
virus that causes gastrointestinal disease in dogs and gastrointestinal and systemic disease in cats that is usually spread through contaminated feces; in cats is known as feline infectious peritonitis (FIP); abbreviated CCV in dogs

feline calicivirus (kah-lē-sē-vī-ruhs)
picornavirus infection in felines that is associated with upper respiratory and ocular infections

feline chlamydia (klah-mihd-dē-ah)
bacterial infection in felines that is associated with upper respiratory and ocular infections

feline leukemia (loo-kē-mē-ah)
feline retrovirus that may produce increased numbers of abnormal leukocytes, immune suppression, cancer, and illness associated with immune suppression; abbreviated FeLV

feline panleukopenia (pahn-loo-kō-pē-nē-ah)
parvovirus infection of felines that is associated with fever, vomiting, diarrhea, and a decrease in all types of white blood cells; abbreviated FPV; commonly called feline distemper

feline rhinotracheitis (rī-nō-trā-kē-ī-tihs)
herpesvirus infection in felines that is associated with upper respiratory and ocular infections

leptospirosis (lehp-tō-spī-rō-sihs)
bacterial disease caused by various serotypes of *Leptospira;* signs include renal failure, jaundice, fever, and abortion

Lyme disease (līm)
bacterial disease caused by the bacteria *Borrelia burgdorferi* in which a tick vector transports the bacteria; associated with fever, anorexia, joint disorders, and occasionally neurologic signs; also called borreliosis

rabies (rā-bēz)
fatal zoonotic rhabdovirus infection of all warm-blooded animals that causes neurologic signs; transmitted by a bite or infected body fluid; abbreviated RV

REVIEW EXERCISES

Case Studies—Define the underlined terms in each case study.

A 3 yr old F/S black Labrador retriever was presented to the clinic for removal of a round bone from the mandible. Hx: the dog had been chewing on the bone during the day and had gotten the bone stuck on its mandible. The dog has been pawing at the bone for the past 45 min. On PE the dog was tachycardic, anxious, and tachypnic. The rostral end of the mandible was swollen. The dog was sedated with an IV sedative so that the bone could be removed. The bone was situated caudal to the lower canine teeth. Gigli wire was threaded through the hole in the center of the bone and the bone was cut in two places to allow for its removal. While the bone was being sawed tissue trauma occurred to the skin of the mandible. The dog was sent home on antibiotics 1 T bid PO × 7d.

yr_____

F/S_____

mandible_____

Hx_____

PE_____

tachycardic_____

tachypnic_____

rostral_____

IV_____

caudal_____

canine teeth_____

antibiotics_____

T_____

bid_____

PO_____

d_____

A 9 wk old ♀ DSH kitten was presented to the clinic for inappetence. On PE it was noted that the kitten had bilateral yellow-green mucopurulent ocular and nasal discharge. T = 103.8 °F, HR = 170 BPM, RR = 40 breaths/min, mm = pink, CRT = 2 sec. The kitten was alert. An audible wheeze was heard on thoracic auscultation; lungs had increased bronchial sounds and referred URT sounds. The conjunctiva was reddened and edematous. The abdomen palpated normally. Dx: URI; DDx: 1) Rhinotracheitis virus, 2) calicivirus, 3) chlamydia

wk_____

♀_____

DSH_____

inappetence_____

PE_____

bilateral_____

mucopurulent_____

ocular_____

nasal_____

T _____

°F _____

HR _____

BPM _____

RR _____

min _____

mm _____

CRT _____

sec _____

wheeze _____

thoracic _____

auscultation _____

bronchial _____

URT _____

conjunctiva _____

edematous _____

abdomen _____

palpated _____

Dx _____

DDx _____

URI _____

rhinotracheitis _____

A 5 yr old F/S DSH cat was presented to the clinic with stranguria and hematuria. T = 102.4 °F, HR = 180 BPM, RR = 35 breaths/min, mm = pink and moist, CRT = 1 sec. Heart and lungs ausculted normally. Oral exam revealed mild tartar with grade II gingivitis. Abdominal palpation yielded normal kidneys, normal intestinal loops, a tense and painful caudal abdomen, and turgid urinary bladder. Dx: cystitis; DDx: 1) FUS, 2) crystalluria

yr _____

F/S _____

DSH _____

stranguria _____

hematuria _____

ausculted _____

oral _____

tartar _____

gingivitis _____

abdominal _____

palpation _____

caudal _____

turgid _____

Dx _____

cystitis _____

DDx _____

FUS _____

crystalluria _____

A 2 yr old <u>intact</u> male Golden retriever was presented with a <u>4″</u> <u>laceration</u> with extensive <u>hemorrhage</u> on his right <u>carpus</u>. Pressure bandages were immediately applied for <u>hemostasis</u>. When the bleeding was under control, the dog was <u>anesthetized</u>, so that the blood vessels could be <u>ligated</u> and the wound <u>sutured</u>.

intact _____

4″ _____

laceration _____

hemorrhage _____

carpus _____

hemostasis _____

anesthetized _____

ligated _____

sutured _____

A 6 mo old F black Labrador retriever was presented to the clinic for <u>OHE</u>. A <u>preanesthetic</u> <u>blood screen</u> (<u>PCV, ALT, BUN, GLU</u>) and <u>IV</u> fluid line were done prior to <u>sx</u>. The animal was <u>anesthetized</u>, clipped, and prepped for surgery. A <u>ventral midline incision</u> was made, and the reproductive tract was identified. The ovaries, uterine horns, and uterus were removed after proper <u>ligation</u>. When the abdominal incision was being closed, the veterinarian noted pooling of blood in the abdomen. The ligatures were rechecked and still in place. A large amount of blood was coming from the abdominal incision and the veterinarian had the technician reassess the animal. The <u>CRT</u> was prolonged, the <u>mm</u> were pale, and the animal was <u>tachycardic</u> and <u>hypothermic</u>. Blood was taken for another PCV and the fluid rate was increased. The PCV was low normal. The owner was called to see if the dog had been sick recently, and the owner stated that the dog was seen eating rat bait about 3 days ago. Additional blood was collected in a <u>heparin</u> tube for assessment of clotting times, and the dog was given vitamin K. The incision was closed, and the dog was closely monitored during recovery. The dog made a slow recovery and was hospitalized an additional night for observation. Clotting times from the lab demonstrated prolonged clotting times.

OHE _____

preanesthetic _____

blood screen _____

PCV _____

ALT _____

BUN _____

GLU _____

IV _____

anesthetized _____

ventral midline incision _____

ligation _____

CRT_____

mm _____

tachycardic _____

hypothermic _____

heparin _____

A 10 yr old M/N cock-a-poo was presented to the clinic for scooting (dog assumes a sitting position and drags the anal region along the ground) and licking the perianal region. On PE it was noted that the dog was obese, had dermatitis of the tail head region, and oily skin. The TPR were normal. The dog had a hx of tenesmus and reluctance to stand. Rectal palpation of the anal sacs revealed moderately enlarged sacs. Both anal sacs were expressed and inspissated material was expressed. Both anal glands were flushed with an antiseptic. The dog was discharged with antibiotics and an appointment was made to reassess the anal sacs in 7 days.

perianal_____

dermatitis _____

TPR_____

hx _____

tenesmus_____

rectal palpation_____

anal sacs _____

inspissated_____

antiseptic_____

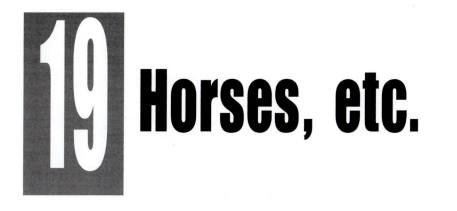

19 Horses, etc.

Objectives *In this chapter, you should learn to:*

▶ Recognize, define, spell, and pronounce terms related to horses, donkeys, mules, and ponies
▶ Understand case studies and medical terminology in a practical setting

HORSES, DONKEYS, MULES, AND PONIES

Equine animals have been used for transportation, field work, pack work, and recreation. Horses, donkeys, mules, and ponies have been used as companion animals as well.

Many of the anatomy and physiology concepts, as well as medical terms, related to equine species have been covered in previous chapters. The lists below contain words that apply more specifically to the care and treatment of equine species.

MODE OF MOVEMENT

See Figure 19–1.

amble (ahm-buhl)
    lateral gait that is different from the pace by being slower and more broken in cadence

back
    trotting in reverse

beat (bēt)
    time when the foot (or feet if simultaneous) touch the ground

canter (kahn-tər)
    slow, restrained, three-beat gait in which the two diagonal legs are paired

dressage (druh-sahzh)
    method of riding in which a rider guides (rather than using hands, feet, or legs) a trained horse through natural maneuvers

fox trot (fohcks troht)
    slow, short, broken type of gait in which the head usually nods

gallop (gahl-ohp)
    fast, four-beat gait where the feet strike the ground separately (1st = one hind foot, 2nd = other hind foot, 3rd = front foot on the same side as 1st step, 4th = other front foot on the same side as 2nd step); also called **run**

jog (johg)
    slow trot

pace (pās)
    fast, two-beat gait in which the front and hind feet on the same side start and stop at the same time

pointing (poyn-tihng)
    stride in which extension is more pronounced than flexion

rack (rahck)
    fast, flashy, unnatural four beat gait in which each foot meets the ground separately at equal intervals; also called **single-foot**

249

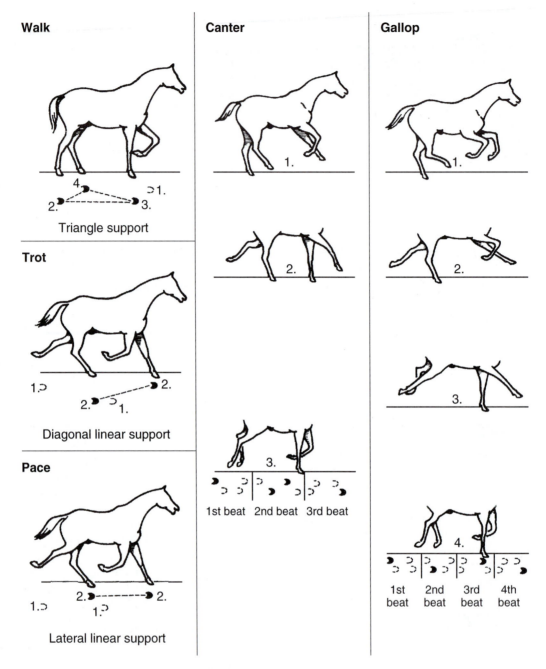

FIGURE 19–1 Basic gaits of equine

rolling (rō-lihng)

    excessive side-to-side shoulder motion

running (ruhn-ihng) **walk**

    slow, four-beat gait intermediate in
    speed between a walk and rack

stride (strīd)

    distance between successive imprints
    of the same foot

suspension (suh-spehn-shuhn)

    time in which none of the feet are in
    contact with the ground

swing (swihng)

    non-weight-bearing phase of a stride
    short, quick, choppy stride

trappy (trahp-pē)

    short, quick, choppy stride

trot (troht)

    natural, rapid, two-beat, diagonal gait
    in which the front foot and the oppo-
    site hind foot take off and hit the
    ground at the same time

walk
> natural, slow, flat-footed, four-beat gait in which each foot takes off and strikes the ground at separate intervals

ANATOMY, PHYSIOLOGY, AND DISEASE TERMS

bad mouth
> malocclusion where the top and bottom teeth do not meet (Figure 19–2)

bag up
> development of mammary glands or udder near parturition; also called **bagging up**

bars (bahrz)
> support structure that angles forward from the hoof wall to keep it from overexpanding; also the gap between a horse's incisors and molars; also the side points on the tree of a saddle

bishoping (bihsh-ohp-ihng)
> artificial altering of teeth of an older horse to sell it as a younger horse

check ligament (chehck lihg-ah-mehnt)
> one of two ligaments to the digital flexors of equine; function to maintain the limbs in extended position during standing

chestnut (chehs-nuht)
> horny growths on the medial surface of the equine leg either above the knee in the front limb or toward the caudal area of the hock in the rear limb

cracks (krahkz)
> hoof wall defects that form because the hoof is too long and not trimmed frequently enough

croup (krūp)
> top part of equine rump

cups (kuhpz)
> deep indentations of the incisors in the center of the occlusal surface in young permanent teeth

curb (kərb)
> enlargement on the caudal aspect of the hind leg below the hock

dental star (dehn-tahl stahr)
> marks on the occlusal surface of the incisor teeth appearing first as narrow, yellow lines then as dark circles near the center of the tooth

flexor tendon (flehck-ər tehn-dohn)
> tendon that cause the fetlock joint to bend

foal heat (fōl hēt)
> estrus that occurs directly after parturition (usually not fertile)

full-mouthed (fuhl mouthd)
> horse having all the permanent teeth and cups present

Galvayne's groove (gahl-vānz groov)
> mark on labial surface of the equine tooth; used to determine age; usually appears around 11 years of age (Figure 19–3)

guttural pouch (guht-ər-ahl powch)
> large, air-filled ventral outpouching of the eustachian tube in equine

hindgut (hihnd-guht)
> collective term for the cecum, small colon, and large colon

in wear
> condition when a tooth has risen to the masticatory level

lamina (lah-mih-nah)
> tissue that attaches hoof to the underlying foot structures

laminitis (lahm-ihn-ī-tihs)
> inflammation of the sensitive laminae under the horny wall of the hoof; also called **founder** (fownd-ər) (Figure 19–4)

milk teeth
> first teeth that the animal develops

nippers (nihp-pərz)
> central incisors of equine; also a tool to remove excess hoof wall

overshot jaw
> condition where the maxilla is longer than the mandible; also called parrot mouth

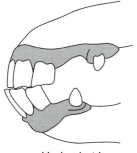
Undershot jaw or monkey mouth

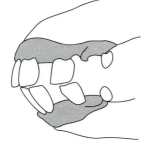
Overshot jaw or parrot mouth

FIGURE 19–2 Bad mouth

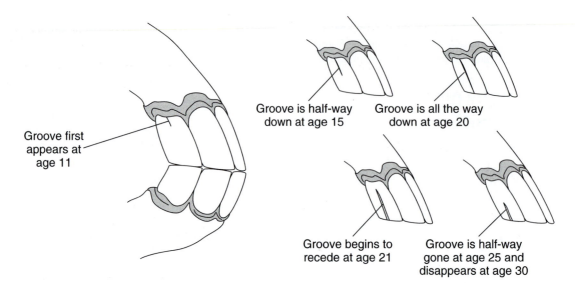

Groove first appears at age 11

Groove is half-way down at age 15

Groove is all the way down at age 20

Groove begins to recede at age 21

Groove is half-way gone at age 25 and disappears at age 30

FIGURE 19–3 Galvayne's groove

periople (pehr-ē-ō-puhl)
> varnish-like coating that holds moisture in the hoof and protects the hoof wall

quidding (kwihd-ihng)
> condition in which a horse drops food from the mouth while chewing

quittor (kwihd-ər)
> festering of the foot anywhere along the border of the coronet

scratches (skrahtch-ihz)
> low-grade infection or scab in the skin follicles around the fetlock; also called **grease heel**

smooth mouth
> condition where no cups are present in the permanent teeth

stay apparatus (stā ahp-ahr-ah-tuhs)
> anatomical mechanism of the equine limb that allows the animal to stand with little muscular effort; includes many muscles, ligaments, and tendons

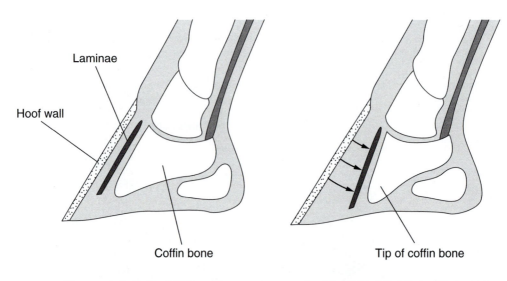

Laminae

Hoof wall

Coffin bone

Tip of coffin bone

The normal relation of the coffin to the laminae and hoof wall.

In chronic founder, the coffin bone is detached and rotated toward the sole.

FIGURE 19–4 Laminitis

undershot jaw

condition where the mandible is longer than the maxilla

winking (wihnk-ihng)

opening of the labia to expose the clitoris while the female assumes a mating position

wolf teeth

rudimentary first upper premolar in equine that is usually shed in maturity

MARKINGS

See Figure 19–5.

bald face

wide white marking that extends beyond both eyes and nostrils

banding (bahn-dihng)

style of mane that is sectioned and fastened with rubber bands; seen in Western show horses

blaze (blāz)

broad white stripe on the face of a horse

blemish (blehm-ihsh)

unattractive defect that does not interfere with performance

bloom (bloom)

shiny coat for show horses

distal spots (dihs-tahl)

dark circles on a white coronet band

half-stocking (hahlf stohk-ihng)

white marking from the coronet to the middle of the cannon

points (poyntz)

black coloration from the knees and hocks down in bays and browns (may include the ear tips)

stocking (stohck-ihng)

white marking from the coronet to the knee

EQUIPMENT

aids (ādz)

means by which a rider communicates with a horse (voice, hands, legs, seat, etc.)

bit (biht)

part of the bridle that is put in the horse's mouth to control the animal

breeching (brē-ching)

part of a harness that passes around the rump of a harnessed horse

bridle (brī-duhl)

part of a harness that includes the bit, reins, and headstall (Figure 19–6)

 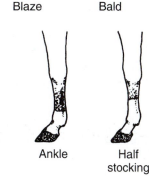 

| Star and snip | Stripe | Blaze | Bald | Spot | Race |

 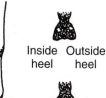

Inside heel   Outside heel

| Coronet | Half pastern | Pastern | Ankle | Half stocking | Full stocking | Both heels |

FIGURE 19–5 Face and leg markings

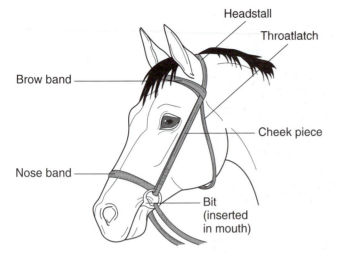

FIGURE 19–6 Parts of a bridle

Headstall

Throatlatch

Brow band

Cheek piece

Nose band

Bit
(inserted
in mouth)

calks (kawkz)

grips on the heels and the outside of the front shoes of horses

cinch (sihnch)

part of a saddle used to hold it onto the horse; placed around the girth area

clinch cutter (klihnch kuht-tər)

tool used to remove horseshoes

cradle (krā-duhl)

device used to prevent an animal from licking or biting an injured area

halter (hahl-tər)

device used to lead and tie a horse; also called a **head collar**

hobble (hohb-uhl)

type of restraint in which either the front feet or hind feet are placed in straps to keep them from moving

hoof pick (huhf pihck)

instrument used to clean the sole, frog, and hoof wall

hoof testers (huhf tehs-tərs)

instrument used to test sensitivity in the equine foot

pincher (pihn-shər)

tool used to remove horseshoes; pincers are central incisors

puller (puhl-ər)

tool used to remove worn horseshoes

rasp (rahsp)

tool used for leveling the foot (Figure 19–7)

shoe (shū)

plate or rim of metal nailed to the ventral surface of an equine hoof to protect the hoof from injury or to aid in hoof disease management

tack (tahck)

equipment used in riding and driving horses

throatlatch (thrōt-lahtch)

bridle part that connects the bridle to the head located under the horse's throat; also area under throat where head and neck are joined and where the harness throat latch fits (refer to Figure 4–1)

MANAGEMENT TERMS

as-fed basis (ahs-fehd bā-sihs)

amount of nutrients in a diet expressed in the form in which it is fed

bedding (behd-ihng)

material used to cushion the animal's shelter

birth date (bərth dāt)

for racing or showing, a foal's birthday is considered as January 1 (regardless of the actual month it was born)

blistering (blihs-tər-ihng)

application of an irritating substance to treat a blemish

board (bōrd)

to house

bolt (bōlt)

to eat rapidly or startle

bots (bohtz)

larvae of the bot fly, *Gastrophilus*; occur in the stomach

bow-line knot (bō-līn noht)

type of nonslippable knot

box stall (bohcks stahl)

enclosure where a horse can move freely

cast (kahst)

to be caught in a recumbent position and unable to rise

cribbing (krihb-ihng)

vice of equine in which an object is grasped between the teeth and pressure is applied (Figure 19–8)

cross tying (krohs tī-ihng)

method of using two ropes to secure a horse so that the head is level

(a)

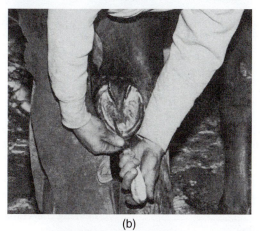

(b)

(c)

FIGURE 19–7 Trimming a horse's hoof. (a) Use of nippers to cut the horny wall to a proper length and angle; (b) Use of hoof knife to pare dead and flaky tissue from sole; (c) Use of rasp to eliminate jagged and sharp corners

FIGURE 19–8 Cribbing

diluters (dī-loo-tərs)
>type of fluid that is used to increase the volume of semen (thus diluting the sample)

driving (drī-vihng)
>horses harnessed and controlled from behind

dry-matter basis (drī-mah-tər bā-sihs)
>method of expressing concentration of a nutrient based on absence of water in the feed

extenders (ehcks-tehn-dərs)
>additive used to extend the lifespan of sperm cells

farrier (fār-ē-ər)
>person who cares for equine feet including trimming and shoeing

feathering (feh-thər-ihng)
>fringe of hair around an equine foot just above the hoof; also used to describe fringe of hair on caudal aspects of canine limbs

firing (fihr-ihng)
>making a series of skin blisters with a hot needle over an area of lameness

flighty (flī-tē)
>nervous

floating (flō-tihng)
>filing off the sharp edges of equine teeth (Figure 19–9)

get (geht)
>offspring

grade (grād)
>animal that is not registered with a specific breed registry

hand
>unit used to measure an equine that is equal to 4 inches

heaving (hē-vihng)
>extra contraction of the flank muscles during respiration; caused by loss of lung elasticity

hunters (huhn-tərz)
>horses that are judged while jumping fences or hunting fox

jumpers (juhm-pərz)
>horses that compete at shows by jumping and are judged on height and time and faults

lather (lah-thər)
>accumulation of sweat on a horse's body

leg cues (lehg kūz)
>signals given to the horse through movement of the rider's legs

longe (luhng)
>act of exercising a horse on the end of a long rope, usually in a circle (Figure 19–10)

near side
>left side of horse

off side
>right side of horse; also called **far side**

paddock (pah-dohck)
>small fenced in area; also called **corral**

pasture (pahs-chər)

area for grazing animals; also means grass or other forage that grazing animals eat

pasture mating (pahs-tər-mā-tihng)

natural breeding; also called **natural cover**

plumb line (pluhm)

line formed when a weight is placed on the end of a string to measure the perpendicularity of something (used to detect straightness of a horse leg)

quick-release knot

knot that breaks loose easily

saddle (sahd-uhl)

piece of tack placed over the back of an equine for riding, draft or pack (Figure 19–11)

settle (seht-uhl)

breeding successfully; said of a mare when she becomes pregnant

shod (shohd)

equine with horseshoes

strike (strīk)

defensive or aggressive movement of a horse in which the front leg is moved quickly and cranially

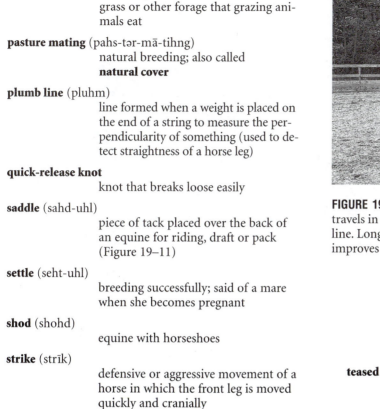

FIGURE 19–10 **Longeing a horse.** Longeing is when a horse travels in a large circle around the handler on a long strap or line. Longeing helps train young horses, exercise horses, and improves balance and development of stride.

teased (tēzd)

act of determining whether a mare is in heat (estrus) by presenting a stallion to her

teaser (tē-zər)

stallion used to determine which mares are in heat (estrus)

twitch (twihtch)

mode of restraint in which a device is twisted on the upper lip or muzzle

waxing (wahcks-ihng)

accumulation of colostrum on the nipples of mares usually prior to foaling; also called **waxed teats**

TYPES OF HORSES

draft horse (drahft)

large breed of working horse; usually over 17 hands

light horse

breed of horse that is intermediate in size and stature (usually greater than 14.2 hands)

miniature horse

breed of horse that is small (usually less than 8.2 hands)

pony (pō-nē)

small breed of horse (usually about 14 hands)

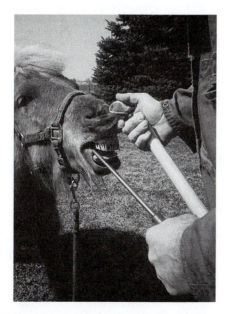

FIGURE 19–9 **Twitch applied to a horse's upper lip to allow floating of teeth.** *Source:* Courtesy of Ron Fabrizius, DVM, Diplomat ACT

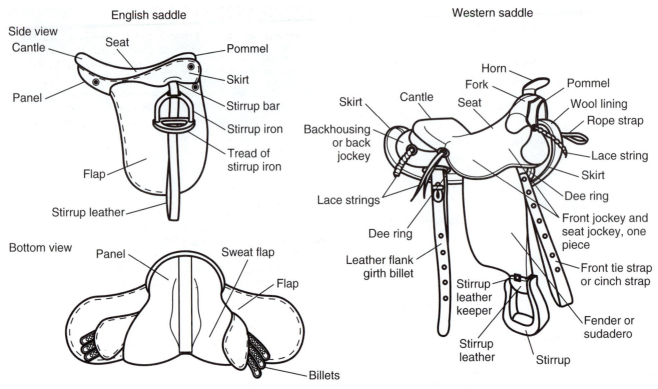

FIGURE 19–11 Parts of the English saddle and Western saddle

TERMS FOR UNSOUNDNESS IN HORSES

See Figure 19–12.

bog spavin (bohg spah-vihn)
  enlargement of proximal hock due to distention of the joint capsule; **spavin** means swelling

bone spavin (bōn spah-vihn)
  bony enlargement at the base and medial surface of the hock

bowed tendons (bōd tehn-dohnz)
  thickening of the caudal surface of the leg proximal to the fetlock

capped hock (kahpd hohck)
  thickening of the skin or large callus at the point of the hock

fistulous withers (fihs-tyoo-luhs wih-thərz)
  inflammation of the withers

grease heel (grēs-hēl)
  infection or scab in the skin around the fetlock; also called **scratches** (skrahch-ehz)

osselets (ohs-eh-lehts)
  soft swellings on the cranial (and sometimes sides) of the fetlock joint

poll evil (pōl ē-vihl)
  fistula on the poll that is difficult to heal

quarter crack, (kwahr-tər krahck) **toe crack,** (tō krahck) or **heel crack** (hēl krahck)
  cracks in toe, quarters, or heel of hoof wall due to poor management

quittor (kwihd-ər)
  festering of the foot along the border of the coronet

ringbone (rihng-bōn)
  bony enlargement on the pastern bones; high ringbone occurs at the pastern joint; low ringbone occurs at the coffin joint

splints (splihntz)
  inflammations of the interosseous ligament that holds the splint bones to the cannon bone

stifled (stī-flehd)
> displaced patella

sweeney (swē-nē)
> atrophy of the shoulder muscle

thoroughpin (thər-ə-pihn)
> fluctuating enlargement located
> in the hollows proximal to the hock;
> throughpins can be pressed from side
> to side, hence the name

EQUINE VACCINATIONS

equine encephalomyelitis (ehn-sehf-ah-lō-mī-ih-lī-tihs)
> mosquito-transmitted infectious
> alphaviral disease of horses that is
> associated with motor irritation,
> paralysis, and altered consciousness;
> there are three types: Eastern, Western,
> and Venezuelan; also known as sleep-
> ing sickness

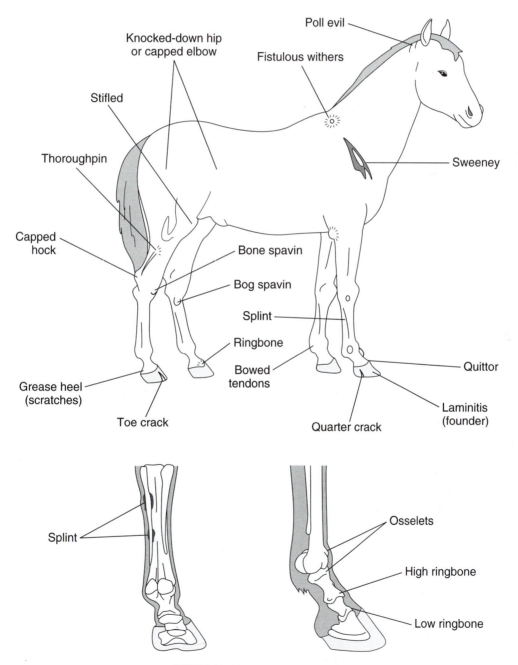

FIGURE 19–12 Unsoundness in horses

equine influenza (ihn-flū-ehn-zah)
>   myxovirus infection of horses that is associated with mild fever, watery eyes, and persistent cough; commonly called **flu**

equine viral arteritis (ahr-tər-ī-tihs)
>   togavirus infection of horses that is associated with upper respiratory disease signs, abortion, and lesions in small arteries; abbreviated EVA

equine viral rhinopneumonitis (rī-nō-nū-mohn-ī-tihs)
>   herpesvirus infection of horses that is associated with upper respiratory disease signs and abortion; abbreviated EVR

Potomac horse fever (pō-tō-mihck)
>   rickettsial bacterial disease of horses that is associated with fever, anorexia, incoordination, diarrhea, and edema of the extremities; also called **equine ehrlichiosis**

rabies (rā-bēz)
>   fatal zoonotic rhabdovirus infection of all warm-blooded animals that causes neurologic signs; transmitted by a bite or infected body fluid; abbreviated RV

strangles (strā-guhlz)
>   contagious bacterial disease of horses caused by the bacteria *Streptococcus equi;* signs include high fever, nasal discharge, anorexia, and swollen and abscessed lymph nodes

tetanus (teht-ah-nuhs)
>   highly fatal, bacterial disease caused by the toxin of *Clostridium tetani* that is associated with hyperesthesia, tetany, and convulsions; also called **lockjaw**

REVIEW EXERCISES

Case Studies—Define the underlined terms in each case study.

A 6 yr old quarter horse <u>mare</u> was presented for signs of pawing at the abdomen, <u>flank</u>-watching, <u>anorexia</u>, and lack of <u>stool</u> production. On PE she was <u>tachycardic and hyperpnic</u>, <u>mm</u> were tacky and discolored, and the <u>CRT</u> was 2 seconds. Her ears and limbs were cold to the touch. The gut was <u>ausculted</u> and <u>borborygmus</u> was noted. The veterinarian did a <u>rectal palpation</u> of the horse and noted gas-filled segments of bowel that felt <u>proximal (oral)</u> to an <u>impaction</u>. <u>Ventral midline abdominocentesis</u> was performed, and blood was collected for a <u>CBC</u>. Laboratory results revealed few <u>peritoneal</u> fluid changes on the abdominocentesis sample, and the CBC was unremarkable. The veterinarian felt the impaction was mild and opted for medical treatment. A <u>nasogastric tube</u> was passed to relieve the gas distention and mineral oil was given via the <u>NG tube</u>. An <u>analgesic</u> was given <u>IV</u> for pain relief. The horse recovered uneventfully.

mare _____

flank _____

anorexia _____

stool _____

tachycardic _____

hyperpnic _____

mm _____

CRT _____

ausculted _____

borborygmus _____

rectal palpation _____

proximal (oral) _____

impaction _____

ventral midline abdominocentesis _____

CBC _____

peritoneal _____

nasogastric tube _____

NG tube _____

analgesic _____

IV _____

A 2 yr old Arabian <u>colt</u> was anorectic and reluctant to stand. The owner called the veterinarian for an examination. The veterinarian walked the horse, and it showed a short-striding gait. The horse was <u>febrile</u> and <u>tachycardic</u>. Upon <u>palpation</u> of the hoof, the area near <u>the coronary band</u> was warm and a <u>pulse</u> could be palpated. The veterinarian suspected <u>laminitis</u> and recommended <u>radiographs</u> of the hoof. Radiographs revealed rotation of the <u>coffin bone</u> from the hoof wall. The <u>dx</u> of laminitis (<u>founder</u>) was made. Treatment consisted of <u>NSAIDs</u> and hoof trimming and reshoeing by a <u>farrier</u>.

colt _____

febrile _____

tachycardic _____

palpation _____

coronary band _____

pulse _____

laminitis _____

radiographs _____

coffin bone _____

dx _____

founder _____

NSAID _____

farrier _____

20 Make Room for the Ruminants

Objectives *In this chapter, you should learn to:*

▶ Recognize, define, spell, and pronounce terms related to cattle, sheep, goats, and llamas
▶ Understand case studies and medical terminology in a practical setting

RUMINANTS

A **ruminant** (roo-mihn-ehnt) is a cud-chewing animal that has a forestomach which allows for fermentation of ingesta. Cattle, sheep, and goats have four stomach compartments. The first three, the rumen, reticulum, and omasum, are actually outpouchings of the esophagus. The abomasum is considered the true or glandular stomach. Llamas have three stomach compartments and are referred to as pseudoruminants.

CATTLE

Cattle provide humans with meat, milk, hides, and other by-products. There are basically two types of cattle: dairy and beef. Dairy cattle are bred for their milk-producing qualities, while beef cattle are bred for meat. Some breeds are considered dual purpose, which means they have both dairy and beef traits.

Many of the anatomy and physiology concepts, as well as medical terms, related to cattle have been covered in previous chapters. The lists below contain words that apply more specifically to the care and treatment of cattle.

INDUSTRY TERMS

artificial insemination
(ahr-tih-fih-shahl ihn-sehm-ihn-ā-shuhn)
breeding method in which semen is collected, stored, and deposited in the uterus or vagina without copulation; abbreviated AI

balling gun (bahl-ihng)
tool used to administer pills or magnets to livestock (Figure 20–1)

barren (bār-ehn)
not able to reproduce

body capacity (boh-dē kah-pah-siht-ē)
the heart girth and barrel

brand (brahnd)
method of permanently identifying animal by scarring the skin with heat, extreme cold, or chemicals

bred (brehd)
animal that is mated and is pregnant

breed (brēd)
group of animals that are genetically similar in color and conformation that when mated to each other they produce young identical to themselves

calving interval (kahv-ihng ihn-tər-vahl)
amount of time between the birth of a calf and birth of the next calf from the same cow

carcass (kahr-kuhs)
body of animal after it has been slaughtered; usually has head, hide, blood, and offal removed

cattle (kah-tuhl)
greater than one member of the genus *Bos*

263

FIGURE 20–1 Balling gun used to administer a magnet to a Holstein cow. *Source:* Courtesy of Ron Fabrizius, DVM, Diplomat ACT

chute (shoot)
mechanical device that is used to restrain cattle (Figure 20–2)

cleaning a cow
common term for removal of a retained placenta; also called **cleansing a cow**

cod (kohd)
remnants of steer scrotum

colostrum (kō-lah-struhm)
first milk-like substance produced by the female after parturition, which is thick, yellow, and high in protein and antibodies

conformation (kohn-fōr-mā-shuhn)
shape and body type of an animal

corium (kōr-ē-uhm)
specialized, highly vascular cells that nourish the hoof and horn

crossbred (krohs-brehd)
offspring resulting from mating two different breeds within the same species

cull (kuhl)
removal of an animal from the rest because it does not meet a specific standard or is unproductive

dehorn (dē-hōrn)
mechanical, heat, or chemical removal of horns or horn buds

dual purpose (dool pər-puhs)
animals that are bred and used for both meat and milk production

ear tagging (ēr tahg-ihng)
placement of identification device in the ear

embryo transfer (ehm-brē-ō trahnz-fər)
removal of an embryo from a female of superior genetics and placing it in the reproductive tract of another female

F1 generation (F-1 jehn-ər-ā-shuhn)
first offspring from purebred parents of different breeds or varieties; F1 stands for first filial

feeder (fē-dər)
beef cattle that are placed in a feedlot based on age and weight

feedlot (fēd-loht)
confined area where an animal is fed until it is slaughtered (Figure 20–3)

flushing (fluhsh-ihng)
act of increasing feed before breeding or embryo transfer to increase the number of ova released

FIGURE 20–2 Headgate chute

(a)

(b)

(c)

FIGURE 20–3 Cattle housing. (a) Feedlot; (b) Stanchion; (c) Hutch

fly strike (flī-strīk)
  infestation with maggots

gomer bull (gō-mər)
  bull used to detect female bovines in heat; bull may have penis surgically deviated to the side, may be treated with androgens, or may be vasectomized so as not to impregnant female; also called **teaser bull**

halter (hahl-tər)
  head harness worn by animals for restraint that extends behind the head and over the nose

heart girth (hahrt gərth)
  circumference around the thoracic cavity used to estimate an animal's weight and capacity of the heart and lung

hutch (huhch)
  individual housing pen for calves (and other small animals like rabbits)

hybrid (hī-brihd)
  offspring resulting from mating of two different species

inbred (ihn-brehd)
  offspring resulting from mating two closely related animals, i.e., son to dam, sire to daughter; also called **close breeding**

lead rope (lēd rōp)
  piece of rope, leather, or nylon that is attached via a clasp to a halter

magnet (māg-neht)
  charged metal device that is used to prevent hardware disease (traumatic reticuloperitonitis); it is given orally and placed in the reticulum

malpresentation (mahl-prē-sehn-tā-shuhn)
  abnormal position of a fetus just before parturition

marbling (mahr-blihng)
  streaks of fat interdispersed throughout meat to enhance its tenderness

offal (aw-fuhl)
  inedible visceral organs and unusable tissues removed from the carcass of a slaughtered animal

parturient paresis (pahr-too-rē-ahnt pahr-ē-sihs)
  hypocalcemic metabolic bolic disorder of ruminants seen in late pregnancy or early lactation; also called **milk fever**

pinch (pihnch)
  common term for a bloodless castration using an emasculatome

proved (proovd)

    animal whose ability to pass on specific traits is known and predictable

rectal palpation (rehck-tahl pahl-pā-shuhn)

    method of determining pregnancy, phase of the estrus cycle, or disease process via insertion of a gloved arm into the rectum of the animal and feeling for a specific structure (Figure 20–4)

render (rehn-dər)

    to melt down fat by heat

replacement (rē-plās-mehnt)

    animal that is raised for addition to the herd (one that replaces a less desirable animal)

scurs (skərz)

    underdeveloped horns that are not attached to the skull

somatic cell count (sō-mah-tihck)

    determination of number of cells (leukocytes, epithelial cells, etc.) in milk to test for mastitis; abbreviated SCC

spotter bull (spoh-tər)

    vasectomized male bovine used to find and mark female bovines in estrus

springing (sprihng-ihng)

    anatomic changes in a ruminant that indicate parturition is near

FIGURE 20–5 Tail jack or tailing a cow. *Source:* Courtesy of Ron Fabrizius, DVM, Diplomat ACT

stanchion (stahn-chuhn)

    restraint device that secures cattle around the neck to allow accessibility for milking, feeding, and examining

standing heat

    female bovine in the phase of estrus in which she will stand to be mounted

switch (swihtch)

    distal part of a bovine tail that consists of long, coarse hairs

tailing (tā-lihng)

    restraint technique used in cattle in which the tailhead is grasped and raised vertically; also called a **tail jack** (Figure 20–5)

tankage (tahnk-ahj)

    animal residues left after rendering fat in a slaughter house that is used for feed or fertilizer

tattoo (taht-too)

    permanent identification of an animal using indelible ink that is injected under the skin

veal (vēl)

    confined young dairy calf that is fed only milk or milk replacer to produce pale, soft, and tender meat

wean (wēn)

    act of removing young from its mother so that it can no longer nurse

FIGURE 20–4 Rectal palpation in a cow. *Source:* Courtesy of Ron Fabrizius, DVM, Diplomat ACT

windbreak (wihnd-brāk)
> shelter in which an animal can stand and be protected from the wind

MILK-RELATED TERMS

alveoli (ahl-vē-ō-lī)
> milk-secreting sacs of mammary gland; also used to describe gas exchange sac of respiratory system

dry (drī)
> animal that is not lactating

drying off
> ending the production of milk when milk yield is low or before freshening

gland cistern (sihs-tərn)
> area of udder where milk collects before entering the teat cistern (Figure 20–6)

milking
> process of drawing milk from the mammary glands

milk solids
> portion of milk that is left after water is removed; includes protein and fat

milk veins
> veins found near the ventral midline of a cow; also called **mammary veins**

milk well
> depression in the cow's ventral underline where milk veins enter the body

milk yield
> amount of milk produced in a given period

streak canal (strēck kah-nahl)
> passageway that takes milk from the teat cistern to the outside; also called the **papillary duct** or **teat canal**

strip cup
> metal cup with a lid that is used for detecting mastitis

supernumerary teats (soo-pər-nū-mahr-ē tētz)
> more than the normal amount of nipples

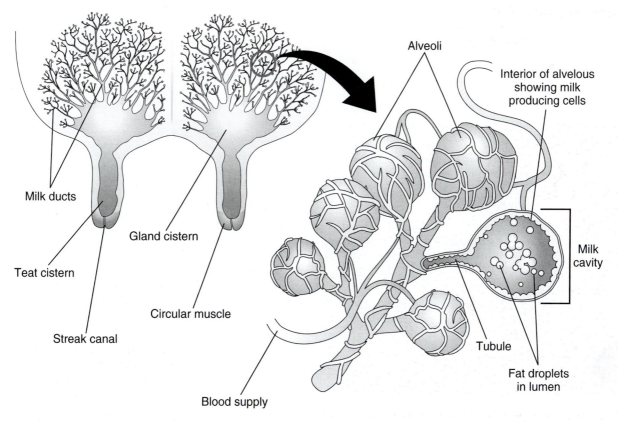

FIGURE 20–6 Parts of the udder

teat (tēt)

nipple, especially the large nipples of ruminants and equine

teat cannula (tēt kahn-yoo-lah)

short and narrow, round-pointed metal or plastic tube used to pass from the exterior through the teat canal and into the teat cistern to relieve teat obstructions

teat cistern

cavity in udder where milk is secreted before leaving the teat

teat dipping (tēt)

submerging or spraying nipple with disinfectant to prevent the development of mastitis

teat stripping (tēt)

removal of the last milk in the teat after milking by occluding the teat at the top between the thumb and forefinger and then pulling downward to express all the milk; also called **stripping**

udder (uh-dər)

milk production organ in ruminants and equine species

FEED ME

by-pass protein (bī-pahs prō-tēn)

protein that is heat or chemically treated so that it does not get altered in the ruminant stomach

concentrate (kohn-sehn-trāt)

type of feed that is high in total digestible nutrients and low in fiber

creep feed (krēp fēd)

high-energy feed that is fed to young animals in special feed devices so that adult animals cannot gain access to the feed

cudding (kuhd-ihng)

act of chewing cud; **cud** (kuhd) is regurgitated food particles, fiber, rumen fluid, and rumen microorganisms

dental pad (dehn-tahl pahd)

hard surface of the upper mouth of cattle that serves in place of upper teeth

ensiling (ehn-sī-lihng)

process in which a forage is chopped, placed in a storage unit that excludes oxygen, and ferments to allow longer preservation of feed

feedstuff (fēd-stuhf)

any dietary component that provides a nutrient; also called **feed**

finishing (fihn-ihsh-ihng)

act of feeding beef cattle high-quality feed before slaughter to increase carcass quality and yield

graze (grāz)

eating grasses and plants that grow close to the ground

legume (lehg-yoom)

roughage plants that have nitrogen-fixing nodules on their roots; examples include alfalfa and clover

premix (prē-mihx)

ration mixed with various feedstuffs at the feedmill

ration (rah-shuhn)

amount of food consumed by animal in a 24-hour period

roughage (ruhf-ahj)

type of feed that is high in fiber and low in total digestible nutrients; examples include pasture and hay; also called **forage**

silage (sī-lahj)

type of roughage feed that is produced by fermenting chopped corn, grasses, or plant parts under specific moisture conditions to ensure preservation of feed without spoilage

supplement (suhp-lah-mehnt)

additional feed product that improves and balances a poorer ration; also called **additive**

sweetfeed (swēt-fēd)

food that consists of grains and pellets mixed with molasses to increase palatability

SHEEP

Sheep are raised for wool, meat, and research models. As in cattle, there are breeds that are better known for their wool production and others that are better known for their meat quality. Sheep usually give birth to twins rather than single lambs (Figure 20–7).

Many of the anatomy and physiology concepts, as well as medical terms, related to sheep have been covered in previous chapters. The list below contains words used more specifically for sheep and sheep production.

Sheep Terminology

band (bahnd)
> large group of range sheep

carding (kahr-dihng)
> process of separating wool fibers

clip (klihp)
> one season's wool yield

crimp (krihmp)
> amount of wave in wool

crutching (kruhtch-ihng)
> process of clipping wool from dock, udder, and vulva of sheep before lambing; also called **tagging**

docking (dohck-ihng)
> removal of the distal portion of the tail; also means to reduce in value (Figure 20–8)

felting (fehl-tihng)
> property of wool fibers to interlock when rubbed together under heat, moisture, or pressure

fleece (flēs)
> another term for wool

grease wool
> wool that has been shorn from a sheep and that has not been cleaned

lanolin (lahn-ō-lihn)
> fatlike substance secreted by the sebaceous glands of sheep

mutton (muh-tihn)
> adult sheep meat

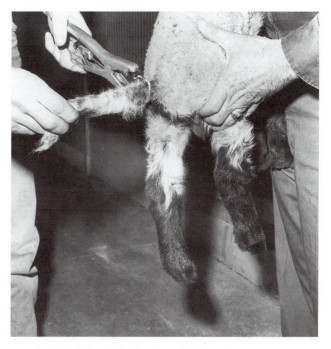

FIGURE 20–8 Docking. Tail docking of a lamb with an elastrator

rumping (ruhm-phing)
> method of restraining sheep by placing them in a sitting position with the front legs elevated; also called **tipping**

scouring (skow-ər-ihng)
> cleaning wool

shear (shēr)
> to shave off wool, hair, or fur

GOATS

Goats are raised for milk, meat, wool, and research models, and as pets.

Many of the anatomy and physiology concepts, as well as medical terms, related to goats have been covered in previous chapters. The list below contains words that apply more specifically to the care and treatment of goats. Some additional goat terms are under the cattle section, since they apply to both animals.

Goat Terminology

disbud (dihs-buhd)
> removal of horn growth in kids or calves by use of a hot iron or caustic substance; also called **debudding**

FIGURE 20–7 Twins are common in sheep. *Source:* Courtesy of Ron Fabrizius, DVM, Diplomat ACT

— Wattle

FIGURE 20–9 Wattle in a goat. *Source:* Courtesy of Ron Fabrizius, DVM, Diplomat ACT

wattle (waht-tuhl)

appendages suspended from the head (usually the chin) in chickens, turkeys, and goats (Figure 20–9)

LLAMAS

Llamas are becoming popular pets as well as being used as pack animals and for fiber production. Llamas, alpacas, vicunas, and guancos are all part of the Camelid family. Llamas are the larger species with less desirable wool, alpacas are smaller than llamas and have high quality wool, vicunas are the smallest and rarest with fine, high quality wool, and guanacos are slightly smaller than llamas and are undomesticated.

Many of the anatomy and physiology concepts, as well as medical terms, related to llamas have been covered in previous chapters. The list below contains words that apply more specifically to the care and treatment of llamas.

Llama Terminology

cushing (kuhsh-ihng)

common term for copulation of llamas

fighting teeth

set of six teeth in llamas that include an upper vestigial incisor and an upper and lower canine on each side (Figure 20–10)

VACCINATIONS OF RUMINANTS

bovine respiratory

paramyxovirus infection

syncytial virus (sihn-sihsh-ahl)

of bovine that is associated with fatal pneumonia; abbreviated BRSV or RSV

bovine viral diarrhea

togavirus infection of bovine that is associated with acute stomatitis, gastroenteritis, and diarrhea; abbreviated BVD

clostridial disease (klohs-trihd-ē-ahl)

group of bacterial infectious conditions of ruminants caused by various species of *Clostridium*, which includes diseases like blackleg, malignant edema, pulpy kidney, and overeating disease

infectious bovine rhinotracheitis (rī-nō-trā-kē-ī-tihs)

herpesvirus infection of bovine that is associated with fever, anorexia, tachypnea, and cough; abbreviated IBR

leptospirosis (lehp-tō-spī-rō-sihs)

bacterial disease caused by various serotypes of *Leptospira;* signs include renal failure, jaundice, fever, and abortion.

parainfluenza (pār-ah-ihn-flū-ehn-zah)

paramyxovirus infection of ruminants that is associated with fever, cough, and diarrhea; one part of the shipping fever complex; abbreviated PI-3

tetanus (teht-ah-nuhs)

highly fatal, bacterial disease caused by the toxin of *Clostridium tetani* that is associated with hyperesthesia, tetany, and convulsions; also called lockjaw

vibriosis (vihb-rē-ō-sihs)

Campylobacter fetus bacterial infection that is associated with infertility and irregular estrus cycles; bulls are vaccinated

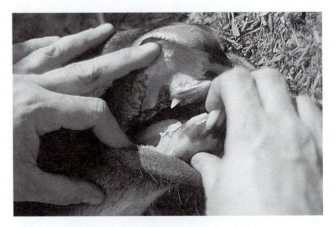

FIGURE 20–10 Fighting teeth of a llama

REVIEW EXERCISES

Case Studies—Define the underlined terms in each case study.

A 2 yr old French Alpine <u>doe</u> was presented with recurrent abdominal distention, decreased milk production, and <u>dyspnea</u>. On PE the <u>mm</u> were pink, <u>CRT</u> was normal, <u>TPR</u> was normal, breathing was labored but no abnormal respiratory sounds were <u>ausculted</u>, and the mammary glands were normal. The herd that this doe is in is closed, and the rest of the animals were normal. Urine was tested and revealed a trace amount of glucose and negative ketones. Rumen fluid had a <u>pH</u> of 6.9 and a healthy population of small and medium <u>protozoal</u> organisms. The majority of the large protozoal population was dead. A <u>rumenostomy</u> was performed and administration of 0.5 <u>L</u> warm water was initiated.

doe _____

dyspnea _____

mm _____

CRT _____

TPR _____

ausculted _____

pH _____

protozoal _____

rumenostomy _____

L _____

A 2 yr old Suffolk <u>ram</u> was examined for <u>lethargy, anorexia</u>, and weight loss. The <u>PE</u> was <u>WNL</u> except for pale <u>mucous membranes</u> and a distended abdominal area. The owner was questioned about grazing, nutritional, and deworming practices on the farm. The veterinarian took a blood sample for an <u>hematocrit</u> and a stool sample for parasite testing. The hematocrit revealed that the ram was mildly <u>anemic</u> and the fecal exam revealed *Haemonchus* eggs. The owner was advised to administer a broad-spectrum <u>anthelmintic</u>, to rotate pastures, and monitor animals for signs of anemia.

ram _____

lethargy _____

anorexia _____

PE _____

WNL _____

mucous membranes _____

hematocrit _____

anemic _____

anthelmintic _____

A <u>herd</u> of Hereford cattle were showing signs of <u>alopecia</u> and crusty skin <u>lesions</u> on the head and neck. Some of the more severely affected cattle had lesions over most of their bodies. The <u>PE</u> was unremarkable except for the skin lesions. The veterinarian took several <u>skin scrapings</u> to identify possible mite infestation. <u>Microscopic examination</u> revealed *Psoroptes* <u>mites</u>. The veterinarian diagnosed <u>scabies</u> (Figure 20–11), recommended pour-on medication, and advised the owner to observe withdrawal times. The cattle recovered uneventfully.

FIGURE 20–11 Scabies in a Hereford cow

herd_____

alopecia_____

lesions _____

PE _____

skin scraping_____

microscopic examination _____

mites _____

scabies_____

21 Hog Heaven

Objectives *In this chapter, you should learn to:*

► Recognize, define, spell, and pronounce terms related to swine
► Understand case studies and medical terminology in a practical setting

PIGS

Pigs are used for meat, hide, research models, and pharmaceutical production. Some pigs, such as potbellied pigs, have more recently been housed as pets.

Many of the anatomy and physiology concepts, as well as medical terms, related to pigs have been covered in previous chapters. The charts below and Figure 21–1 contain words that apply more specifically to the care and treatment of pigs.

EQUIPMENT AND INDUSTRY TERMS

casting (kahs-ting)
restraint method using ropes to place animal in lateral recumbency

dunging pattern (duhn-jihng pah-tərn)
tendency for animals to eliminate wastes in a particular location

ear notching (ēr nohtch-ihng)
identification method used in swine in which notches of various patterns are cut in the ear (Figure 21–2)

farrowing crate (făr-ō-ihng krāt)
holding pen that limits sow movement before and during parturition (Figure 21–3)

farrowing pen (făr-ō-ihng pehn)
sow holding area that has guardrails and floor junctures that allow young pigs to escape; used before and during parturition; pen is larger than a crate

finish (fihn-ihsh)
degree of fat on an animal that is ready for slaughter

hog hurdle (hohg hər-duhl)
portable partition used to move swine by blocking the area in which the pig should not go (Figure 21–4a)

hog snare (hohg snār)
restraint method in which the pig's snout is secured via a loop tie that is attached to a long handle; also called **snare** (Figure 21–4b)

hog-tight (hohg-tīt)
fencing that prevents animal escape

lard (lahrd)
soft, white fat that is the product of rendering pig fat

needle teeth
eight temporary incisors and canine teeth of young swine (Figure 21–5)

piles
common term for a prolapsed rectum in swine

ringing
act of implanting a wire ring through a pig's nose to discourage rooting

sling (slihng)
swine restraint device with four leg holes and an additional hole under the neck for blood collection; device looks like a hammock

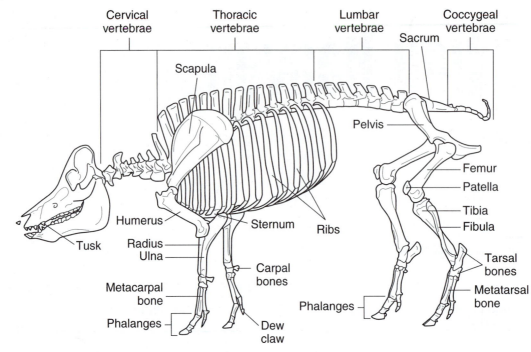

FIGURE 21-1 Pig skeleton

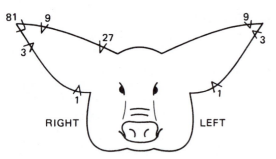

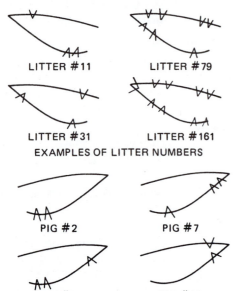

KEY TO STANDARD EAR NOTCHING SYSTEM

LITTER #11   LITTER #79

LITTER #31   LITTER #161
EXAMPLES OF LITTER NUMBERS

PIG #2   PIG #7

PIG #5   PIG #12
EXAMPLES OF INDIVIDUAL PIG NUMBERS

FIGURE 21-2 Ear notching identification in swine

snout (snowt)
upper lip and apex of nose of swine

specific pathogen free (speh-sihf-ihck pahth-ō-jehn frē) management system in which animals are obtained from cesarean section and raised in isolation to prevent certain infectious diseases; abbreviated SPF; not disease free

tusk (tuhsk)
overgrown canine tooth of boar

wallow (wahl-ō)
natural or artificial wading area to cool swine

SWINE MANAGEMENT TERMINOLOGY

closed herd
group of animals that restricts entrance of new animals

farrow-to-finish (fār-ō too fihn-ihsh)
full-service swine operation that houses breeders, newborns, weanlings, and feeder stock

farrow-to-wean (fār-ō too wēn)
swine operation that houses breeding sows and offspring until they reach weaning age or weight

feeder-to-finish (fē-dər too fihn-ihsh)
swine operation that raises weanling pigs to market weight

(a)

(b)

(c)

FIGURE 21–3 Housing of swine. (a) Farrowing or gestation crate restrains sows during farrowing. (b) Farrowing pen allows baby pigs to move away from the sow to prevent being crushed. (c) Larger pigs are housed away from the sow. *Source:* Courtesy of Ron Fabrizius, DVM, Diplomat ACT

(a)

(b)

FIGURE 21–4 Restraint of pigs. (a) hog hurdle; (b) hog snare. *Source:* Courtesy of Ron Fabrizius, DVM, Diplomat ACT

finisher pig (fihn-ihsh-ər pihg)
> swine over 100 pounds to slaughter

grower pig (grō-ər pihg)
> swine from about 40 to 100 pounds

open herd
> group of animals in which animals from other groups are allowed to join the existing group

starter pig (stahr-tər pihg)
> swine from about 10 to 40 pounds

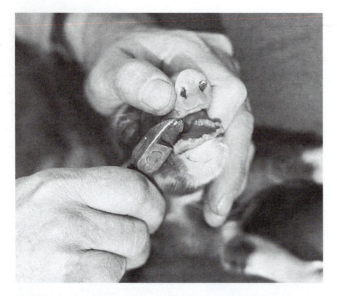

FIGURE 21–5 Needle teeth. Needle teeth of baby pigs are clipped to prevent injury to the sow during nursing.

SWINE VACCINATIONS

Bordetella (bōr-dih-tehl-ah)
   bacterial causing respiratory disease; atrophic rhinitis pathogen

erysipelas (ehr-ih-sihp-eh-lahs)
   bacteria disease of swine causing acute septicemia, skin lesions, chronic arthritis, and endocarditis

Haemophilus (hē-mohf-ih-luhs)
   bacteria causing respiratory disease associated with acute onset, pyrexia, and reluctance to move

leptospirosis (lehp-tō-spī-rō-sihs)
   bacteria disease of swine associated with pyrexia, anorexia, neurologic signs, and abortion

Mycoplasma (mī-kō-plahz-mah)
   bacteria causing respiratory disease seen largely in young pigs with a severe cough

parvovirus (pahr-vō-vī-ruhs)
   parvovirus infecting mainly affecting gilts associated with abortion

Pasteurella (pahs-too-rehl-ah)
   bacteria causing respiratory disease that sometimes leads to pericarditis and pleuritis

pseudorabies (soo-dō-rā-bēz)
   herpesvirus infection associated with pyrexia and neurologic signs

rotavirus (rō-tō-vī-ruhs)
   rotavirus associated with villous destruction in the intestine, malabsorption, and diarrhea

transmissible gastroenteritis
(trahnz-mihs-ih-buhl gahs-trō-ehn-tehr-ī-tihs)
   coronaviral disease of swine characterized by villous destruction of jejunum and ileum, malabsorption, diarrhea, and dehydration; abbreviated TGE

REVIEW EXERCISES

Case Studies—Define the underlined terms in each case study.

A group of 4 mo old barrows were presented with clinical signs of sneezing, purulent nasal discharge, and decreased weight gain. Most of the barrows had a mild to moderate deviation of the snout. The farmer sacrificed one pig for necropsy and found that the nasal turbinates were atrophied and asymmetrical. Dx was atrophic rhinitis, which is a common disease of pigs caused by two types of bacteria. Control measures such as better ventilation and improved hygiene were discussed with the farmer (Figure 21–6).

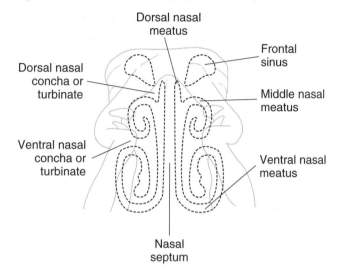

FIGURE 21–6 Atrophic rhinitis. Atrophic rhinitis causes atrophy of the turbinates and distortion of the nasal septum. The turbinates soften and erode due to infection by bacteria.

mo _____

barrows _____

clinical_____

signs _____

purulent _____

nasal _____

deviation_____

snout_____

necropsy _____

turbinates _____

atrophied _____

asymmetrical _____

Dx _____

rhinitis _____

A 3 yr old <u>sow</u> was seen by the area veterinarian for <u>dystocia</u>. She had <u>farrowed</u> two litters normally prior to this, but had problems with <u>mastitis</u> and <u>agalactia</u>. Because the pig had <u>uterine inertia</u>, the veterinarian had to perform an emergency <u>C-section</u> to retrieve the remaining pigs.

yr _____

sow _____

dystocia _____

farrowed _____

mastitis _____

agalactia _____

uterine inertia _____

C-section _____

Provide the medical term for the underlined definitions.

A litter of 2 wk old pigs were presented with <u>vomiting</u>, <u>abnormal frequency and liquidity of fecal material</u>, <u>incoordination</u>, and <u>elevated body temperature</u>. A blood sample was taken and the pigs tested positive for pseudorabies, also called Aujesky's disease, which is a herpesvirus infection. In adult pigs, pseudorabies can cause respiratory disease and <u>termination of pregnancy</u>. All positive animals were isolated and kept under observation. Pseudorabies control measures, such as only adding serologically negative animals to the herd, avoiding visiting of infected premises, keeping <u>wild</u> animals away from swine, and providing separate equipment for each group of animals was discussed with the owner.

vomiting _____

abnormal frequency and liquidity of fecal material _____

incoordination _____

elevated body temperature _____

termination of pregnancy _____

wild _____

Birds of a Feather

Objectives *In this chapter, you should learn to:*

▶ Recognize, define, spell, and pronounce terms related to birds
▶ Understand case studies and medical terminology in a practical setting

BIRDS

Birds and the terms associated with them vary greatly due to the two different types of birds kept by people. Some birds, such as the psittacines, are popular as pets. Other birds, such as poultry, are used more like live-stock. The terms related to birds usually refer to one or the other of these uses for birds.

Birds have unique anatomy because of their ability to fly. Their respiratory and skeletal systems vary greatly from those of other vertebrates to allow for flight (Figure 22–1). The integumentary system also varies in the fact that avian species have feathers and beaks. Other systems, such as the gastrointestinal, uro-genital, and circulatory systems, also vary from other vertebrates.

Some anatomy and physiology concepts and medical terms related to avian species have been covered in previous chapters. Additional anatomy and physiology concepts will be covered in this chapter.

ANATOMY AND PHYSIOLOGY TERMINOLOGY

Respiratory System

air sacs
    spaces in the respiratory tract of birds that store air and provide buoyancy for flight

choana (or **choanal space**) (kō-ā-nah)
    posterior naris; the cleft in the hard palate of birds (Figure 22–2)

nasal gland
    gland that allows sea birds to drink salt water; found in the anterior portion of the beak

syrinx (sehr-ihncks)
    voice organ of birds

Integumentary System

apterium (ahp-tehr-ē-uhm)
    areas or tracts of skin without feathers or down; plural is **apteria** (ahp-tehr-ē-ah)

barb (bahrb)
    one of the parallel filaments projecting from the main feather shaft (rachis); forms the feather vane

barbule (bahr-byoo-uhl)
    one of the small projections fringing the edges of the barbs of feathers; attach to adjacent barbules to give the vane rigidity

beak (bēk)
    modified epidermal structure that covers the rostral part of the maxilla and mandible

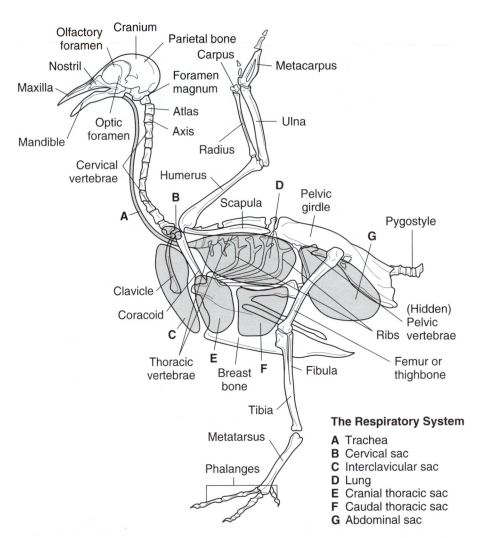

The Respiratory System

A Trachea
B Cervical sac
C Interclavicular sac
D Lung
E Cranial thoracic sac
F Caudal thoracic sac
G Abdominal sac

FIGURE 22–1 Skeletal and respiratory system of birds. Air flow through the lungs and air sacs. (1) first inspiration: air flows into the trachea, through the primary bronchi, and goes to the caudal and abdominal air sacs. Air already in the caudal air sacs moves to the cranial air sacs. (2) First expiration: air travels back to the parabronchi and gas exchange occurs. (3) Second inspiration: air moves from the parabronchi to the cranial air sacs. (4) Second expiration: air moves out of the cranial air sacs, into the parabronchi, and out of the trachea.

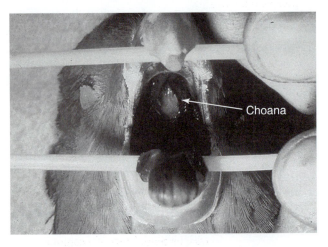

FIGURE 22–2 Choana of a parrot

calamus (kahl-ah-muhs)
hollow shaft at the proximal end of the feather shaft

cere (sēr)
thickened skin at base of external nares of birds; may be different colors in some birds to denote sex (Figure 22–3)

contour feather (kohn-tər fehth-ər)
body or flight feather and are arranged in rows

feather (fehth-ər)
epidermal structure analogous to hair; used for insulation, and thermoregulation, in courtship displays, and for flight (Figure 22–4)

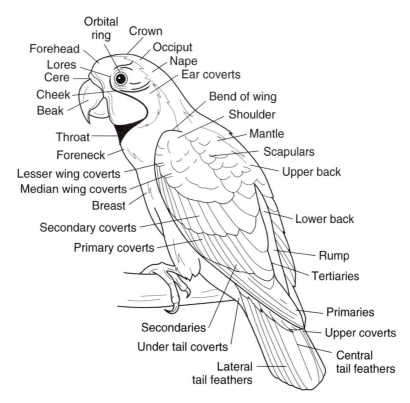

FIGURE 22–3 External parts of a bird

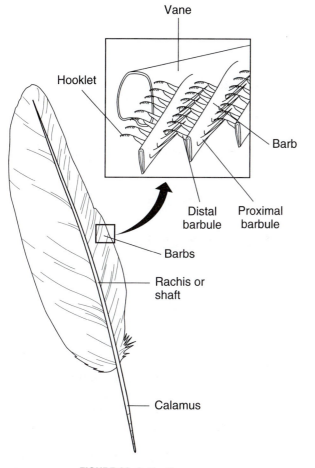

FIGURE 22–4 Feather parts

filoplume (fihl-ō-ploom)
> type of feather that resembles a bristle feather topped by a down feather

molt (mōlt)
> process of casting off feathers before replacement feathers appear

plume (ploom)
> down feathers

pteryla (tehr-ih-lah)
> feather tract of birds; plural is **pterylae** (tehr-ih-lā)

rachis (rā-kuhs)
> distal end of the feather shaft

rectrices (rehck-trih-sehz)
> tail flight feathers

remiges (rehm-ih-jehz)
> primary wing feathers; singular is **remix**

setae (shē-tā)
> sensitive bristles that grow on the heads of many birds; also called **bristles**

shaft (shahft)
> quill or central part of a contour feather; also called **scapus** (skā-puhs)

snood (snowd)
> long, fleshy extension at the base of a turkey's beak

spurs (spərz)
> projecting body (as from a bone) or a sharp, horn-covered, bony projection from the shank of male birds of some species

uropygial gland (yoor-ō-pihg-ih-ahl)
> gland located laterally to the tail feather attachment that secretes oil used to waterproof or preen feathers; also called the **preen gland** (prēn)

wattle (waht-tuhl)
> appendages suspended from the head (usually the chin) in chickens, turkeys, and goats

Gastrointestinal System

cloaca (klō-ā-kah)
> common passage for fecal, urinary, and reproductive systems in birds and lower vertebrates (Figure 22–5)

coprodeum (kōp-rō-dē-uhm)
> rectal opening into the cloaca

crop (krohp)
> esophageal enlargement that stores, moistens, and softens food in some birds (Figure 22–6)

droppings
> composite of feces and urine in birds

Meckel's diverticulum
(mehck-ehlz dī-vər-tihck-yoo-luhm)
> structure at the terminal end of the jejunum that functions as a lymphatic organ.

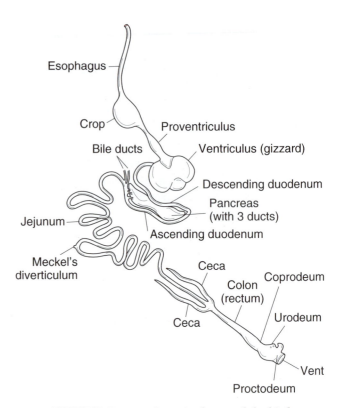

FIGURE 22–6 Gastrointestinal tract of the bird

proventriculus (prō-vehn-trihck-yoo-luhs)
> elongated, spindle shaped, glandular stomach of birds

vent (vehnt)
> external opening of the cloaca of birds

ventriculus (vehn-trihck-yoo-luhs)
> muscular stomach of birds; also called the gizzard

Musculoskeletal System

columella (kohl-uhm-eh-lah)
> bony structure which replaces the malleus, incus, and stapes within the middle ear

keel (kēl)
> sterum or breastbone of birds

pygostyle (pihg-ō-stī-uhl)
> bony termination of the vertebral column in birds where tail feathers attach; also called **rump post**

scleral ring (skleh-rahl)
> overlapping bony plates encircling the eye at the corneal-scleral junction

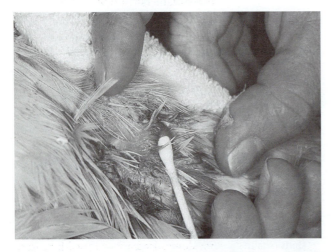

FIGURE 22–5 Cloacal swab with the normal mucosa everting

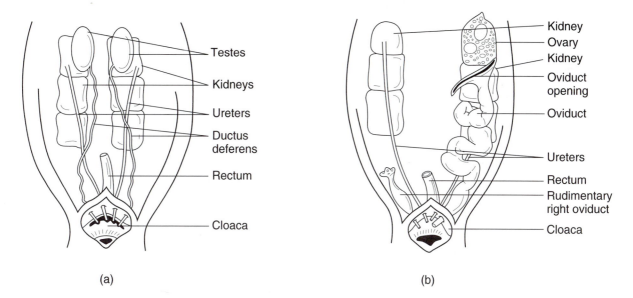

(a)   (b)

FIGURE 22–7 Urogenital system of the bird. (a) Male bird; (b) Female bird. Only the left ovary and oviduct are functional in female birds.

Urogenital System

See Figure 22–7 a and b.

isthmus (ihs-muhs)
  portion of the oviduct that adds the shell membranes

magnum (mahg-nuhm)
  portion of the oviduct that separates the albumin and chalaza over the egg yolk and sperm

sperm nests
  clusters of spermatozoon held in readiness in the infundibulum to fertilize the egg as it comes from the ovary

urodeum (yoo-rō-dē-uhm)
  area of the cloaca in which the ureters and vagina open

uterus (yoo-tər-uhs)
  portion of the oviduct in birds that produces the shell and shell pigments; also the area that the egg turns around so that it is layed blunt end first

vagina (vah-jī-nah)
  portion of the oviduct in birds that directs the egg to the cloaca

POULTRY TERMS

broiler (broy-lər)
  young chicken approximately 8 weeks old weighing 1.5 kg or greater; also called fryer or young chicken

brooder (brū-dər)
  housing unit for rearing birds after hatching

cage operation
  method of raising chickens in which the hens are kept in confinement as they produce eggs

candling (kahn-dlihng)
  process of shining a light through an egg to check embryo development

chalaza (kahl-ā-zah)
  rope-like structure that holds the yolk to the center of the egg

comb (kōm)
  in domestic fowl, the vascular, red cutaneous structure attached in a sagittal plane to the dorsum of the skull (Figure 22–8)

debeaking (dē-bēk-ihng)
  removal of about one-half of the upper beak and a small portion of the lower beak in poultry to prevent feather picking, cannibalism, and fighting; also called **beak trimming** in poultry (Figure 22–9)

incubation (ihn-kyoo-bā-shuhn)
  process of a fertilized poultry egg developing into a newly hatched bird

layer (lā-ər)
  chicken raised for egg production

poultry (pōl-trē)
  any domesticated fowl raised for meat, eggs, or feathers

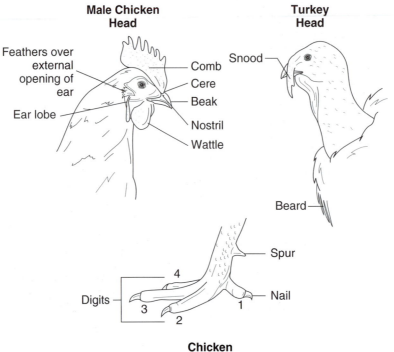

FIGURE 22–8 Head and feet of poultry

yolk (yōk)

yellow part of the egg that contains the germinal cells

PET BIRD TERMINOLOGY

clipping (klihp-ihng)

trimming wings of birds to alter their flight; also referred to as to **pinion** (pihn-yehn) (Figures 22–10 and 22–11)

columbiformes (kō-luhm-bih-fōrmz)

group of dove like birds that includes doves and partridges

fledgling (flehdg-lihng)

young bird that has recently acquired its flight feathers

passeriformes (pahs-ər-ih-fōrmz)

group of perching birds that includes finches, sparrows, mynahs, and canaries

psittacine (siht-ah-sēn)

group of parrot-like birds that includes parrots, macaws, budgerigars, and parakeets

FIGURE 22–9 Debeaking a chicken

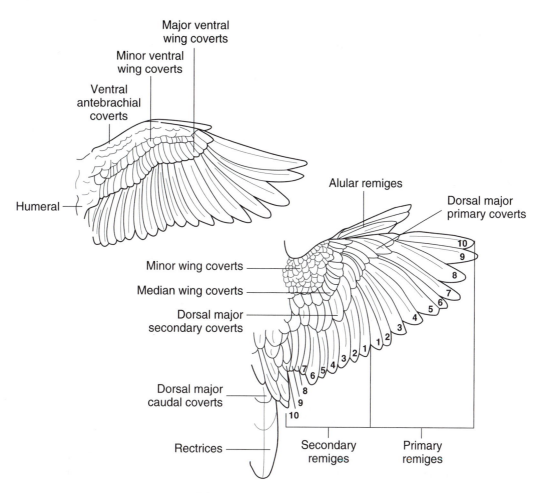

Major ventral
wing coverts

Minor ventral
wing coverts

Ventral
antebrachial
coverts

Humeral

Alular remiges

Dorsal major
primary coverts

10
9
8
7
6
5
4
3
2
1

Minor wing coverts

Median wing coverts

Dorsal major
secondary coverts

1 2 3 4 5 6 7 8 9 10

Dorsal major
caudal coverts

Rectrices

Secondary
remiges

Primary
remiges

FIGURE 22–10 Parts of a wing

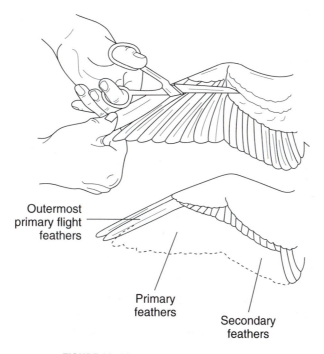

Outermost
primary flight
feathers

Primary
feathers

Secondary
feathers

FIGURE 22–11 Wing clipping in a pet bird

REVIEW EXERCISES

Case Studies—Define the underlined terms in each case study.

A <u>neonatal</u> greater sulfur-crested cockatoo was presented with a history of being tube fed with a red rubber tube. The tube became dislodged from the feeding syringe and was swallowed by the chick. A <u>radiograph</u> was taken which confirmed that the tube was swallowed and located extending from the <u>crop</u> to the <u>proventriculus</u>. An <u>ingluviotomy</u> was considered. As it would be a less traumatic procedure, it was decided to retrieve the feeding tube with an <u>endoscope</u>. Under general <u>anesthesia</u> with isoflurane, the tube was successfully removed. The bird recovered uneventfully, and the owner was counseled on hand feeding with a spoon instead of a feeding tube.

neonatal _____

radiograph _____

crop _____

proventriculus _____

ingluviotomy (also known as cropotomy) _____

endoscope _____

anesthesia _____

A 3 yr old, male budgerigar (parakeet) was presented for a growth on its <u>thorax</u> near the <u>thoracic inlet</u>. The bird had been kept singly and was fed a predominantly seed diet. On physical examination, the bird was noted to have a 2.5 <u>cm</u> diameter mass located <u>subcutaneously</u> on the <u>rostral</u> portion of the thorax. Differential <u>diagnoses</u> included <u>neoplasia</u> resulting in <u>lipoma</u> or a <u>xanthoma</u>. A <u>fine needle</u> <u>aspirate</u> (FNA) was performed, which confirmed the mass to be a lipoma. The owner declined surgical removal. The owner did elect to administer Lugol's iodine in the drinking water as this condition may be related to <u>hypothyroidism</u> (caused by iodine deficiency relating to an all seed diet).

thorax _____

thoracic inlet _____

cm _____

subcutaneously _____

rostral _____

diagnoses _____

neoplasia _____

lipoma _____

xanthoma _____

fine needle aspirate _____

hypothyroidism _____

23 All the Rest

Objectives *In this chapter, you should learn to:*

▶ Recognize, define, spell, and pronounce terms related to laboratory animals
▶ Recognize, define, spell, and pronounce terms related to exotic animals or pocket pets
▶ Understand case studies and medical terminology in a practical setting

LABORATORY ANIMALS AND POCKET PETS

Laboratory animals include a wide range of species, even species some consider pets or livestock. Pocket pets are small, "non-traditional" pets that were once thought of solely as laboratory animals or exotic to many areas of the country. Many terms used in the laboratory animal field have already been covered under other species chapters. New terminology related to laboratory animals includes terms related to scientific studies and the facilities in which they occur and terms used in the care of pocket pets (Table 23–1).

RODENTS

Rodents are gnawing animals that have continuously growing upper and lower incisors. The most common examples of rodents are hamsters, gerbils, mice, rats, guinea pigs, and chinchillas. The order Rodentia is divided into three suborders based on the musculature of the jaw and other skull structures. These suborders are the Sciuromorphas (squirrel-like rodents), the Myomorpha (rat-like rodents), and the Hystricomorpha (porcupine-like rodents). The two suborders of primary importance to the veterinary field are the Hystricomorpha, which include chinchillas and guinea pigs, and the Myomorpha, which include rats, mice, hamsters, and gerbils (Figure 23–1).

Mice and Rats

Mice and rats are rodents and are classified in the subfamily murine. Rats are larger than mice and have more rows of scales on their tales. Rats and mice were once mainly used as research animals, but are becoming more common as pets. Both are relatively clean and quiet animals to keep. Both rats and mice like to be housed with others of their species.

There are many species of rats; however, only some species, like the black rat, the brown rat (Norway rat) and the albino rat (Norway rat), have been domesticated and used for research and as pets. Rats have the ability to adapt to many different habitats, environments, and food sources, which explains why rats are found in all parts of the world. Rats are agile climbers, excellent swimmers, and very curious. They do best when kept with other rats and are primarily nocturnal. Rats lack a gallbladder or tonsils (Figure 23–4). Rats have 16 teeth and the dental formula is 2(1I/1I, 0C/0C, 0P/0P, 3M/3M).

Mice are believed to have originated from Asia and have spread throughout the world. Mice have a pointed nose and slit upper lip. Mice have 16 teeth and

TABLE 23-1

Common Laboratory and/or Pocket Pet Terms

Term	Pronunciation	Definition
acclimatization	(ahck-lih-mah-tih-zā-shuhn)	adjustment of an animal to a new environment
agouti	(ah-goo-tē)	naturally occurring coat color pattern, which consists of dark-colored hair bands at the base of the hair and lighter increments of hair color toward the tip
albino	(ahl-bī-nō)	animal with a white coat and pink eyes; devoid of melanin
ambient	(ahm-bē-ahnt)	surrounding
analogous	(ahn-ahl-oh-guhs)	refers to structures that differ anatomically but have similar functions
anogenital distance	(ā-nō-jehn-ih-tahl)	area between the anus and genitalia; females have a shorter anogenital distance than males, which is used to determine the sex of animals (Figure 23–2)
antivivisectionist	(ahnt-ih-vihv-ih-sehck-shuhn-ihst)	person who opposes surgery on live animals for research or educational purposes
autosome	(aw-tō-zōm)	non-sex-determining chromosome
axenic	(ā-zehn-ihck)	germ free
barbering	(bahr-bər-ihng)	behavioral disorder in animals where dominant animals bite or chew the fur from subordinate animals
barrier sustained	(bār-ē-ər suh-stānd)	gnotobiotic animals that are maintained under sterile conditions in a barrier unit
biohazard	(bī-ō-hahz-ahrd)	substance that is dangerous to life
calvarium	(kahl-vār-ē-uhm)	top of the skull
cannibalism	(kahn-ih-bahl-ihz-uhm)	devouring one's own species
cesarean derived	(sē-sā-rē-ahn)	animal is delivered via cesarean section into a sterile environment to avoid possible contamination
cheek pouch		space in oral cavity of hamsters that carries food and bedding
chromodachryorrhea	(krō-mō-dahck-rē-ō-rē-ah)	shedding of colored (blood-colored) tears
contact bedding		substrate with which animal comes into direct contact; also called **direct bedding**
control		standard "normal" against which experimental results are compared; also called **experimental control**
crepuscular	(krē-puhs-kuh-lahr)	becoming active at twilight or before sunrise
data	(dah-tah)	mass of accumulated information or results of an experiment
dusting	(duhs-tihng)	cleaning method in which chinchillas roll in dust
emission	(ē-mihsh-uhn)	discharge
estivate	(ehs-tih-vāt)	dormant state in which an animal reduces its body temperature, heart, and respiration rates and metabolism in summer
exsanguination	(ehcks-sahn-gwih-nā-shuhn)	removal or blood loss from the body
fomite	(fō-mīt)	inanimate carriers of disease
fur-slip	(fər-slihp)	shedding of hair patches from rough handling in chinchillas
genotype	(jē-nō-tīp)	genetic make-up of an individual for a particular trait
gnotobiotic	(nōt-ō-bī-ah-tihck)	germ-free animals that have been introduced to one or two known nonpathogenic microorganisms
heterozygous	(heht-ər-ō-zī-guhs)	condition of having two different genes for a given gentic trait; usually one gene is dominant over the other
hibernate	(hī-bər-nāt)	dormant state in which an animal reduces its body temperature, heart, and respiration rates and metabolism in winter
homologous	(hō-mohl-ō-guhs)	having a common origin but different function(s) in different species
homozygous	(hō-mō-zī-guhs)	condition of having two identical genes for a given trait
hooded	(huhd-ehd)	rats that have a white coat with a black "hood" over their head and shoulders and pigmented eyes (Figure 23–3)
horizontal transmission	(hōr-ih-zohn-tahl trahnz-mihs-shuhn)	disease transfer from one animal to the other
hybrid	(hī-brihd)	strain resulting from mating two inbred strains
hypothesis	(hī-pohth-eh-sihs)	statement of research supposition

TABLE 23–1
Common Laboratory and/or Pocket Pet Terms *continued*

hystricomorph	(hihs-trihck-ō-mōrf)	type of rodent that includes guinea pigs, chinchillas, and porcupines
inbred	(ihn-brehd)	strain resulting from at least 20 brother-sister or parent-offspring matings
in situ	(ihn sih-too)	at the normal site
in vitro	(ihn vē-trō)	outside living organisms; in test tubes or other laboratory glassware
in vivo	(ihn vē-vō)	inside living organisms
latent infection	(lā-tehnt ihn-fehck-shuhn)	condition that may not be clinically noticed, but under stress or poor health will develop into a recognizable disease state
macroenvironment	(mahck-rō-ehn-vī-ərn-mehnt)	surroundings above the cellular level
metanephric	(meht-ah-nehf-rihck)	embryonic-like kidney
microenvironment	(mīk-rō-ehn-vī-ərn-mehnt)	surroundings at the cellular level
monogamous	(moh-noh-goh-muhs)	pairing with one mate
murine	(moo-rēn)	mice and rats
outbred	(owt-brehd)	offspring from unrelated parents; also called **random bred**
pithing	(pihth-ihng)	destruction of brain and spinal cord by thrusting a blunt needle into the cranium or vertebral column
phenotype	(fē-nō-tīp)	physical characteristics of an individual
phylogeny	(fī-lohj-eh-nē)	developmental history of a species
polygamous	(poh-lihg-ah-muhs)	having multiple mates
polytocous	(poh-liht-ō-kuhs)	giving birth to multiple offspring
prehensile	(prē-hehn-sihl)	adapted for grasp and seizing
progenitor	(prō-jehn-ih-tōr)	parent or ancestor
progeny	(proh-jehn-ē)	offspring or descendents
propagate	(proh-pah-gāt)	to reproduce
protocol	(prō-tō-kawl)	written procedure for carrying out experiments
rack	(rahck)	metal device that supports caging units
reduction	(rē-duhck-shuhn)	theory of using the minimal number of animals for a project that will yield valid results; one of the three R's principle of Russell and Burch
refinement	(rē-fīn-mehnt)	theory of inflicting minimal stress and pain to animals in research; one of the three R's principle of Russell and Burch
replacement	(rē-plās-mehnt)	theory of using cell/tissue culture or mathematical models instead of animals in research if possible; one of the three R's principle of Russell and Burch
ringtail	(rihng-tā-uhl)	abnormal condition in which anular lesions form on the tails of rats housed in low humidity environments
rodent	(rō-dehnt)	class of animal that have chisel-shaped incisor teeth
rosette	(rō-seht)	swirl hairgrowth pattern in Abyssinian guinea pig
rudimentary	(roo-dih-mehn-tār-ē)	incompletely developed
sable	(sā-buhl)	color pattern that has cream-colored undercoat and black guard hairs on feet, tail, and mask
scurvy	(scər-vē)	common term for vitamin C deficiency
sexual dimorphism	(sehcks-yoo-ahl dī-mōrf-ihzm)	physical or behavioral differences between females and males of a given species
shoebox	(shū-bohcks)	caging that has a solid bottom flooring
suspended cage	(suh-spehnd-ehd)	caging that hangs from a metal rack and has wire flooring
teratology	(tehr-aht-ohl-ō-jē)	study of embryo development
test group		collection of animals used for experimental manipulation
transgenic	(trahnz-gehn-ihck)	removing or synthesizing specific genes from one strain and injecting them into the cells of another strain
vector	(vehck-tər)	something that carries disease from one animal to another
vertical transmission	(vər-tih-kahl trahnz-mihs-shuhn)	disease transfer from mother to fetus
vestigial	(vehs-ti-jē-ahl)	structure that has lost a function it previously had

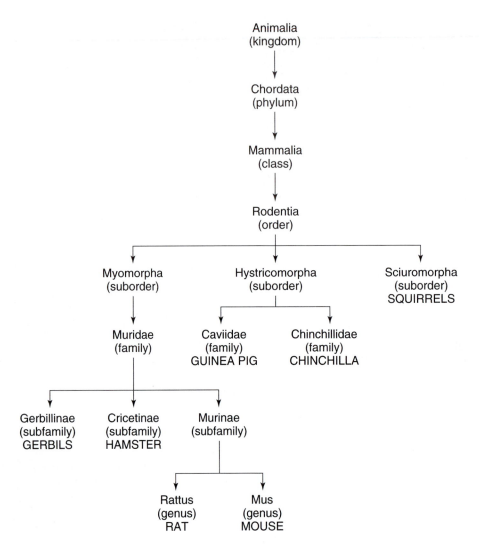

Animalia
(kingdom)

↓

Chordata
(phylum)

↓

Mammalia
(class)

↓

Rodentia
(order)

Myomorpha
(suborder)

Hystricomorpha
(suborder)

Sciuromorpha
(suborder)
SQUIRRELS

Muridae
(family)

Caviidae
(family)
GUINEA PIG

Chinchillidae
(family)
CHINCHILLA

Gerbillinae
(subfamily)
GERBILS

Cricetinae
(subfamily)
HAMSTER

Murinae
(subfamily)

Rattus
(genus)
RAT

Mus
(genus)
MOUSE

FIGURE 23–1 Classification scheme of rodents

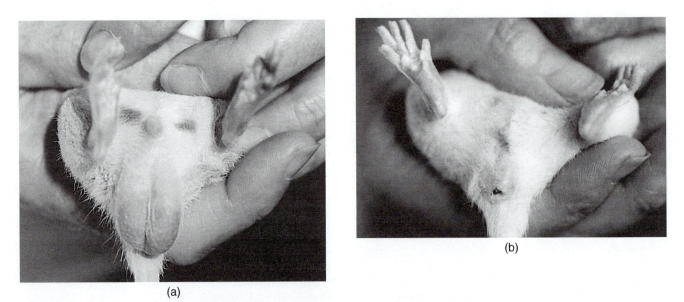

(a)

(b)

FIGURE 23–2 Anogenital distance of a rat. (a) Genital area of a male rat; (b) Genital area of a female rat. *Source:* Photos by Dean Warren

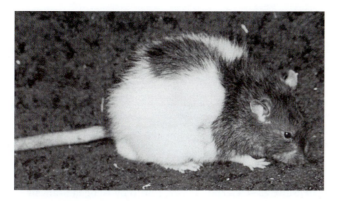

FIGURE 23–3 A rat with hooded markings

the dental formula of the mouse is 2(1I/1I, 0C/0C, 0P/0P, 3M/3M). Mice have a perfect visual field due to the placement and shape of their eyes; however, their detailed vision is poor. To compensate, mice have large ears and a highly developed sense of hearing. Mice have a highly developed sense of smell as well. Mice do best when kept with other mice and are primarily nocturnal. A mouse colony is lead by one head male. The head male is the only one allowed to mate with the females. Fighting between males for the head male status is common in mouse colonies.

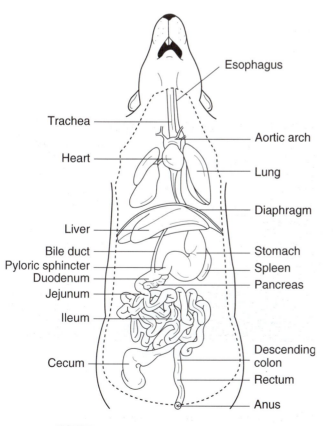

Esophagus
Trachea
Aortic arch
Heart
Lung
Diaphragm
Liver
Bile duct
Stomach
Pyloric sphincter
Spleen
Duodenum
Pancreas
Jejunum
Ileum
Cecum
Descending colon
Rectum
Anus

FIGURE 23–4 Internal structures of a rat

Guinea Pigs

Guinea pigs, also called cavies, are rodents that have short bodies, stocky legs, and short sharp claws. Guinea pigs tend to be gentle, easy to handle rodents. There are several common varieties of guinea pigs (Figure 23–5).

(a)

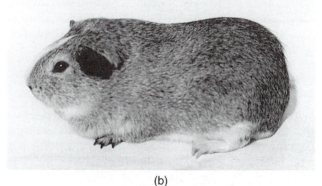

(b)

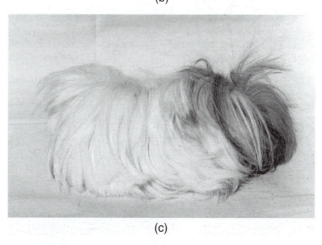

(c)

FIGURE 23–5 Examples of guinea pig varieties. (a) Abyssinian; (b) American short-haired; (c) Peruvian. *Source:* Photos (a and b) by Isabella Francais; photo (c) by Dean Warren

The male guinea pig is larger than the female guinea pig. Guinea pigs have very sensitive senses of hearing and smell.

Guinea pigs cannot synthesize vitamin C like most other mammals can; therefore, vitamin C must be supplied in the diet. Feeding fresh pellets with supplemental fresh produce can provide the guinea pig with enough vitamin C. The dental formula of the guinea pig is 2(1I/1I, 0C/0C, 1P/1P, 3M/3M). The guinea pig has a large cecum with numerous pouches; therefore, feeding adequate roughage is important.

Hamsters

Hamsters are used as pets and research animal. The Syrian or Golden hamster (Figure 23–6) was bred from wild hamsters in the 1930s and all domesticated hamsters sold are descendants of this original breeding. Hamsters are rodents that are noctural and solitary. A few species of hamster exist; however, the Golden hamster is the most common. Hamsters have virtually no tails. Hamster skin is abundant, loose, and pliable.

Hamsters prefer a temperature around 70 °F and will go into a deep estivation if the temperatures rise above 80 °F or so. Hamsters have prominent scent glands on their flanks. The secretions from these glands are used to mark territory. Hamsters also have cheek pouches to store and transport food. Hamsters have 16 teeth and the dental formula is 2(1I/1I, 0C/0C, 0P/0P, 3M/3M).

Gerbils

Gerbils, also known as jirds, are burrowing rodents that have bodies that are short and have a hunched appearance (Figure 23–7). The tail is covered with fur, has a bushy tip, and is used for support while standing. The male gerbil has a scent gland on its stomach. This scent gland allows the adult male to leave his scent by sliding his stomach across an object.

The Mongolian gerbil is the most common type of gerbil kept as pets. Gerbils are also used as research animals. Gerbils are quiet animals; however, they are

FIGURE 23–7 Gerbil. *Source:* Photo by Isabella Francais

quite active. Gerbils are hardy and relatively resistant to disease. Gerbils have 16 teeth and the dental formula is 2(1I/1I, 0C/0C, 0P/0P, 3M/3M). The adrenal gland of the gerbil is large and contributes to the gerbil's ability to conserve water.

Chinchillas

Chinchillas are rodents that were originally used for their fur, but are becoming increasingly popular as pets and research animals. Chinchillas resemble small rabbits with shorter ears and bushy tails. Their fur is thick and soft due to the fact that they have fewer guard hairs than some of the other pocket pets (Figure 23–8).

Chinchillas are nocturnal and live in groups. Chinchillas clean their fur by rolling in dust (Figure 23–9). Chinchillas can also release their fur as a defense mechanism. If grabbed too roughly, either by a predator or handler, the chinchilla will leave a patch of fur behind. Chinchillas are hindgut fermenters with a relatively large stomach, jejunum, and cecum. Chinchillas have 20 teeth and the dental formula is 2(1I/1I, 0C/0C, 1P/1P, 3M/3M). Male chinchillas have open inguinal rings. The testes are located in the inguinal canal without a true scrotal sac.

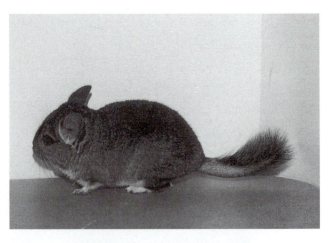

FIGURE 23–8 A standard chinchilla

FIGURE 23–6 Hamsters. Long-haired Golden hamster; (left) and short-haired Golden hamster (right)

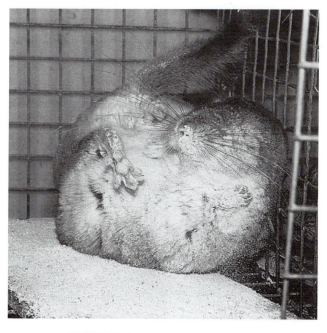

FIGURE 23–9 Dusting in a chinchilla

FERRETS

Ferrets (Figure 23–10) are mammals that belong to the family Mustelidae which includes weasels, mink, and polecats. Ferrets have been in the United States for more than 300 years and were originally used for rodent control. Today ferrets are used primarily as pets; however, they are curious and like to get into small spaces and eat just about anything lying around. Ferrets are also used in research and as working animals because of their ability to reach difficult places.

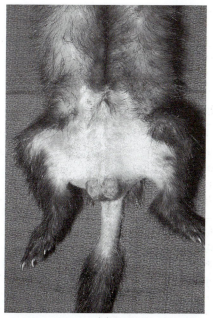

(a)

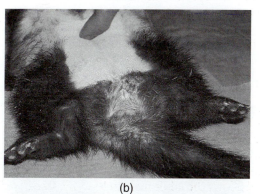

(b)

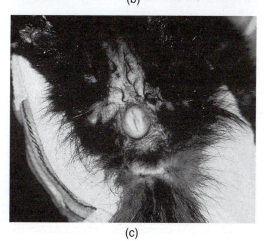

(c)

FIGURE 23–11 External genitalia of ferrets. (a) Male ferret; (b) Nonestrus female ferret; (c) Estrus female ferret

FIGURE 23–10 Fitch ferrets Ferrets may be restrained by scruffing the skin of the dorsal neck. Male ferrets are larger than female ferrets of similar age.

Ferrets have long slender bodies and long tails. Ferret tails are about one-half the length of the head and body. Ferrets have short legs and small, rounded ears. Ferrets are primarily nocturnal; hence, they do not see well in bright light. Ferrets have highly developed senses of hearing, smell, and touch. Ferrets have 40 permanent teeth and the dental formula is 2(3I/3I, 2C/2C, 4P/3P, 1M/2M). Female ferrets are induced ovulators, which has caused some health problems if the female does not come out of heat. **Hyperestrogenism** (hī-pər-ehs-trō-jehn-ihz-uhm) is elevated blood estrogen levels seen in intact cycling female ferrets if not bred (Figure 23–11).

RABBITS

Rabbits are mammals classified as lagomorphs. Rabbits have four upper incisor teeth versus rodents which have two incisor teeth. Rabbits may be dwarf, mini, standard, giant in size and/or lop eared (Figure 23–12). Hares are in the same family as rabbits and are usually larger than rabbits and have longer ears. Hares do not build nests as rabbits do, and young hares are born fully furred with their eyes open. Hares live above ground and do not dig tunnels as rabbits do.

FIGURE 23–13 Snuffles in a rabbit. *Source:* USDA

Rabbits are kept indoors or outdoors as pets, are raised for pelts and food, and are used for research. Proper ventilation is important in keeping rabbits disease free. Rabbits are hindgut fermenters, so adequate roughage in the diet is important. Rabbits have a simple, glandular stomach and a large cecum. Rabbits cannot vomit. Rabbits also need to gnaw to keep the length of their incisors in check. Rabbits have 28 permanent teeth and the dental formula of the rabbit is 2(2I/1I, 0C/0C, 3P/2P, 3M/3M).

Rabbits have powerful hindquarters for jumping. If not restrained properly, a rabbit can kick its rear legs and fracture its spine. In male rabbits, the inguinal canals remain open for life, the scrotum is hairless, and the testes descend at about 12 weeks of age.

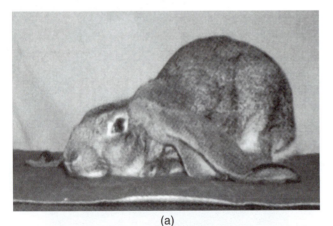

(a)

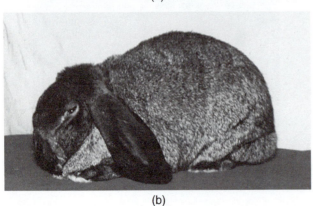

(b)

FIGURE 23–12 Examples of lop eared rabbits (a) English lop; (b) French lop

FIGURE 23–14 Torticollis in a rabbit. *Source:* USDA

night stool (nīt-stool)
> rabbit noctural feces that is looser than normal and contains vitamins and nutrients that the rabbit consumes

snuffles (snuhf-uhlz)
> common term for upper respiratory disease of rabbits caused by *Pasteurella multocida* (Figure 23–13)

sore hocks
> ulceration of the foot pads and foot area caused by the animal's body weight pressing down on the foot; commonly seen in rabbits housed on wire cage floors

torticollis (tōr-tih-kō-luhs)
> contracted state of the cervical muscles producing torsion of the neck; also called **wry neck** (Figure 23–14)

REPTILES

Reptiles are cold-blooded vertebrates that have lungs and breathe air. Reptiles have a body covering (bony skeleton, scales, or horny plates) and a heart that has two auricles and in most species one ventricle.

There are four different types of reptiles: Chelonia (turtles, tortoises, and terrapins), Serpentes (snakes, pythons, and boas), Squamata (iguanas and lizards), and Crocodilia (crocodiles and alligators).

Modern-day reptiles have three body types. One type have long bodies and clearly defined tails such as lizards. The second type have long bodies that taper into tails such as snakes. The third type have short, thick bodies encased in shells such as turtles.

carapace (kahr-ah-pāc)
> dorsal region of a turtle shell (Figure 23–15)

chin glands
> secretory organs located on the throats of turtles; also called **mental glands**

dysecdysis (dihs-ehck-dī-sihs)
> difficult or abnormal shedding

ecdysis (ehck-dī-sihs)
> shedding or molting

femoral pores (fehm-ōr-ahl poorz)
> sexually dimorphic glands prominent in mature lizards that are located on the ventral surface of lizard thighs; also called **femoral glands**

head gland
> small secretory organ located on the head of snakes

musk glands
> four secretory organs that open lateroventrally near the carapace edge of turtles

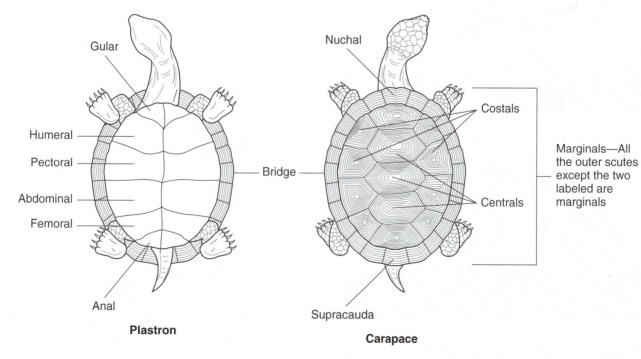

FIGURE 23–15 Scutes (scale-like structures) on a turtle shell

plastron (plahs-trohn)

> ventral region of a turtle shell

scent gland

> sac-like secretory organ located at the base of the tail in snakes

spectacle (spehck-tah-kuhl)

> transparent, highly vascular, unshed, abnormal covering over the cornea of some reptiles; also called an **eyecap** (Figure 23–16)

tail autotomy (tā-uhl aw-toh-tō-mē)

> lizard's ability to lose its tail

urostyle (yoor-ō-stīl)

> long caudal vertebra of some amphibians

FIGURE 23–16 Spectacle or eyecap in a snake

REVIEW EXERCISES

Case Studies—Define the underlined terms in the following case studies.

A 6 mo old, female, green iguana was presented for examination for lethargy, ataxia, and limb twitching. It was kept in an aquarium with food, water, a hot rock, and a white light source. The diet consisted mainly of head lettuce and grapes. Upon PE, the iguana was noted to be underweight and undersized for its age and limb tremors could be induced while handling. Based on the history, it was likely that this animal had secondary nutritional hyperparathyroidism, otherwise known as metabolic bone disease. Radiographs were taken, and blood was drawn for a serum chemistry profile. Results indicated poor bone density, hypocalcemia, and hypoproteinemia. The iguana was treated with calcium and vitamin D3 supplementation. The owners were counseled on proper husbandry and nutrition of green iguanas (Figure 23–17).

FIGURE 23–17 Green iguana

lethargy _____

ataxia _____

PE _____

hyperparathyroidism _____

metabolic _____

radiographs _____

serum _____

chemistry profile _____

hypocalcemia _____

hypoproteinemia _____

A 5 yr old, three-toed box turtle was presented with a swelling on the right side of its head. The turtle was housed in an aquarium with dirt substrate, a water dish, and a hide box. On PE, the right tympanic membrane was noted to bulge out laterally from the head. Oral examination revealed stomatitis characterized by yellow plaques on the oral mucosa. A diagnosis of otitis media and potential hypovitaminosis was made. The abscess in the ear was lanced and debrided. The oral plaques were swabbed for cytology and were found to demonstrate squamous metaplasia consistent with hypovitaminosis A. The turtle was treated with antibiotics and vitamin A supplementation. Husbandry and nutrition were discussed with the owner.

tympanic membrane _____

laterally _____

oral _____

stomatitis _____

plaques _____

mucosa _____

diagnosis _____

otitis media _____

hypovitaminosis _____

abscess _____

lanced _____

debrided _____

cytology _____

squamous _____

metaplasia _____

A 5 yr old M/N ferret was presented for signs of <u>alopecia</u> on its <u>dorsal lumbar and tail regions</u> and <u>dysuria</u>. Although differential diagnoses of <u>ectoparasites</u>, <u>dermatophytosis, atopy, cystitits, and urolithiasis</u> are considered, the most likely <u>dx</u> is <u>hyperadrenocorticism</u>. This endocrine disease is caused by adrenal gland <u>hyperplasia</u> or <u>neoplasia</u>. Production of excessive androgen hormones results in <u>cessation</u> of the normal hair growth cycle, and may cause <u>prostatic hyperplasia</u>, which in turn produces the clinical signs described. Treatment options include surgical <u>excision</u> of the affected adrenal glands and/or medical therapy aimed at reducing the functional tissue of the adrenal glands. Many ferrets respond very well to the therapeutic options.

alopecia _____

dorsal lumbar and tail regions _____

dysuria _____

ectoparasites _____

dermatophytosis _____

atopy _____

cystitis _____

urolithiasis _____

dx _____

hyperadrenocorticism _____

hyperplasia _____

neoplasia _____

cessation _____

prostatic hyperplasia _____

excision _____

A 3 yr old chinchilla was noted to be limping on its left rear leg. He had taken a fall from the top of a bedroom dresser. Upon presentation to the veterinarian, a <u>midshaft, spiral femoral Fx</u> was <u>diagnosed</u> with <u>radiographs</u>. Options included <u>coaptation</u> by <u>splinting</u> the leg, surgical repair of the <u>Fx</u>, or restricted cage rest. The owner chose to have the leg splinted. The chinchilla was anesthetized with isoflurane, an <u>inhalant anesthetic</u>, via a face mask. A splint was fashioned with fiberglass casting and bandage materials. One week later, it was noted that the splint had shifted, so the animal was returned to the veterinarian. Upon removal of the splint, it was discovered that the leg was no longer <u>viable</u> due to the <u>vascular</u> and soft tissue damage that had occurred. The chinchilla was again anesthetized for surgical <u>amputation</u> of the leg. He recovered well over the next few weeks and <u>ambulated</u> well as a three-legged chinchilla.

midshaft _____

spiral _____

femoral Fx _____

diagnosed _____

radiographs _____

coaptation _____

splinting _____

Fx _____

inhalant anesthetic _____

viable _____

vascular _____

amputation _____

ambulated _____

A 2 yr old M guinea pig was noted to have <u>hematuria</u> and <u>pollakiuria</u> from the owner's <u>Hx</u>. On PE, the guinea pig was <u>BAR</u>. All systems examined were <u>WNL</u> other than <u>palpation</u> of the urinary bladder, which elicited <u>vocalization</u> as a result of assumed pain. The guinea pig was restrained for <u>cystocentesis</u>, and an <u>UA</u> was performed. The urine contained numerous <u>erythrocytes</u> and <u>leukocytes</u>, as well as calcium oxalate <u>crystals</u>. A <u>urolith</u> was suspected. Whole body <u>radiographs</u> confirmed the presence of a <u>radiopaque</u> density in the urinary bladder. A <u>cystotomy</u> was performed to remove the calcium oxalate stone. The owners were instructed to feed timothy hay instead of alfalfa hay to decrease dietary calcium intake.

hematuria _____

pollakiuria _____

Hx _____

BAR _____

WNL _____

palpation _____

vocalization _____

cystocentesis _____

UA _____

erythrocytes _____

leukocytes _____

crystals _____

urolith _____

radiographs _____

radiopaque _____

cystotomy _____

A 1 yr old F domestic mouse was presented for <u>pruritus</u> and a powdery substance on the <u>dorsal</u> aspect of the neck. PE revealed a pruritic mouse with white granular material on the <u>dorsal cervical region</u>. A cellophane tape preparation was performed on the neck <u>debris</u>. <u>Ectoparasitic</u> ascariasis was diagnosed, the most likely agent being *Myobia musculi.* The mouse was treated <u>topically</u> with the <u>parasiticide</u> Ivermectin, which was placed on the <u>interscapular</u> skin. The owner was instructed to return for a second treatment in 10 days.

pruritus _____

dorsal _____

dorsal cervical region _____

debris _____

ectoparasitic _____

topically _____

parasiticide _____

interscapular _____

Socrates, a $1^1/_2$ yr old hooded rat was examined for what the owner described as a cold. <u>Conjunctivitis</u> and <u>sinusitis</u> were noted on PE. The owner was concerned that there was bleeding from the <u>nares</u>. He was informed that when rats become ill, they groom themselves less, which allows nasal discharge to accumulate on the fur. The red tinge that is being seen is not blood, but a prophyrin pigment, which is a normal component of the <u>nasal</u> secretions. Various viral and bacterial <u>etiologic</u> agents are responsible for respiratory disease in rats, including *Mycoplasma.* The <u>antibiotic</u> oxytetracycline was prescribed <u>PO</u> for 2 weeks. The owner was informed that this may be a <u>chronic</u>, recurrent disease.

conjunctivitis _____

sinusitis _____

nares _____

nasal _____

etiologic _____

antibiotic _____

PO _____

chronic _____

APPENDIX A

Abbreviations

IMAGING ABBREVIATIONS

a or amp	ampere
A/P	anterior/posterior
Ba	barium
CT or CAT	computed (axial) tomography
D/V	dorsal/ventral
f	frequency
Hz	Hertz (a unit of frequency)
λ	wavelength
IVP	intravenous pyelogram
kv	kilovolt
kVp	kilovolts peak
kw	kilowatt
mA	milliamperage
mAs	milliamperage in seconds
mc	millicurie
MRI	magnetic resonance imaging
P/A	posterior/anterior
rad	unit of measurement of absorbed dose of ionizing radiation
RT	radiation therapy
T	period of time (used in ultrasound)
v	velocity or volt
V/D	ventral/dorsal
w	watt
X-ray	roentgen ray

SPECIALIST OR TITLE ABBREVIATIONS

The letter "D" may appear in front of these specialties, which means that the person is a diplomat of that particular college. For example, DAVCIM after someone's name means that the person is a diplomat in the American College of Internal Medicine.

ABVP	American Board of Veterinary Practitioners
ABVT	American Board of Veterinary Toxicology

ACLAM	American College of Laboratory Animal Medicine
ACPV	American College of Poultry Veterinarians
ACT	American College of Theriogenologists
ACVA	American College of Veterinary Anesthesiologists
ACVB	American College of Veterinary Behaviorists
ACVCP	American College of Veterinary Clinical Pharmacology
ACVD	American College of Veterinary Dermatology
ACVECC	American College of Veterinary Emergency and Critical Care
ACVIM	American College of Veterinary Internal Medicine
ACVM	American College of Veterinary Microbiologists
ACVN	American College of Veterinary Nutrition
ACVO	American College of Veterinary Ophthalmologists
ACVP	American College of Veterinary Pathologists
ACVPM	American College of Veterinary Preventive Medicine
ACVR	American College of Veterinary Radiology
ACVS	American College of Veterinary Surgeons
ACZM	American College of Zoological Medicine
AHT	Animal Health Technician
AVDC	American Veterinary Dental College
ALAT	Assistant Laboratory Animal Technician
CVT	Certified Veterinary Technician
DVM	Doctor of Veterinary Medicine
LAT	Laboratory Animal Technician

LATG	Laboratory Animal Technologist		ED_{50}	median effective dose
RVT	Registered Veterinary Technician		fl oz	fluid ounce
VMD	Veterinary Medical Doctor (Veterinariae Medicinae Doctor)		g	gram

ASSOCIATION ABBREVIATIONS

AAHA	American Animal Hospital Association
AALAS	American Association of Laboratory Animal Science
AKC	American Kennel Club
AO	arbeitsgemeinschaft fur osterosynteses-fragen (association for the study of fracture treatment in man founded by a group of Swiss surgeons); used to describe specialized bone plates and instruments used in orthopedic repair
ASIF	Association of the Study of Internal Fixation
ASPCA	American Society for the Prevention of Cruelty to Animals
AVMA	American Veterinary Medical Association
CDC	Centers for Disease Control and Prevention
DEA	Drug Enforcement Agency
DHIA	Dairy Herd Improvement Association
DOT	Department of Transportation (used for OSHA regulation of hazardous material transfer)
FDA	Food and Drug Administration
NAVTA	North American Veterinary Technician Association
OFA	Orthopedic Foundation for Animals
OSHA	Occupational Safety and Health Administration
USDA	United States Department of Agriculture
USP	United States Pharmacopeia

PHARMACOLOGY ABBREVIATIONS

ac	before meals (ante cibum)
ad lib	as much as desired (ad libitum)
bid	twice daily (bis in die)
BSA	body surface area
cal	calorie
cap	capsule
cc	cubic centimeter (same as ml)
cm	centimeter
conc	concentration
dr	dram; equal to $^1/_8$ oz or 4 ml
D_5W	5% dextrose in water
ED	effective dose

ED_{50}	median effective dose
fl oz	fluid ounce
g	gram
gal	gallon
gr	grain; unit of weight approximately 65 mg
gt	drop (gutta)
gtt	drops (guttae)
hr	hour
IA	intraarterial
IC	intracardiac
ID	intradermal
IM	intramuscular
IP	intraperitoneal
IT	intrathecal
IU	international units or intrauterine
IV	intravenous
kg	kilogram
km	kilometer
L or l	liter
lb or # behind a number	pound(s)
LD	lethal dose
LRS	lactated Ringer's solution
m	meter
mcg or μg	microgram
MED	minimal effective dose
mEq	milliequivalent
mg	milligram
MID	minimum infective dose
MIC	minimum inhibitory concentration
ml	milliliter (same as cc)
MLD	minimum lethal dose
mm	millimeter (also used for muscles)
NPO/npo	nothing by mouth (non per os)
NS	normal saline
OTC	over the counter
oz	ounce
%	percent
pc	after meals (post cibum)
PDR	*Physician's Desk Reference*
VPB	*Veterinary Pharmaceuticals and Biologicals*
pH	hydrogen ion concentration (acidity and alkalinity measurement)
PO/po	orally (per os)
ppm	parts per million
PR	per rectum

prn	as needed
pt	pint
PZI	protamine zinc insulin
q	every
qd	every day
qh	every hour
q4h	every four hours
q6h	every six hours
q8h	every eight hours
q12h	every twelve hours
qid	four times daily (quarter in die)
qn	every night
qod or eod	every other day
qp	as much as desired
qt	quart
®	registered trade name when superscript by drug name
sid	once daily
sig	let it be written as (used when writing prescription)
sol'n or soln	solution
SQ, SC, subQ, or subc	subcutaneous
T	tablespoon or tablet (or temperature)
tab	tablet
tid	three times daily (ter in die)
tsp or t	teaspoon
vol	volume

LABORATORY ABBREVIATIONS

ab	antibody
ABO	human blood groups
ag	antigen
alb	albumin
ALT	alanine aminotransferase (formerly SGPT)
AST	aspartate aminotransferase (formerly SGOT)
BP	blood pressure
BUN	blood urea nitrogen
CBC	complete blood count
CFT	complement fixation test
CMT	California mastitis test
diff	differential blood count
EDTA	ethylenediaminetetraacetic acid; type of anticoagulant
ESR	erythrocyte sedimentation rate
GTT	glucose tolerance test
H&E	hematoxylin and eosin stain

Hb or Hgb	hemoglobin
Hct or crit	hematocrit
HDL	high-density lipoprotein
HPF	high power field
HW	heartworm
LDL	low-density lipoprotein
LPF	low power field
MCH	mean corpuscular hemoglobin
MCHC	mean corpuscular hemoglobin concentration
MCV	mean corpuscular volume
ME ratio	myeloid-erythroid ration
NRBC	nucleated red blood cell
PCV	packed cell volume
PMN	polymorphonuclear neutrophil leukocyte
PT	prothrombin time
PTT	partial thromboplastin time
qns	quantity not sufficient
qs	quantity sufficient
RBC	red blood cell
rpm	revolutions per minute
SAP	serum alkaline phosphatase
SCC	somatic cell count
sed rate or SR	sedimentation rate
SGOT	serum glutamic oxaloacetic transaminase; now abbreviated AST
SGPT	serum glutamic pyruvic transaminase; now abbreviated ALT
SPF	specific pathogen free
sp. Gr.	specific gravity
Staph	*Staphylococcus* bacteria
Strep	*Streptococcus* bacteria
T_3	triiodothyronine (one type of thyroid hormone)
T_4	thyroxine (one type of thyroid hormone)
TB	tuberculin
TNTC	too numerous to count
UA	urinalysis
WBC	white blood cell
WMT	Wisconsin mastitis test

VACCINATION ABBREVIATIONS

BVD	bovine viral diarrhea
DHLPP	distemper, hepatitis, leptospirosis, parainfluenza, and parvovirus
DHLPP-CV	distemper, hepatitis, leptospirosis, parainfluenza, parvovirus, and coronavirus

EEE	Eastern equine encephalitis
EIA	equine infectious anemia
FeLV or	feline leukemia virus
FeLeuk or FeLuk	feline leukemia
FIP	feline infectious peritonitis
FIV	feline immunodeficiency virus
FVRCP	feline viral rhinotracheitis, calicivirus, and panleukopenia
FVRCP-C	feline viral rhinotracheitis, calicivirus, panleukopenia, and chlamydia
IBR	infectious bovine rhinotracheitis
MLV	modified live vaccine
PHF	Potomac horse fever
PI-3	parainfluenza 3 virus
RV	rabies vaccine
TE	tetanus
TGE	transmissible gastroenteritis
WEE	Western equine encephalitis

PHYSICAL EXAMINATION, PHYSIOLOGY, AND PATHOLOGY ABBREVIATIONS

ACh	acetylcholine
ACH	adrenocortical hormone
ACTH	adrenocorticotrophic hormone
ADH	antidiuretic hormone
AI	artificial insemination
ANS	autonomic nervous system
ASAP	as soon as possible
BAR	bright, alert, responsive
BM	bowel movement
BPM	beats or breaths per minute
c̄	with
C	castrated
cath	catheter
C-section	cesarean section
CHF	congestive heart failure
ChE	cholinesterase
CNS	central nervous system
CO	carbon monoxide
CO_2	carbon dioxide
CP	conscious proprioception
CPR	cardiopulmonary resuscitation
CRT	capillary refill time
CSF	cerebrospinal fluid
CSM	carotid sinus massage
CVP	central venous pressure
DA	displaced abomasum

DIC	disseminated intravascular coagulation
DLH	domestic long hair (feline)
DNA	deoxyribonucleic acid
DOA	dead on arrival
DSH	domestic short hair (feline)
ECG or EKG	electrocardiogram/graph
EEG	electroencephalogram/graph
EMG	electromyogram
F	Fahrenheit or female
FA	fatty acid
FLUTD	feline lower urinary tract disease
F/S	female spayed
FSH	follicle-stimulating hormone
FUO	fever of unknown origin
FUS	feline urological syndrome
GFR	glomerular filtration rate
GH	growth hormone
GI	gastrointestinal
HBC	hit by car
hCG	human chorionic gonadotropin
HR	heart rate
ICSH	interstitial cell-stimulating hormone
ICU	intensive care unit
IVDD	intervertebral disc disease
K-9	canine
Ⓛ	left
Ⓡ	right
LA	large animal
LDA	left displaced abomasum
LE	lupus erythematosus
lg	large
LH	luteinizing hormone
LOC	level of consciousness
LV	left ventricle
M	male
M/C	male castrated
MDB	minimum data base
mm	mucous membrane
mm Hg	millimeters of mercury
M/N	male neutered
MS	mitral stenosis
MSDS	material data safety sheet
N	neutered (or normal on physical examination)
NA or N/A	not applicable
NPN	nonprotein nitrogen
OB	obstetrics
OHE or OVH	ovariohysterectomy

OR	operating room
$\bar{p}$	after
P	pulse
PD	polydipsia
PDA	patent ductus arteriosus
PE	physical examination
pg	pregnant
PM	post mortem; also abbreviation for evening
PMI	point of maximal intensity
PNS	peripheral nervous system
POVMR	problem-oriented veterinary medical records
PU	polyuria
PVC	premature ventricular complex
R	respirations
RDA	right displaced abomasum
RNA	ribonucleic acid
R/O	rule out
RP	retained placenta
RR	respiration rate
$\bar{s}$	without
S	spayed
SA	sinoatrial or small animal
SOAP	subjective, objective, assessment, plan (record-keeping acronym)
stat	immediately (statim)
T	temperature (tablespoon or tablet)
TLC	tender loving care
TPN	total parenteral nutrition
TPO	triple pelvic osteotomy
TPR	temperature, pulse, and respiration
TSH	thyroid-stimulating hormone
TTA	transtracheal aspiration
TTW	transtracheal wash
TVT	transmissible venereal tumor
URI	upper respiratory infection
UTI	urinary tract infection
VM	vagal maneuver
VSD	ventricular septal defect
WNL	within normal limits
wt	weight

SYMBOLS

$\cong$	approximately equal to
$\uparrow$	increased
$\downarrow$	decreased
$+$	positive (used to describe test results); may have multiple +'s to indicate degree
$-$	negative (used to describe test results); may have multiple $-$'s to indicate degree
$\times$	times or multiplication sign
$\sqrt{}$	check
$=$	equal to
#	number (in front of number; for example, #1) pound (behind number; for example, 50#)
$\neq$	not equal to
$<$	less than
$>$	greater than
@	at or each
%	percent
♂	male
♀	female

THE X'S

Bx	biopsy
Dx	diagnosis
Fx	fracture
Hx	history
Rx	prescription
Sx	surgery
Tx	treatment

EYE AND EAR ABBREVIATIONS

AD	right ear
AS	left ear
AU	both ears
IOP	intraocular pressure
OD	right eye (also abbreviation for overdose)
OS	left eye
OU	both eyes

CHEMICAL ABBREVIATIONS

Fe	iron
H	hydrogen
H_2O	water
H_2O_2	hydrogen peroxide
HCl	hydrochloric acid
I	iodine
K	potassium
KCl	potassium chloride
N	nitrogen (also abbreviation for normal)
Na	sodium
NH_3	ammonium
O_2	oxygen

Plural Forms of Medical Terms

Many plural word forms are produced by adding an "s" to the singular term. This is true for medical terms as well. The plural of laceration is lacerations, the plural of bone is bones, etc. However, there are some rules to follow when using plural forms of medical terms. These rules are presented in the table below.

If the singular ending is:	Change or deletion from singular form:	Add plural ending:	Examples (singular)	Plural Form
s, ch, or h		es	abscess	abscesses
			stitch	stitches
			cough	coughes
y	delete y	ies	capillary	capillaries
is	delete is	es	diagnosis	diagnoses
um	delete um	a	bacterium	bacteria
us*	delete us	i	alveolus	alveoli
a	delete a	ae	vertebra	vertebrae
ix	delete ix	ices	cervix	cervices
ex	delete ex	ices	cortex	cortices
ax	delete ax	aces	thorax	thoraces
ma		s	carcinoma	carcinomas
ma	delete ma	mata	stoma	stomata
nx	delete nx	nges	phalanx	phalanges
on†	delete on	a	spermatozoon	spermatozoa

*except plural of virus is viruses, and plural of sinus is sinuses
†except plural of chorion is chorions

Index

Page numbers followed by "t" indicate tables; page numbers followed
by "f" indicate figures.

NEW COMPUTERIZED STUDY GUIDE

Flash!® for Windows

Flash!® is a computerized study guide with an interactive question and answer association program.

Flash!® allows you to quickly and easily master the information and improve test-taking skills.

Flash!® is always ready to help you throughout the course-when you want it.

How does Flash!® work?

Flash!® displays the question, then, when you click the mouse button on the screen, displays the answer. You can set **Flash!**® to learning mode to display both the question and answer at the same time or you can opt to display the answer, and guess the question. Features in **Flash!**® allow you to:

► select which questions to display
► see a timer to automatically advance from item-to-item
► print out questions and answers

What's in Flash!®?

Each Chapter has a computerized flashcard review of all key terms in the book. The order of presentation is the same as in the book and includes:

► flashcard for each chapter
► abbreviations
► prefixes, combining forms, and suffixes alphabetically listed

Minimum hardware requirements: 386 or better personal computer running Windows 3.1 or later; at least 4 MB of RAM; a mouse is required

For technical support call 1-800-477-3692 (9:30 am to 4:30 pm EST) or fax 1-800-464-7000 (24 hours a day).